This volume includes contributions by leading workers in the field given at the workshop on Numerical Relativity held in Southampton in December 1991. Numerical relativity, or the numerical solution of astrophysical problems using powerful computers to solve Einstein's equations, has grown rapidly over the last 15 years. It is now an important route to understanding the structure of the universe, and is the only route currently available for approaching certain important astrophysical scenarios. The Southampton meeting was directed at providing a dialogue between theoreticians in general relativity and practitioners in numerical relativity. It was also notable for the first full report of the new 2+2 approach and the related null or characteristic approaches, as well as for updates on the established 3+1 approach, including both Newtonian and fully relativistic codes. The contributions range from theoretical (formalisms, existence theorems) to the computational (moving grids, multiquadrics and spectral methods).

This book will be of value to relativists, cosmologists, astrophysicists and applied mathematicians interested in the increasing contribution made to this subject using numerical techniques.

Approaches to Numerical Relativity

Approaches to Numerical Relativity

Proceedings of the International Workshop on Numerical Relativity,
Southampton, December 1991

Edited by
Ray d'Inverno
Faculty of Mathematical Studies, Southampton University

CAMBRIDGE
UNIVERSITY PRESS

CAMBRIDGE UNIVERSITY PRESS
Cambridge, New York, Melbourne, Madrid, Cape Town, Singapore, São Paulo

Cambridge University Press
The Edinburgh Building, Cambridge CB2 2RU, UK

Published in the United States of America by Cambridge University Press, New York

www.cambridge.org
Information on this title: www.cambridge.org/9780521439763

First published 1992
This digitally printed first paperback version 2005

A catalogue record for this publication is available from the British Library

ISBN-13 978-0-521-43976-3 hardback
ISBN-10 0-521-43976-0 hardback

ISBN-13 978-0-521-01735-0 paperback
ISBN-10 0-521-01735-1 paperback

TABLE OF CONTENTS

CONTRIBUTORS

Miguel Alcubierre,
Department of Physics and Astronomy,
University of Wales College of Cardiff,
Cardiff,
CF1 3TH,
Wales,
UK.

Gabrielle D. Allen,
Department of Physics and Astronomy,
University of Wales College of Cardiff,
Cardiff,
CF1 3TH,
Wales,
UK.

John Barrett,
Department of Applied Mathematics
and Theoretical Physics,
University of Cambridge,
Silver Street,
Cambridge,
CB3 9EW,
UK.

Nigel Bishop,
Department of Mathematics, Applied Mathematics
and Astronomy,
University of South Africa,
0001 Pretoria,
South Africa.

Carles Bona,
Department of Physics,
University of the Balearic Isles,
E-07071 Palma de Mallorca,
Spain.

Silvano Bonazzola,
D.A.R.C.,
UPR176 du C.R.N.S.,
Observatoire de Meudon,
92195 Meudon,
France.

Joan M. Centrella,
Department of Physics and Atmospheric Science,
Drexel University,
Philadelphia,
Pennsylvania 19104,
USA.

Matt Choptuik,
Center for Relativity,
Department of Physics,
University of Texas,
Austin,
Texas,
TX 78712-1081,
USA.

Chris J. S. Clarke,
Faculty of Mathematical Studies,
Southampton University,
Southampton,
SO1 5BR,
UK.

Ray A. d'Inverno,
Faculty of Mathematical Studies,
Southampton University,
Southampton,
SO1 5BR,
UK.

Mark Dubal,
Center for Relativity,
Department of Physics,
University of Texas,
Austin,
TX 78712-1081,
USA.

Jörg Frauendiener,
Max Planck Institute for Astrophysics,
Karl Schwarzschild Strasse 1,
D-8046 Garching bei München,
Germany.

Helmut Friedrich,
Max Planck Institute for Astrophysics,
Karl Schwarzschild Strasse 1,
D-8046 Garching bei München,
Germany.

Roberto O. Gómez,
Department of Physics and Astronomy,
University of Pittsburgh,
Pittsburgh,
PA 15260,
USA.

Eric Gourgoulhon,
D.A.R.C.,
UPR176 du C.R.N.S.,
Observatoire de Meudon,
92195 Meudon,
France.

Jerry Griffiths,
Department of Mathematical Sciences,
Loughborough University of Technology,
Loughborough,
LE11 3TU,
UK.

Pawel Haensel,
D.A.R.C.,
UPR176 du C.R.N.S.,
Observatoire de Meudon,
92195 Meudon,
France.

Sean Hayward,
Max Planck Institute for Astrophysics,
Karl Schwarzschild Strasse 1,
D-8046 Garching bei München,
Germany.

José Maria Ibáñez,
Department of Theoretical Physics,
46100-Burjassot,
Valencia,
Spain.

Fjodor V. Kusmartsev,
Department of Physics,
University of Oulu,
Linnanmaa,
SF 90570,
Finland.

Antonio Lanza,
International School of Advanced Studies (SISSA),
Strada Costiera 11,
I-34014 Trieste,
Italy.

Kei-Ichi Maeda,
Department of Physics,
Waseda University,
Shinjuku-ku,
Tokyo 169,
Japan.

Jean-Alain Marck,
D.A.R.C.,
UPR176 du C.R.N.S.,
Observatoire de Meudon,
92195 Meudon,
France.

José Maria Martí,
Department of Theoretical Physics,
46100-Burjassot,
Valencia,
Spain.

J. Massó,
Department of Physics,
University of the Balearic Isles,
E-07071 Palma de Mallorca,
Spain.

Richard A. Matzner,
Center for Relativity,
Department of Physics,
University of Texas,
Austin,
TX 78712-1081,
USA.

Eckehard W. Mielke,
Faculty of Mathematics,
University of Kiel,
Kiel,
Gemany.

John C. Miller,
Astronomical Observatory of Trieste,
Trieste,
Italy.

Juan A. Miralles,
Department of Theoretical Physics,
46100-Burjassot,
Valencia,
Spain.

Takashi Nakamura,
Yukawa Institute for Theoretical Physics,
Kyoto University,
Kyoto 606,
Japan.

Ken-Ichi Nakao,
Department of Physics,
Kyoto University,
Sakyo-ku,
Kyoto 606,
Japan.

S. R. Oliveira,
Department of Physics,
University of Brasilia,
Brazil.

Niall Ó Murchadha,
Physics Department,
University College Cork,
Cork,
Ireland.

Ken-Ichi Oohara,
National Laboratory for High
Energy Physics (KEK),
1-1 Oho,
Tsukuba,
Ibaraki 305,
Japan.

O. Pantano,
Department of Physics,
University of Padua,
Padua,
Italy.

Alan Rendall,
Max Planck Institute for Astrophysics,
Karl Schwarzschild Strasse 1,
D-8046 Garching bei München,
Germany.

J. V. Romero,
Department of Theoretical Physics,
46100-Burjassot,
Valencia,
Spain.

Franz Schunck,
Institute for Theoretical Physics,
University of Cologne,
Zülpicher Str. 77,
5000 Köln 41,
Germany.

Bernard F. Schutz,
Department of Physics and Astronomy,
University of Wales College of Cardiff,
Cardiff,
CF1 3TH,
Wales,
UK.

Paul Shellard,
Depart of Applied Mathematics
and Theoretical Physics,
University of Cambridge,
Silver Street,
Cambridge,
CB3 9EW.
UK.

Scott C. Smith,
Department of Physics and Atmospheric Science,
Drexel University,
Philadelphia,
Pennsylvania 19104,
USA.

John Stewart,
Department of Applied Mathematics
and Theoretical Physics,
University of Cambridge,
Silver Street,
Cambridge,
CB3 9EW,
UK.

James A. G. Vickers,
Faculty of Mathematical Studies,
Southampton University,
Southampton,
SO1 5BR,
UK.

Jeff Winicour,
Department of Physics and Astronomy,
University of Pittsburgh,
Pittsburgh,
PA 15260,
USA.

INTRODUCTION

Ray d'Inverno

Faculty of Mathematical Studies, University of Southampton, Southampton, UK

This volume derives from a workshop entitled "Approaches to Numerical Relativity" which was held in the week 16-20th December, 1991, in the Faculty of Mathematical Studies at Southampton University, England. It was held principally because it was thought that the time was opportune to begin a dialogue between theorists in classical general relativity and practitioners in numerical relativity. Numerical relativity - the numerical solution of Einstein's equations by computer - is a young field, being possibly only some fifteen years old, and yet it has already established an impressive track record, despite the relatively small number of people working in the field. Part of this dialogue involved bringing participants up to date with the most recent advances. To this end, international experts in the field were invited to attend and give presentations, including Joan Centrella, Matt Choptiuk, John Miller, Ken-Ichi Oohara, Paul Shellard and Jeff Winicour. In addition, a significant number of European scientists, both theoreticians and practitioners in numerical relativity, were invited, the majority of whom attended. In the event, there were some 35 participants, most of whom gave presentations. This volume is largely comprised of the written versions of these presentations (their length being roughly proportional to the time requested by the authors for their presentations).

In an attempt to highlight the distinctive nature of the workshop, I have divided the contributions into Part A, Theoretical Approaches and Part B, Practical Approaches. This is to a large extent somewhat arbitrary, since several of the theoretical contributions involve a significant element of computing, and all of the practical contributions involve theoretical aspects. Again, where possible, I have tried to locate related contributions together. I hope that my arbitrary division and ordering will not offend anyone, but rather help to point out the dialogue nature of the workshop. Indeed, this is exemplified further by the inclusion of an edited version of the final Panel Discussion.

Most numerical relativity makes considerable demands on computer processing power. As a consequence, much of the work has been carried out in the US and Japan, although there have been significant contributions from Europe, especially in Germany, France, Italy and Spain. More surprising, perhaps, is the fact that the field has attracted a significant number of British scientists, and there are a growing number beginning to take an interest in it. Thus, an additional reason for holding the workshop was to hold it in the UK as a mark of this involvement and growing interest and, in so doing, to announce the setting up of three centres for numerical relativity, namely at Cambridge (under John Stewart), Cardiff (under Bernie Schutz) and Southampton (under Chris Clarke).

There was also a notable difference in the emphasis of this workshop as compared with other recent meetings in numerical relativity. Previously, most attention has been paid to the 3+1 approach and, to a lesser extent, the Regge calculus approach. This workshop, for the first time, gave greater attention to the newer null or characteristic approach and the related 2+2 approach. As someone who was involved in developing the 2+2 approach, I have a particular interest in seeing its application in numerical relativity. And so, as editor of this volume, I decided to reflect this change in emphasis by including theses approaches first in each part, and would wish to draw particular attention to the pioneering work of the groups of Jeff Winicour and John Stewart. Part A starts off with my own contribution, partly because it was the first presentation at the workshop, and partly because it includes a short (personally conceived) introduction to numerical relativity.

I would like to express my thanks for their co-operation to the authors of the articles and, in particular, to the head of our group, Chris Clarke, for agreeing to add a preface. I would like to thank Chris Clarke additionally, together with James Vickers, for their considerable help in the preparation of this volume. Finally, the three of us as co-organisers, would wish to express our gratitude for financial support for the workshop from the Science and Engineering Research Council, the London Mathematical Society and the Institute of Physics.

PREFACE

C. J. S. Clarke

Faculty of Mathematical Studies, University of Southampton, Southampton, UK

General relativity was for too long the ugly duckling of science. In the 50s and 60s the dominant impression was of the difficulty of the equations, solvable only by arcane techniques inapplicable elsewhere; of the scarcety of significant experimental tests; of the prohibitive cost of computational solutions, compounded by a lack of rigorous approximation techniques; and of the isolation of the subject from the physics of the other fundamental forces. This led to a situation where, even in the 70s, much theoretical work was becoming increasing irrelevant to physics. Exact solutions proliferated but (with the exception of cosmology) attempts at physical interpretation were few and unconvincing. Mathematical investigations in the wake of the singularity theorems became increasingly sophisticated, but few were applied to actual physical models. In the 70s and 80s, however, all this changed, with the growth of experimental relativity, the trend to geometrical methods in high energy physics, and the inception of numerical relativity. The workshop reported in this book marks the complete clearing of this last hurdle, as reliable and practical computational techniques are established.

It brought together numerical and classical relativists, and showed that the cultural gap between them was closing fast. Dramatically increased standards of reliability and accuracy had been set, and were being achieved in many cases, so that numerical work can no longer be seen merely as providing a rough indication for the 'proper' work of analysis. In increasingly many areas numerical simulation will clearly be the decisive component in answering the questions being posed by the theorists. On the astronomical side, the sophistication of hydrodynamic codes is now enabling them to cope with real astrophysical problems.

A precise account of some of the main technical points is contained in the first review section of d'Inverno's paper, to which I refer the reader for an expansion of what follows. Here I shall try to give an overview of the place of the workshop papers in the subject as a whole. The papers cover the two main classes of numerical techniques: those that use Cauchy surfaces and those that use null surfaces. (This is in itself interesting. As Miller points out in the concluding discussion, the field is no longer totally dominated by the former method.) They include expositions of numerical techniques, both generally applicable ones and those specific to relativity, discussions of different formulations of relativity appropriate to numerical work, presentations of new physically interesting results, and reviews of areas where problems seem ripe for a numerical approach. To give a flavour of the many dialogues that enlivened the workshop, and to indicate the main directions opened up for future work, we include at the end an edited transcript of the concluding round-table discussion.

Einstein's equations are perhaps unique among physical models in the variety of different choices of coordinates and representations that can be made, all having some claim to simplicity or efficiency in some circumstances. The main subdivision, already referred to, is between those in which evolution is governed by a coordinate t with the surfaces $t = $ constant spacelike (3+1 or Cauchy approaches) and those which instead use a u with $u = $ constant null (2+2 or characteristic approaches). d'Inverno describes the geometric features of the methods, while Winicour's article begins with a useful survey of the relative merits of the two, concluding that Cauchy approaches are best suited to matter-dominated regions where trapped surfaces may be forming, while characteristic methods are suited to the intermediate zones round a system emitting radiation, where they can efficently handle the self-interaction and back-scattering of radiation, which, as his article shows, could crucially effect the appearance of the object to gravitational detectors.

While most papers used a traditional approach to the characteristic method, an illuminating alternative was provided by Hayward, who showed that, just as the Cauchy formalism can be seen as a Hamiltonian dynamical system, so the characteristic formalism can be seen as generalised '2-time' Hamiltonian system.

Within the characteristic methods there is further subdivision into those based on the use of tetrads and the Newman-Penrose or Geroch-Held-Penrose formalisms (discussed by Frauendiener, Vickers and Stewart) and those based directly on the metric (discussed by Winicour, Bishop and d'Inverno). The great strength of all these methods is the simplification of the equations: equations lying in the surfaces $u = $ const. become ordinary differential equations in contrast to the elliptic equations specifying the constraints in the Cauchy approach. In addition, some of the methods give a dramatic reduction in the total size of the equation set. These together can lead to very fast and efficient numerical codes.

The majority of the papers use the Cauchy approach, however. Here also there are many different choices of coordinates, particularly since the choice of the hypersurfaces t = const. (the slicing condition) now introduces much more freedom. There seems to be a lull in the debate as to the best approach here, with different groups being content to be judged by their fruits. The issue is far from dead, however, as witnessed by the recent controversy over whether or not a naked singularity can be produced by a collapsing cloud of particles (Wald and Iyer, 1991). The range of work on Cauchy methods, and particularly the elegant work of Bona's group, certainly shows that there is as much scope as in the characteristic methods for judicious simplification of the structure of equations.

Not surprisingly, in view of the complementarity of the two approaches there is interest in combining them. Thus Bishop presents a well articulated (but as yet untested) numerical scheme for joining together the two methods in different regions. Friedrich, on the other hand, shows that it is possible to carry out an intriguing scheme using hyperboloidal slices that have all the advantages of the Cauchy method in the interior, but which asymptote to null hypersurfaces, enabling one to compactify the entire space-time just as for the characteristic method.

In addition to the two above divisions of the full equations of relativity, there is a broad class of approaches which first approximate the equations and then solve numerically – typically used, as by Mark, Bonazzola and Nakamura, in collapsing stellar problems where the full equations are very complex. The status of approximation methods in relativity is an uncertain area; there is no approximation scheme known to be convergent and no effective estimates on the errors associated with existing schemes. It is therefore all the more interesting to see both exact and approximated methods developing numerically so that more can be learnt about the validity of these methods.

The most striking feature of the workshop, for a classical relativist like myself, was a repeated emphasis on the importance of discriminating fine structure. Underlying this is the fact that, even without general relativity, the hydrodynamics of a collapsing star is a difficult numerical problem: shock waves form, which, in the absence of high symmetry, can rapidly lead to great complexity. Relativity also has its tendency to generate fine structure, even with a system as simple as the spherically symmetric scalar field. Christodoulou had already demonstrated that the field evolved to a step function, and Choptuik's paper traces this evolution through the emergence of progressively finer structure. A repeated theme, here and with Lanza, for example, is the use of multi-grids for covering adequately the region with fine structure, while Centrella and others use adaptive grids.

It is clear from this that considerable sophistication is needed in the choice of numerical methods. Centrella, for example, uses a discretisation of advection that ensures local angular momentum conservation, while Ibañez and his group, in a widely applauded paper, used Godunov methods for hyperbolic equations to capture the shocks occuring in realistic models of white dwarfs with quite remarkable accuracy. Oohara, on the other hand, achieved good accuracy with a more classical TVD limiter approach. This well illustrates the power of modern numerical techniques, which at present are only applicable to the Cauchy implementation of general relativity.

At the level of numerical technique, several alternatives to standard finite differences were presented. Bonazzola and Marck used pseudo-spectral methods with great success, in combination with multi-domains to capture shocks; while Dubal successfully used multi-quadrics, an impressive technique, but one that seemed to require considerable artistry in its use. Regretably, the relativistic equivalent of finite elements – Regge calculus – was under-represented, with only one paper, by Barrett. This analysed the combinatorics of developing the triangulation of hypersurfaces, a subject with ramifications both for low dimensional topology (new invariants have recently been discovered by this method) and quantum theory, as well as for mesh-generation in Regge calculus. The direct use of the calculus for general relativity seems to have been somewhat sidelined, however.

A potentially vital numerical method, classical in essence but brought to the fore by the particular needs of relativity, was the ADI method of the Cardiff group (Allen and Alcubiere) adapted to grids moving with a super-characteristic velocity, as is needed for black hole problems. A simple idea with the benefit of hindsight, it was applied to achieve stability over a remarkably wide range.

A further feature of general relativity is the non-triviality of the problem of establishing initial data. In the Cauchy approach this involves solving the elliptic constraint equations, while in the characteristic approach, though it is numerically easier, there are subtle problems concerning the specification of 'no incoming radiation'. Several papers concerned the Cauchy problem here, including some on equilibrium configurations, such as Lanza's useful analysis of thin-disc equilibria. An interesting contribution by O'Murchadha showed how the multipole structure of the source was reflected in the details of the boundary conditions to be imposed on the constraint equations.

Perhaps the work most characteristic of the mood of the meeting was that by Choptuik, in both his own paper and in the closing discussion. While previous meetings had stressed the importance of calibration and test-bed calculations, comparing with known exact solutions (themes that were still present here), now the emphasis was on the construction of self-validating programmes: codes that could be run with a wide range of parameters governing step-sizes so as to produce estimates of their own error. This was controversial material, many participants worrying that such procedures could only check convergence to something, not necessarily convergence to Einstein's equations. But it was countered that, providing all the equations are monitored, properly designed code should be able to verify this, giving an estimate of the actual discrepancy from the exact result. This is clearly an area where valuable work could be done in collaboration with analytic theory. It is also an area where psychological issues are as important as mathematical ones, in convincing the relativistic community that numerical results can now be produced that are as firm as analytic results.

Though my own prejudices have placed mathematics first, the papers also marked a considerable advance in physical understanding. Miller argued powerfully that a careful analysis of numerical data could be used to develop our conceptual understanding of the role of angular momentum and so on in the relativistic regime, despite the difficulty of defining many of these concepts in the abstract. And the presentation of Oohara, on the interaction of neutron stars, showed distinct physical structures emerging in different parameter ranges – a valuable conceptual adjunct to the data.

The range of problems addressed covered the gamut of astrophysics and relativity. Particularly exciting was the number of 3 (spatial) dimensional calculations presented. Most used approximations, but Bona demonstrated an exact three dimensional vacuum code (available by e-mail, so we can all try it!) Cosmology was represented by Nakao's account of horizon formation in inflation, a new application of numerical techniques, as was the analysis of boson stars (possible candidates for dark matter) by Schunck, interestingly using catastrophe theory to explore genericity ideas that were also predominant in Choptuik's related paper. Cosmic strings were presented by Shellard, who not only showed that the subject was still very active, but also broadened the context of numerical relativity by linking it with numerical work in this other area.

The majority of numerical papers, however, were either directly concerned with gravitational wave generation, or (as with Centrella and Miller) studied related problems of stability in relativistic stellar physics – very appropriately so in view of the rapidly developing experimental situation. The central problem here is the efficiency of generation from likely astrophysical sources of gravitational radiation, exploring the conjecture that 3-dimensional configurations should be much more efficient than axisymmetric ones. This was well illustrated by Oohara's finding of a 30-fold increase in efficiency compared with axisymmetry, but this had to be qualified by doubts concerning more global relativistic effects in the wave-zone, whose importance was stressed by Winicour, and concerning the effect of red shifts on the radiation.

The interface between standard relativity problems and numerical work was also addressed, most provocatively in the paper by Rendal which bore on the work already referred to, by Shapiro and Teukolsky (1991), on naked singularities. Colliding waves, reviewed by Griffiths, also offered an arena where current techniques for 3-D vacuum codes should already be able to shed light on the outcome of generic interactions, where the occurrence or otherwise of singularities is very uncertain.

Though hardware was not singled out for discussion, some interesting trends emerged. Not all problems required the power of super-computers; the simplifications achieved by Stewart, for instance, enabled his code to run on a workstation. And, while high power was often necessary (for hydrodynamics in particular), this did not now involve prohibitive cost, as illustrated by the use of transputer networks by the Southampton group (d'Inverno and Bishop). Thus not only was numerical relativity important, but it was also accessible.

The conclusion of the meeting was that numerical work had emerged as focally important in general relativity as well as astrophysics, with almost no topic that was not decisively affected by it. General relativity now emerges as no longer the ugly duckling, but a swan-like subject combining both elegance and power.

REFERENCES

Shapiro, S. L. and Teukolsky, S. A. (1991). *Phys. Rev. Lett.*, **66**, 994–997.
Wald, R. M. and Iyer, V. (1991). *Phys. Rev. D*, **44**.

PART A

THEORETICAL   APPROACHES

NUMERICAL RELATIVITY ON A TRANSPUTER ARRAY

Ray d'Inverno

Faculty of Mathematical Studies, University of Southampton, Southampton, UK

Abstract. The area of numerical relativity is briefly reviewed and its status in general relativity is considered. The 3+1 and 2+2 approaches to the initial value problem in general relativity are described and compared. A 2+2 approach based on null cones emanating from a central timelike geodesic, together with an implementation on a transputer array is discussed.

1 EXACT SOLUTIONS OF EINSTEIN'S FIELD EQUATIONS

In 1915 Einstein proposed his field equations for the gravitational field

$$G_{ab} = \kappa T_{ab} \tag{1}$$

where G_{ab} is the Einstein tensor for the gravitational field with metric g_{ab}, κ is a coupling constant and T_{ab} is the energy-momentum tensor for any matter field present (d'Inverno (1992)). In the absence of matter fields, the equations reduce to the vacuum equations

$$G_{ab} = 0. \tag{2}$$

Einstein always considered the vacuum equations as being more fundamental in character. They may be viewed as second order non-linear partial differential equations for the metric potentials g_{ab}. The non-linearity means that the equations are difficult to solve, indeed Einstein originally thought that it would not be possible to solve them exactly. For example, they do not satisfy a superposition principle, and so complicated configurations cannot be analysed in terms of simpler constituent ones. It came as something of a surprise when Schwarzschild discovered an exact solution in 1916.

In the ensuing decades there were relatively few exact solutions discovered. However, the invariant techniques of the Petrov classification, optical scalars and Killing vectors led, in the 1960's, to the discovery of numerous exact solutions. It is difficult to count the number of known exact solutions, because many depend on parameters, or on solutions of subsidiary ordinary or partial differential equations. However, the number of authors involved in the discovery of exact solutions is certainly well into four figures. The area of exact solutions was for a long time a confused one until, in 1980, considerable progress was made in attempting to put known solutions into some sort of systematic framework through the publication of the exact solutions book (Kramer et al (1980)).

There are two major problems associated with this field, one practical and the other theoretical. The first concerns the problems associated with the horrendous algebraic calculations involved in work with exact solutions, especially when carried out by hand. The second involves the well-known equivalence problem: given two geometries g_1 and g_2, are they in fact different or is there a coordinate transformation which (locally) transforms one into the other? Significant advances with both these problems were made with the advent of computer algebra systems, some of which were specifically designed for the metric calculations involved in general relativity. Perhaps the best known and most used system in general relativity is the system Sheep (Frick (1977), d'Inverno and Frick (1982)). The power of a system like Sheep is that calculations can be undertaken which would have taken lifetimes to complete by hand. Moreover, the results are error-free.

The theoretical advance came with the discovery of the Karlhede algorithm for classifying a geometry (Karlhede (1980)). This is essentially achieved by introducing a frame in which the Riemann tensor and its covariant derivatives take on canonical forms. This classification can be undertaken, essentially automatically, in the extension of Sheep called Classi. Then, given two geometries, if the classifications are different then so are the geometries; if they are the same them they are candidates for identification. A search is then made for a coordinate transformation which can map one geometry into the other. This last procedure reduces, in general, to solving four algebraic equations, a process which is not algorithmic but which is often manageable in practise. The Karlhede classification program has led to the establishment of the computer database project, a joint research effort aimed at classifying and documenting all known exact solutions (Åman et al (1985)). The first aspiration of the project is to put all the solutions of the exact solutions book into the database. At present, several hundred solutions have been classified. The ultimate hope is that the database will be freely accessible to the scientific community and continuously kept up to date. Then any newly discovered solution can be checked out against the data base which can be updated if the solution is genuinely new.

The database project holds out the prospect of placing the field of exact solutions onto a much more coherent basis. Unfortunately, although this large number of exact solutions exist, very few would appear to be physically realistic or even approximately so. As is well known, partial differential equations admit large classes of solutions, many of which are pathological in nature. One usually has to apply boundary conditions or initial conditions to pick out the solutions which are of physical interest.

Apart from black hole, cosmological and plane wave solutions, the likelihood is that the remaining solutions are indeed pathological in nature. More importantly, we do not possess exact solutions corresponding to or approximating to important physical scenarios such as a 2-body system, an n-body system, a radiative source, the interior of a rotating object undergoing gravitational collapse, and so on. Yet these are precisely the objects that are of interest to us, especially on an astrophysical scale. This is where, I believe, numerical relativity comes in.

2 NUMERICAL RELATIVITY

Numerical Relativity consists of solving Einstein's equations numerically on a computer. The standard scenario is to specify the 3-metric $^{(3)}g$ — the intrinsic geometry — of some spacelike slice ($t = t_1 =$ constant, say), and use the field equations to compute the 3-metric at some future time ($t = t_2 > t_1$). The significance of being able to do this is that we can thereby model physically interesting scenarios. Indeed, given the freedom to vary the initial configuration, we can consider the resulting numerical simulations as being in the arena of experimental relativity. This significance will become more pronounced when the long awaited detection of gravitational waves is at last reported and we move into the era of gravitational astronomy. The need will then arise of finding theoretical justifications for actual observations, and this need will likely push numerical relativity into the forefront of general relativity.

There are, in essence, three distinct approaches to numerical relativity: the 3+1 approach, the 2+2 approach and the Regge calculus. These proceedings are largely concerned with the first two approaches, and so we shall overview them briefly in turn.

3 THE 3+1 APPROACH

The basis of this approach is to decompose 4-dimensional space-time into families of 3-dimensional spacelike hypersurfaces and 1-dimensional timelike lines (see, for example, article of York in Smarr (1979)). In more geometrical language, space-time is decomposed into a spacelike foliation and a (transvecting) timelike fibration (figure 1). We can introduce a constructive procedure for generating the decomposition if we start off with a 4-dimensional manifold possessing no metrical or affine structure on it — a so-called bare manifold — and prescribe on it a vector field which transvects some 3-dimensional submanifold i.e. the vector field nowhere lies in the submanifold (figure 2). We then use the vector field to propagate the initial submanifold or hypersurface into a family of hypersurfaces (technically by Lie dragging). The standard initial value problem (or IVP for short), sometimes called the Cauchy IVP, consists of specifying a positive definite 3-metric on the initial hypersurface Σ_0 and then using the vacuum field equations to determine the 3-geometries on successive hypersurfaces Σ_t, say.

There is an analogous initial value problem when matter fields are present.

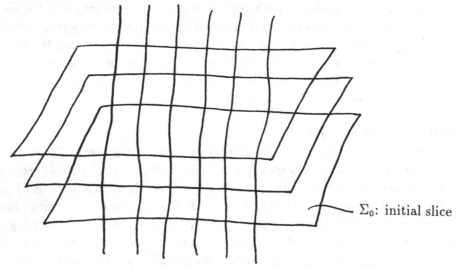

Figure 1. Spacelike foliation and timelike fibration.

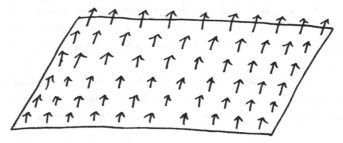

Figure 2. 3-dimensional submanifold and transvecting vector field.

The next step is to introduce some formalism which is adapted to the 3+1 decomposition. The most elegant method is to work with a 4-dimensional formalism, that is one which is manifestly covariant, coupled with projection operators to accomplish the decomposition and Lie derivatives to accomplish the propagation. In this overview, we shall simply use adapted coordinates in which the hypersurfaces have equation $x^0 = t =$ constant, and possess intrinsic coordinates x^α (where latin indices run from 0 to 3 and greek from 1 to 3). Then Lichnerowicz has shown that the vacuum field equations are equivalent to six evolution equations (d'Inverno (1992))

$$R_{\alpha\beta} = 0 \qquad (3)$$

and four constraint equations

$$G^0{}_a = 0. \qquad (4)$$

If these constraint equations hold everywhere on Σ_0 and the evolution equations hold everywhere, then the contracted Bianchi identities reveal that the constraint equations are automatically satisfied everywhere. A typical numerical scheme consists of taking the dynamical variables to be the 12 variables consisting of $g_{\alpha\beta}$, the 3-metric, and $K_{\alpha\beta}$, its extrinsic curvature, which is defined as (apart from an unimportant numerical factor)

$$K_{\alpha\beta} = \dot{g}_{\alpha\beta} \tag{5}$$

where a dot denotes differentiation with respect to time. The evolution equations then reduce to the first order propagation equations

$$\dot{g}_{\alpha\beta} = \text{known} \tag{6}$$

$$\dot{K}_{\alpha\beta} = \text{known} \tag{7}$$

where the right hand sides are known functions of $g_{\alpha\beta}$, $K_{\alpha\beta}$ and spatial derivatives. The initial data consists of prescribing $g_{\alpha\beta}$ and $K_{\alpha\beta}$ on Σ_0, and use of the propagation equations means that both these quantitites are known on the next neighbouring hypersurface. By taking the time derivatives of equations (6) and (7) we can repeat the process on the next neighbouring hypersurface. Proceeding in this way, we obtain an iterative procedure for generating a solution forward in time. In a numerical regime the derivatives are obtained by a finite difference procedure. There are many variants to this approach, but this short description should serve to illustrate the essential characteristics of the standard approach.

There are two main problems associated with this approach. First of all, the initial data is not freely specifiable but must satisfy the constraints initially. These can be decomposed into the Hamiltonian constraint

$$^{(3)}R - K^{\alpha\beta}K_{\alpha\beta} + (K^{\alpha}_{\alpha})^2 = 0 \tag{8}$$

and the momentum constraint

$$^{(3)}\nabla_{\alpha}(K^{\alpha}_{\beta} - K^{\gamma}_{\gamma}\delta^{\alpha}_{\beta}) = 0. \tag{9}$$

This problem can be resolved by extracting a conformal factor from the 3-geometry and investigating the resulting elliptic partial differential equations. This reveals that the gravitational field possesses two true degrees of freedom, namely, in Hamiltonian language, two coordinates (associated with the $g_{\alpha\beta}$) and two momenta (associated with $K_{\alpha\beta}$). The second problem relates to the extent of the development. Although there are existence theorems which say that a solution can be generated for some finite time to the future of the initial slice, they do not indicate how far this may be. Moreover, the approach fails if the foliation goes null. Yet null foliations are important in there own right as we shall next see.

4 THE 2+2 APPROACH

The basis of this approach is to decompose space-time into two families of space-like 2-surfaces. We can view this as a constructive procedure in which an initial 2-dimensional submanifold S_0 is chosen in a bare manifold, together with two vector fields v_1 and v_2 which transvect the submanifold everywhere (figure 3).

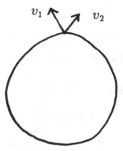

Figure 3. 2-dimensional submanifold and two transvecting vector fields.

The two vector fields can then be used to drag the initial 2-surface out into two foliations of 3-surfaces. The character of these 3-surfaces will depend in turn on the character of the two vector fields. The most important cases are when at least one of the vector fields is taken to be null. For example, if the two vector fields are null, then they give rise to a double-null foliation (indicated schematically in figure 4).

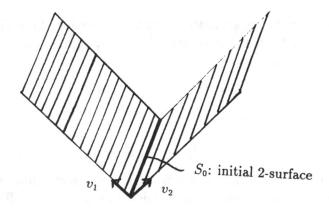

Figure 4. Double-null foliation.

Or if one is null and the other is timelike, this gives rise to a null-timelike foliation (figure 5).

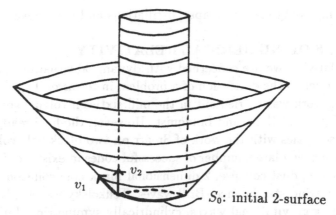

v_2

v_1

S_0: initial 2-surface

Figure 5. Null-timelike foliation.

The most elegant way of proceeding is to introduce a formalism which is manifestly covariant and which uses projection operators and Lie derivatives associated with the two vector fields. The resulting formalism is called the 2+2 formalism (d'Inverno and Smallwood (1980), d'Inverno (1984)). When the vector fields are of a particular geometric character, then this can be refined further into a 2+(1+1) formalism. Finally, one extracts a conformal factor from the spacelike 2-geometries to isolate the gravitational degrees of freedom.

The 2+2 formalism leads to a number of advantages. First of all, it identifies the two gravitational degrees of freedom as the conformal 2-geometry (d'Inverno and Stachel (1978)). Secondly, this data is unconstrained. Thirdly, the data satisfy ordinary differential equations along the vector fields. Most importantly, unlike the 3+1 approach, the formalism applies to situations where the foliations go null. Such IVPs are called null or characteristic IVPs (CIVPs for short). These are the natural vehicles for studying gravitational radiation problems (since gravitational radiation propagates along null geodesics), asymptotics of isolated systems (since I^+ and I^- are null hypersurfaces) and problems in cosmology (since we gain information about the universe along our past null cone).

The null or characteristic approach, however, suffers from one main drawback resulting from the fact that, in general, null hypersurfaces develop caustics. There are then two quite distinct ways of proceeding. One approach is to develop techniques for generating solutions through caustics (Corkhill and Stewart (1983)). The other is to restrict attention to caustic-free scenarios. This is possible by considering models which are close to spherical symmetry and where it can be proved rigorously that caustics will not occur. This assumption still allows the investigation of an important class of models including stellar collapse, oscillations and supernovae.

5 THE STATUS OF NUMERICAL RELATIVITY

In numerical relativity, we are concerned with dynamical situations, and hence it is conventional to refer to the dimension of a problem in terms of its spatial dimension. Some pioneering work was carried out in the mid sixties involving spherical collapse by May and White, and Hahn and Lindquist. However, the field really only came of age in the mid seventies with the work of Smarr on two black hole collisions (Smarr (1977)) There are now a large number of successful codes in existence including codes in spherical collapse, dust collapse, 2-dimensional black hole collision, 2-dimensional axisymmetric neutrino star bounce, Brill waves, Teukolsky waves, planar symmetry solutions, colliding gravitational waves, cylindrically symmetric solutions, accretion disks, shock waves, inflationary cosmolgy, n-body calculations, collapse of massless scalar fields, evolution of 3-dimensional wave packets and 3-dimensional relativistic hydrodynamics. Most of the fully 3-dimensional work undertaken to date has involved Newtonian models of one sort or another. We are just on the verge of 3-dimensional relativistic codes; indeed some of these codes are reported elsewhere in this volume. These codes will make enormous demands (by present standards) on computer time and memory.

There are a number of problems associated with numerical relativity. The main one relates to the role of the constraint equations. The finite difference version of Einstein's equations leads to an overdetermined system in which the constraints are either ignored (free evolution) or artificially imposed. In the latter case, one method involves imposing the constraints after finite intervals of times (chopped evolution) and another is to impose them at every stage of integration (fully constrained evolution). Unfortunately, each method has associated drawbacks. For example, computations with particular exact solutions have demonstrated that a free evolution drifts further away from the true solution as it evolves in time. Similar problems arise with chopped and fully constrained evolutions. Piran has indicated this schematically in figure 6, where the plane represents the subspace of solutions which satisfy the constraint equations.

Other problems relate to the finite difference approximation. Unfortunately, there are an infinite number of possible finite element difference schemes, each with its own solution, of which a large number will bear little resemblance to the exact solution of the original equations. This is because of instabilities which arise due to an incorrect discretization of space-time. Even if one is using a stable scheme, another major source of inaccuracy occurs in truncation errors. These latter errors stem from the fact that one is essentially approximating a function by a finite part of a Taylor series

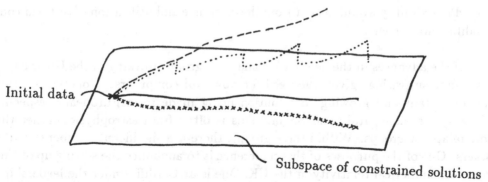

Initial data

Subspace of constrained solutions

Figure 6. Actual trajectory of solution (solid line); free evolution (dashes); chopped evolution (dots); fully constrained (crosses).

expansion. Other difficulties involve applying appropriate coordinate (or gauge) conditions, coordinate singularities and the boundary conditions associated with the use of a finite numerical grid. Then there is the problem of representing and interpreting the solution: in a 3-dimensional code what quantities should be computed and how should they be displayed? Then, as we have indicated before, numerical relativity makes enormous demands on computer time and memory, which produces limits on what is attainable at any one time.

Another issue is that certain formalisms involve long and complicated algebraic computations leading to lengthy expressions which often require conversion into a particular coding format — a process which could well introduce errors. In 1986, Nakamura used the computer algebra system Reduce to generate such algebraic expressions, and then exploited Reduce's ability to convert algebraic expressions into their Fortran equivalent, prior to numerical computation (Nakamura (1986)). Here, computers are being used for both algebraic and numeric work. Nakamura proposes as a name for this combined area CAR — Computer Aided Relativity.

Historically, relativists were originally distrustful of results emanating from computer algebra systems, because they were not convinced that the results were reliable. It was only after very complicated calculations had been checked successfully against each other using algebra systems based on different machines employing different software and design philosophies, that confidence was eventually established in the tool. A similar problem would seem to apply to numerical relativity. The one thing that you can virtually guarantee about a numerical calculation is that it will produce a result; but is the result correct? The field is still a young one, perhaps only some fifteen years old, and it is a small, albeit growing, one. Again, it would appear that the field's credibility would benefit from the existence and utilisation of test bed codes for calibrating systems.

Many of the advances in the field have come from groups working in the US or Japan. This is not surprising, given the need for powerful computers. Nonetheless, there is an important and growing contribution being made from Europeans, especially in France, Germany, Italy and Spain. This is often from astrophysics rather than solely relativity groups. Within these groups, there is a significant number of British workers. One of the purposes of this conference is to announce the setting up of three centres for numerical relativity in the UK. One is at Cardiff, under the leadership of Prof Bernie Schutz, and their work is 3+1 based with a particular interest in implicit numerical integration methods for moving grids (see Alcubierre and Allen in this volume). The second group is at Cambridge under the leadership of Dr John Stewart, where the work is more 2+2 based and has had considerable success, in particular, in modelling colliding gravitational waves and in dealing with caustics. The third group is centred at Southampton and consists of Prof Nigel Bishop, Andrew Garratt and myself under the leadership of Prof Chris Clarke. This group is again 2+2 based and mostly concerned with investigating the viability of using transputer arrays for numerical relativity. Most of the work has been carried out by Bishop, with some recent additions from Garrett, and it will be described in the remainder of this article.

6 TRANSPUTER ARRAYS

An important development in computing machinery in recent years has been the introduction of parallel computer systems. Several different computer architectures have been implemented, and the transputer array is particularly promising because of its flexibility. A single T8-20 transputer chip is a processing unit with a speed of about 1.5 Mflops. The difference between a transputer and other processors is that a transputer is designed to link up with other transputers to form a transputer array. The T8-20 chip has four bidirectional communication channels operating up to 20 Mbits per second, and one may join up the transputers as convenient. There is no shared memory across the array, and the data that a transputer works with are received and sent out along its communication channels. The number of transputers in an array varies up to a present practical limitation of about 1024. The development work at Southampton has been based on a Parsys supernode 32, consisting of a Sun 4.0 workstation for communication, a T414 host computer, up to 32 T8-20 worker transputers, linked through the host to a graphics processor and display (figure 7). The code has been developed in Occam, which is the most efficient parallel programming option, although other versions exist (Bishop et al (1990)).

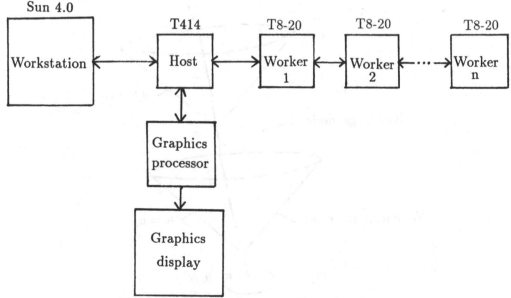

Figure 7. The Southampton Parsys node.

7 THE GEOMETRY OF THE PROBLEM

The approach adopted follows that of Winicour and Isaacson (Isaacson et al (1983), Gomez et al (1986)). We restrict attention to a class of problems which are sufficiently close to spherical symmetry that no caustics develop in the setup we now describe. We introduce Bondi-type coordinates based on a family of outgoing null cones emanating from a central geodesic (figure 8). The proper time, u, along the geodesic labels the null cones. The radial parameter, r, is the luminosity distance measured along null rays emanating from the geodesic. Finally, the angular coordinates θ, ϕ are introduced in the usual way. In the first phase of the development work we have also adopted the assumption of axial symmetry for simplicity. Writing $y = \cos\theta$, for later convenience, then in the coordinates (u, r, y, ϕ) the 4-dimensional line element becomes

$$ds^2 = hu\,du^2 + 2hr\,dudr + 2hy\,dudy + r^2 e^{2q} f^{-1} dy^2 + r^2 e^{-2q} f d\phi^2 \qquad (10)$$

where $f = (1 - y^2)$ and hu, hr, hy and q are all functions of u, r and y. We also allow for the presence of matter fields, but restrict attention to perfect fluids

$$T_{ab} = (\rho + p)w_a w_b - pg_{ab} \qquad (11)$$

with equation of state $p = p(\rho)$ and unit 4-velocity

$$w_a = (wu, wr, wy, 0) \qquad (12)$$

where wu, wr and wy are all functions of u, r and y, and $w_a w^a = 1$.

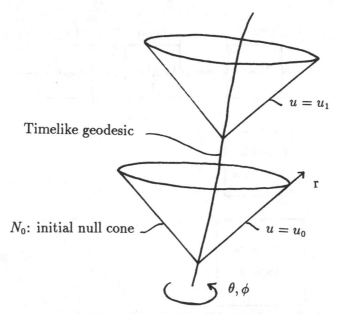

Figure 8. The Bondi-type coordinates.

8 THE CAR EMPLOYED

The computer algebra system Sheep was first employed to compute all the geometric quantities of interest. For example, figure 9 shows part of the output for the zero zero component of the Einstein tensor G_{ab}. Next, the author developed a Lisp program to translate the internal form of Sheep expressions into a form suitable for Occam coding. The program also chopped the output into manageable chunks for processing. An example of part of the resulting output is given in figure 10 (where real variables are held to double precission).

9 THE NUMERICAL ALGORITHM

We now outline the algorithm employed. If we define the quantity E_{ab} by

$$E_{ab} \equiv G_{ab} - \kappa T_{ab} \qquad (13)$$

then, the field equations break up into the following integration hierarchy

$$E_{11} = 0 \qquad (14a)$$

$$E_{12} = 0 \qquad (14b)$$

$$E_{01} = 0 \qquad (14c)$$

$$E_{22} - \frac{1}{2}g_{22}(g^{22}E_{22} + g^{33}E_{33}) = 0 \qquad (14d)$$

$$T_a{}^b{}_{;b} = 0.$$

```
           -2 -2q   2        -2 -2q        -2 -2q                  -2 -2q
G   = 2fhur e   (q )  -fhur e    q    +fr e   hu q   -1/2fr e   hu    +
 00            ,y        ,yy            ,y ,y                    ,yy


       -2 -2q          -2 -2q        -1      -2 -2q
 -2fr e   hy q   +fr e   hy   -4fhr huhyr e   q  q   +
      ,u ,y         ,uy                     ,r ,y


      -1       -2 -2q        -1       -3 -2q        -1      -2 -2q
 +2fhr huhyr e   q   +2fhr huhyr e   q   -2fhr hur e   hr q   +
           ,ry               ,y                   ,y ,y


      -1   -2 -2q        -1   -2 -2q          -1   -2 -2q
 +fhr hur e   hr   +2fhr hur e   hy q   +fhr hur e   hy q   +
          ,yy            ,r ,y                   ,y ,r


      -1   -2 -2q        -1   -3 -2q          -1   -2 -2q
 -fhr hur e   hy   -fhr hur e   hy   +2fhr hyr e   hr q   +
          ,ry            ,y                   ,u ,y


      -1   -2 -2q        -1   -2 -2q          -1   -2 -2q
 -fhr hyr e   hr   -fhr hyr e   hu q   -fhr hyr e   hu q   +
          ,uy            ,r ,y                   ,y ,r


      -1   -2 -2q        -1   -2 -2q          -1   -2 -2q
 +fhr hyr e   hu   +2fhr hyr e   hy q   -fhr hyr e   hy   +
          ,ry            ,u ,r                   ,ur


      -1 -2 -2q          -1 -2 -2q             -1 -2 -2q
 -fhr r e   hr hy   +1/2fhr r e   hr hu   +1/2fhr r e   hu hy   +
        ,u ,y                  ,y ,y                   ,r ,y
```

Figure 9. Sheep output of part of the zero zero component of the Einstein tensor.

```
chuc:=chuc+((-1.0000000E+00(REAL64))*(f*(hr*
(((1.0000000E+00(REAL64))/r)*(e2qa*(hy1*qa2)))))))
chuc:=chuc+((-1.0000000E+00(REAL64))*(f*(hr*
(((1.0000000E+00(REAL64))/r)*(e2qa*(hy2*qa1)))))))
chuc:=chuc+((5.0000000E-01(REAL64))*(f*(hr*
(((1.0000000E+00(REAL64))/r)*(e2qa*hy12)))))
chuc:=chuc+(f*(hr*(((1.0000000E+00(REAL64))/(r*r))*
(e2qa*hy2))))
chuc:=chuc+((-2.0000000E+00(REAL64))*(f*((hy*hy)*
(((1.0000000E+00(REAL64))/r)*(e2qa*(qa1*qa1)))))))
chuc:=chuc+(f*((hy*hy)*(((1.0000000E+00(REAL64))/r)*
(e2qa*qa11))))
chuc:=chuc+((2.0000000E+00(REAL64))*(f*((hy*hy)*
(((1.0000000E+00(REAL64))/(r*r))*(e2qa*qa1)))))
chuc:=chuc+(f*((hy*hy)*(((1.0000000E+00(REAL64))/(r*(r*
r)))*e2qa)))
chuc:=chuc+((5.0000000E-01(REAL64))*(f*(hy*
(((1.0000000E+00(REAL64))/r)*(e2qa*hr12)))))
chuc:=chuc+((2.0000000E+00(REAL64))*(f*(hy*
(((1.0000000E+00(REAL64))/r)*(e2qa*(hy1*qa1)))))))
chuc:=chuc+((-5.0000000E-01(REAL64))*(f*(hy*
(((1.0000000E+00(REAL64))/r)*(e2qa*hy11)))))
chuc:=chuc+((-1.0000000E+00(REAL64))*(f*(hy*
(((1.0000000E+00(REAL64))/(r*r))*(e2qa*hr2)))))
chuc:=chuc+((-1.0000000E+00(REAL64))*(f*(hy*
(((1.0000000E+00(REAL64))/(r*r))*(e2qa*hy1)))))
chuc:=chuc+((2.5000000E-01(REAL64))*(f*
(((1.0000000E+00(REAL64))/r)*(e2qa*(hr2*hr2)))))
chuc:=chuc+((-2.5000000E-01(REAL64))*(f*
(((1.0000000E+00(REAL64))/r)*(e2qa*(hy1*hy1)))))
chuc:=chuc+((-1.0000000E+00(REAL64))*(f*
(((1.0000000E+00(REAL64))/hr)*((hy*hy)*(((1.0000000E+00(REAL64))/
r)*(e2qa*(hr1*qa1)))))))
```

Figure 10. Part of the automatic translation into Occam code.

We choose a null cone, N_0, as an initial null cone, and prescribe as initial data q, ρ, wr and wy on N_0. Then the first three equations in the hierarchy reduce to linear ode's for hr, hy and hu along the null rays in N_0. Equation (14d) then determines $q_{,u}$ on N_0 and the three independent equations in (14e) serve to determine the u-derivatives of the remaining matter variables, namely $\rho_{,u}$, $wr_{,u}$ and $wy_{,u}$. Thus, the u-derivatives of the initial data are known on N_0, and hence the initial data are again known on the next neighbouring null hypersurface, N_1, say. We can now repeat the process on N_1 and continue in this way. Thus the hierarchy of equations constitutes an integration schema for determining the solution to the future of N_0.

It is assumed that the coordinates form a smooth Fermi system along the central geodesic, $r = 0$, which leads to the requirements

$$hu = hr = -1, \qquad hy_{,r} = q = 0 \qquad \text{on} \quad r = 0. \tag{15}$$

The coordinate patch used consists of

$$-1 \leq y \leq 1, \qquad 0 \leq r \leq 1, \qquad 0 \leq u \leq 2 \tag{16}$$

and the coordinate grid for (r, u, y) is $100 \times 500 \times 30$ with typical step sizes for (dr, du, dy) of $(0.01, 0.004, 0.07)$. The solution is arbitrarily cut off at $r = 1$ and a condition (probably representing no incoming radiation) is imposed there. In addition, regularity conditions at the endpoints require

$$q = hy = wy = hr + 1 = 0 \qquad \text{at} \quad y = \pm 1. \tag{17}$$

Numerical instabilities can occur near the origin, where $du > 0.5rdy$. This problem is overcome by fitting the values of the initial data in the region $0 \leq r \leq 0.1$, to an analytic expansion of these functions, by means of a least squares method of fit. Two main methods have been employed to solve the ode's namely, an Adams implicit 1-step and a predictor-corrector method.

There is a natural way to parallelize the algorithm. Each value of y in the grid is placed on a different processor, and the processors are linked up in a linear array. The host processor generates the initial data or reads it from a file. This data is then passed along the chain with each processor retaining the data needed for its particular value of y. Data are passed between processors, when required, to calculate y-derivatives. Also, data are passed along the whole chain to $y = +1$, and then back again, for the least squares calculation with the data near the origin. After a chosen number of iterations the results are displayed graphically or stored on disk.

10 RESULTS

A number of bench marks for the code were first produced. A scalar wave version was tested against an analytic solution and the code was found to be numerically stable, producing a maximum relative error of about 0.1%. Then the code was tested for Minkowski space-time and produced the required result. Two main versions of the code have been developed: one for the matter-free case, and the other for the perfect fluid case. The latter code has been checked against a perfect fluid interior Schwarzschild solution. It was found that error in the initial null cone was quite acceptable, but that stability problems develop as the system evolves in u. It was possible, however, to control these by choosing the step-length for du small enough. To date, most work has been carried out with arbitrarily constructed solutions in which arbitrary initial data is prescribed and the system is allowed to evolve. Examples of the resulting plots for one of the metric functions are given in figures 11 and figures 12.

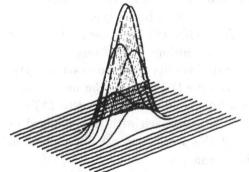

Figure 11. Initial configuration   *Figure 12. Subsequent evolution*
of the metric function q.   *of the metric function q.*

Some timings (given in seconds) for the two codes for a 500×100 (u, r) grid are presented below.

Number of processors	10	20	30
Vacuum	48	50	52
Perfect fluid	162	164	167

It is clear that although increasing the number of transputers involves a small communication overhead, this does not materially affect the overall time. If the number of r-points is increased, then the number of u-points must be increased proportionately. Thus the time is proportional to the square of the number of r-points; which has been confirmed empirically.

The work demonstrates that the idea of using a transputer array for numerical relativity is both a cheap and viable option. We can probably improve on the timings shown by removing a lot of redundancy in the code produced by the Sheep output and the Lisp-generated Occam code. We estimate that if we drop the assumption of axial symmetry, then a fully 3-dimensional relativistic code on an array of 1024 transputers with a $500 \times 100 \times 32 \times 32$ (u, r, y, ϕ) grid would run in about 30 minutes.

11 OTHER DEVELOPMENTS

Future developments planned include improving the efficiency of the code by removing redundancies in the calculation and testing different algorithms for solving ode's. At present the graphical display involves metric components only. We are in the process of providing displays of more physically meaningful quantities such as the Weyl scalars $\Psi_0, \Psi_1, \Psi_2, \Psi_3$ and Ψ_4. The Occam code has been checked in Fortran 77 on both an IBM 3090 mainframe and an Orion 1.5 minicomputer. It has also been implemented in parallel and vectorised Fortran 90 on a Suprenum supercomputer. Work has just finished in developing an 'initial data machine'. This can transform a line element given in an arbitrary coordinate system into Bondi-type coordinates, and extract the corresponding initial data on N_0. This will be a useful device for calibrating the system by test-bedding known exact solutions by evolving them from initial data and comparing the numerically generated results with the exact analytic quantities. Work is reported elsewhere in this volume (Bishop) on the use of compactified coordinates (which enables the use of a finite grid extending out to I^+) and the possibility of combining the 2+2 or characteristic aproach with the standard or Cauchy approach. However, the major thrust of this work it to use the system to evolve fully 3-dimensional physically interesting scenarios.

REFERENCES

Åman, J. E., d'Inverno, R. A., Joly, G. C. and MacCallum, M. A. H. (1985). *Lecture Notes in Computer Science*, **204**, 89–98.

Bishop, N. T., Clarke, C. J. S. and d'Inverno, R. A. (1990). *Class. Quantum Grav.*, **7**, L23–L27.

Corkhill, R. N. and Stewart, J. M. (1983). *Proc. Roy. Soc.*, **A386**, 373–91.

d'Inverno, R. A. and Stachel, J. (1978). *J. Math. Phys.*, **19**, 2447–60.

d'Inverno, R. A. and Smallwood, J. (1980). *Phys. Rev. D.*, **22**, 1233–47.

d'Inverno, R. A. and Frick, I. (1982). *Gen. Rel Grav.*, **14**, 835–63.

d'Inverno, R. A. (1984) in *Problems of collapse and numerical relativity*, D. Bancel and M. Signore (eds.), Reidel, 221-38.

d'Inverno, R. A. (1992). *Introducing Einstein's Relativity*, Oxford University Press.

Frick, I. (1977). The computer algebra system Sheep: What it can and cannot do in general relativity. Inst. of Theoretical Physics, University of Stockholm, report 77–15.

Gomez, R., Isaacson, R. A., Welling, J. S. and Winicour, J. (1986) in *Dynamical spacetimes and numerical relativity*, J. M. Centrella (ed.), Cambridge University Press, 236–55.

Isaacson, R. A., Welling, J. S. and Winicour, J. (1983). *J. Math. Phys.*, **24**, 1824–34.

Karlhede, A. (1980). *Gen. Rel. Grav.*, **12**, 693–707.

Kramer, D., Stephani, H., MacCallum, M. A. H. and Herlt, E. (1980). *Exact solutions of Einstein's field equations*, Cambridge University Press.

Nakamura, T. (1986) in *Proceedings of the 14-th Yamada conference on gravitational collapse and relativity*, H. Sata and T. Nakamura (eds.), World Scientific Publishing, Singapore, 295–312.

Smarr, L. (1977). *Ann. N. Y. Acad. Sci.*, **302**, 569–604.

Smarr, L. (ed.) (1979). *Sources of gravitational radiation*, Cambridge University Press.

SOME ASPECTS OF THE CHARACTERISTIC INITIAL VALUE PROBLEM IN NUMERICAL RELATIVITY

N. T. Bishop

Department of Mathematics, Applied Mathematics and Astronomy, University of South Africa, Pretoria, South Africa

Abstract. This paper is concerned with the axisymmetric characteristic initial value problem (CIVP). Tests on the accuracy and evolution stability of the code are described. The results compare reasonably well with expectations from numerical analysis. It is shown explicitly how to compactify CIVP coordinates so that a finite grid extends to future null infinity. We also investigate the feasibility of interfacing Cauchy algorithms in a central region with CIVP algorithms in the external vacuum.

1 INTRODUCTION

The construction of a new generation of gravitational wave detectors has important implications for numerical relativity. LIGO (Laser Interferometry Gravitational Observatory) is likely to detect gravitational waves from various astrophysical events within the next few years. Numerical relativity will be the main tool for interpreting such data, and will need to be able to calculate waveforms at infinity as accurately as possible. Much work on numerical relativity has been based on the standard $3 + 1$ Cauchy problem, where data is specified on a spacelike hypersurface and then evolved to the future. An alternative approach is the characteristic initial value problem (CIVP) based on a $2 + 2$ decomposition of space-time. It would seem that the CIVP is more appropriate in vacuum, but that it loses this advantage in the presence of matter, whose characteristics do not coincide with those of the gravitational field. Another consideration is the present state of development of numerical codes. For the Cauchy problem many codes incorporate matter, while CIVP codes are less well developed and mainly describe the gravitational field in vacuum.

Previous papers have described [1,2] the development of code for the axisymmetric CIVP. This paper reports on several extensions of that work. First, a summary of the problem is given in section 2 [3,4,13]. The gravitational field is completely described by one function q on an initial null hypersurface. The calculation proceeds in two stages: firstly within the hypersurface, and then q is evolved to the next null hypersurface.

In sections 3 and 4 the accuracy and evolution stability of the code are investigated in a formal way, firstly for the Schwarzschild solution and then for a constructed solution. Computed results are compared with the exact solution, and the behaviour of the error is then discussed in terms of the theoretical behaviour, according to numerical analysis, for the algorithms used.

Many physical situations have matter in a central region and vacuum outside. In these circumstances an interesting approach to numerical relativity may be to combine the CIVP and Cauchy problem, using Cauchy algorithms in the central region and CIVP algorithms outside it, with suitable matching conditions at the boundary (see Figure 1 in section 6). This paper provides a feasibility study of such a combination for the specific case of axial symmetry without rotation and CIVP coordinates based on null cones with vertices on the axis of symmetry [8]. This special case is used because it is the simplest case with interesting structure. The equations of Cauchy and CIVP numerical relativity are such that one would expect the qualitative conclusions obtained for this special case to apply in general non-symmetric circumstances [9].

In section 5 it is shown how to compactify spacetime so that a finite grid of CIVP coordinates extends to future null infinity. (Of course the existence of a suitable compactification is well known, but as will be seen, a numerical implementation is not a trivial matter.) This has two advantages: (1) an outer boundary condition, representing no radiation from past null infinity, is not required; and (2) the prediction for emitted radiation is at future null infinity, which is where it would be measured in a real situation. Section 6 outlines the practicalities of combining Cauchy algorithms in a central region, with CIVP algorithms outside. It is shown how to construct the CIVP coordinates and metric from the Cauchy data, and how CIVP calculations can supply the outer boundary condition for the Cauchy calculation.

2 SUMMARY OF THE CIVP

The numerical formulation of the CIVP of general relativity has been studied by several authors e.g. [1–4], [13], although it has not received as much attention as the Cauchy formulation. The approach and notation used here follow ref. 1 except that, for later convenience we have changed the signs of hu and hr. We summarise the situation for the case of axisymmetry without rotation in vacuum.

The coordinates are based on a family of outgoing null cones with vertices along a timelike geodesic G [1]. The proper time along G is u, and u is constant everywhere on a given null cone. The radial coordinate r is the luminosity distance from the null cone's vertex. Angular coordinates θ, ϕ are defined in the usual way near $r = 0$, and propagated outwards in the null cone by means of radial null geodesics. We assume axial symmetry and so the metric is independent of ϕ. It is convenient to make the substitution

$$y = -\cos(\theta) \tag{1}$$

and the metric is then

$$ds^2 = -hudu^2 - 2hrdu\,dr + r^2 F^{-1}e^{2q}dy^2 + r^2 Fe^{-2q}d\phi^2 \tag{2}$$

where hu, hr, hy and q are functions of u, r and y, and $F = (1-y^2)$. At the origin the coordinates must be a smooth Fermi system; this implies

$$hu = +1\,, \quad hr = +1\,, \quad hy = 0\,, \quad hy_{,r} = 0\,, q = 0\,. \tag{3}$$

Einstein's equations are

$$0 = E_{ab} = G_{ab} - kT_{ab} \tag{4}$$

where T_{ab} is the stress-energy tensor. Here, we are not concerned with the evolution of matter and we regard T_{ab} as either vanishing (i.e. vacuum), or otherwise as somehow given. The remarkable point about Einstein's equations is their simple form as a hierarchy of linear differential equations in r, with dependence on y as a parameter.

The required initial data is q, everywhere on a null cone $u = $ constant. Despite their structural simplicity, Einstein's equations are nevertheless very long and are not reproduced here. We will use $c1, c2, \ldots, c7$ to represent coefficients in the equations that depend on q and on metric variables that have been found earlier in the hierarchy. By "depend" is meant functional dependence on the variable and on its derivatives within the null cone, specifically the derivatives $r, rr, y, yy,$ and ry.

$$E_{11} = 0 \text{ is } 0 = c1\,hr + hr_{,r} \qquad \text{and leads to } hr. \tag{5}$$

$$E_{12} = 0 \text{ is } 0 = c2 + c3hy + c4hy_{,r} + hy_{,rr} \qquad \text{and leads to } hy. \tag{6}$$

$$E_{10} = 0 \text{ is } 0 = c5 + c6hu + hu_{,r} \qquad \text{and leads to } hu. \tag{7}$$

$$EV = E_{22} - \frac{1}{2}g_{22}(g^{22}E_{22} + g^{33}E_{33}) = 0 \text{ is } 0 = c7 + (rq_{,u})_{,r} \qquad \text{and leads to } q_{,u} \tag{8}$$

Initial conditions for the integration are provided by equation (3). Once $q_{,u}$ is found, q may be evolved to the "next" null cone, and the process repeated until the desired region of space-time has been covered.

The numerical procedures used are summarised as they have already been reported [1,2]. The first step is to set up a finite grid. The coordinate patch that will be used is

$$-1 \le y \le 1\,, \quad 0 \le r \le 1\,, \quad 0 \le u \le u_{max}\,. \tag{9}$$

The r-coordinate is cut off arbitrarily at $r = 1$ and a boundary condition on q will be imposed there. As always, the size of finite grid is a matter of balancing the available computer power with accuracy requirements and the smoothness of the initial data. Typical values are given in the following sections. The finite difference representation is straightforward. Derivatives are expressed by the usual central difference formula, and are expected to have second order accuracy. Therefore the equations (5 to 8) are integrated by means of a second-order method, specifically the Adams implicit 1-step [5], because the equations are linear it is possible to use an implicit method. At $y = \pm 1$ pseudo boundary conditions are required on q and hy, and are obtained from analytic expansions about the poles $q = hy = 0$. Where necessary, y derivatives at $y = \pm 1$ are calculated using the end-point formula. Two different algorithms, Adams-Bashforth and Predictor-Corrector, have been used for finding the evolution of q; they are described in section 4.

3 ACCURACY OF THE CIVP CODE

The accuracy of a code is tested against known analytic solutions. One of the simplest solutions is the Schwarzschild interior solution. However, since this solution is spherically symmetric it tests the code in a limited way, and in particular the evolution $(q_{,u})$ is identically zero. The static, constant density, Schwarzschild interior solution [6] is described by two parameters: mass m and radius a, which are taken as

$$m = 0.25 \qquad\qquad a = 1.0. \tag{10}$$

The gravitational field is therefore strong, as m/a is half the value required for a black hole. The metric is independent of y; and q, hy and dq/du vanish identically. The analytic values of the stress-energy tensor T_{ab} are put into Einstein's equations, and the code is used to calculate hr and hu, which have exact values in the range $+1$ at $r = 0$ to approximately $+2$ at $r = 1$. The errors in hr and hu increase from 0 at $r = 0$ to a maximum value at $r = 1$ of approximately (Δr^2); e.g. with 101 r-points, the errors at $r = 1$ are approximately 10^{-4}.

There are no known suitable analytic non-singular axisymmetric solutions. However, one can choose any metric and use Einstein's equations to construct the stress-energy tensor, and then the chosen metric is a solution. Of course, in general the stress-energy tensor is non-physical; but the "solution" is still quite suitable for code testing. The chosen metric is

$$q = FGA, \quad hr = +\exp(2FGA), \quad hy = 0, \quad hu = +\exp(2FGA) \tag{11}$$

where $F = (1 - y^2)$, and $GA = GA(r)$. Using the computer algebra system Sheep on this metric we find

$$G_{11} = -2F^2(GA_{,r})^2 + 4Fr^{-1}GA_{,r} \tag{12a}$$
$$G_{12} = -2FGAF_{,y}GA_{,r} + 2GAr^{-1}F_{,y} + F_{,y}GA_{,r} \tag{12b}$$
$$G_{01} = -F^2(GA_{,r})^2 - FGA^2r^{-2}(F_{,y})^2 + 2Fr^{-1}GA_{,r} + GAr^{-2}(F_{,y})^2 \tag{12c}$$
$$EV = -GA^2(F_{,y})^2 - GAF_{,yy} - r^2GA_{,rr} - 2GA_{,r} \tag{12d}$$

These values are used for the appropriate components of the stress-energy tensor T_{ab}. The function $GA(r)$ needs to be specified. We have taken

$$GA(r) = 0 \text{ if } r < 0.25 \tag{13a}$$
$$GA(r) = 0.1(r - 0.25)^3 \text{ if } r > 0.25. \tag{13b}$$

A boundary condition for d^2q/dr^2 is needed at $r = 1$ and here we have simply used the exact value $(= 0.45F)$.

The code for the problem (11) has been run for various grid-sizes, and the error in $\partial q/\partial u$ is shown in the L_2-norm; in all cases the error is maximum at the grid-point at $r = 1$, $y = 0$. The grid size is shown in terms of IMAX — the number of r-points, and JMAX — the number of y-points. Clearly, $\Delta r = 1/(\text{IMAX} - 1)$ and $\Delta y = 2/(\text{JMAX} - 1)$.

TABLE 1: Error in $\partial q/\partial u$ in the L_2 norm

	IMAX		
JMAX	51	101	201
8	1.1664E-5	4.9132E-6	3.0905E-6
16	9.6247E-6	3.0232 E-6	1.0752E-6
32	8.3568E-6	2.4989E-6	6.7694E-7
64	7.5097E-6	2.2783E-6	5.7955E-7
128	6.9898E-6	2.1659E-6	5.4441E-7

As discussed in section 3, we would expect the errors to be second order, so that approximately

$$\text{error} = a(\Delta r)^2 + b(\Delta y)^2 . \tag{14}$$

The results in Table 1 seem to fit equation (14) well, and this is confirmed by a least squares analysis. We consider the quantity

$$f_i = a(\Delta r_i)^c + b(\Delta y_i)^d - \text{error}_i . \tag{15}$$

Using the IMSL package for nonlinear least squares, UNLSF, the results in Table 1 were fitted to equation (15) with a, b, c, d as unknowns. It was found that $\sum_{i=1}^{15} f_i^2$ is minimised at

$$a = 0.0199 \qquad b = 0.0000661 \qquad c = 1.999 \qquad d = 2.10 \qquad (16)$$

Because b is so small, the value for d is not very accurate. Depending on the details of the calculation, values for d have been found in the range 1.99 to 2.23.

4 EVOLUTION STABILITY OF THE CIVP CODE

Calculations are performed on a grid in (r, y, u) with grid-points denoted by (i, j, k). Let $\bar{q}_k$ denote, at a fixed value of k or u, the matrix

$$\bar{q}_k = q_{ijk}(i \ : \ 0 \text{ to IMAX } -1, j \ : \ 0 \text{ to JMAX } -1). \qquad (17)$$

As described earlier, given $\bar{q}_k$ we can find $q0 = \partial q / \partial u$ at the grid-point i, j. We write

$$q0_{ij} = q0_{ij}(\bar{q}_k) \qquad (18)$$

The Adams-Bashforth algorithm (AB) is

$$q_{ij(k+1)} = q_{ijk} + \frac{du}{2}\left(3q0_{ij}(\bar{q}_k) - q0_{ij}(\bar{q}_{k-1}))\right) \qquad (19)$$

The Predictor-Corrector algorithm (PC) is

$$qp_{ij} = q_{ijk} + \frac{du}{2}\left(3q0_{ij}(\bar{q}_k) - q0_{ij}(\bar{q}_{k-1})\right)$$

$$q_{ij(k+1)} = q_{ijk} + \frac{du}{2}\left(q0_{ij}(\bar{q}p) - (q0_{ij}(\bar{q}_k))\right) \qquad (20)$$

In both cases, the first iteration is performed by means of Euler's method.

The criterion for stability is that the evolution is stable provided the solution is bounded after 500 iterations. The largest value of du for which the evolution is stable is found by a bisection method of two significant figures of accuracy. *E.g.*, if the largest value is given as 0.0021, this means that at $du = 0.0022$ overflow occurred, but at $du = 0.0021$ it did not. Results are given in Table 2. Of course, if more than 500 iterations are performed, the algorithm may become unstable for the value of du given in Table 2. However, for some cases we have evolved the equations stably to several thousand iterations for values of du at half that given in the Table.

It was found that stability was insensitive to the number of r-points (IMAX), perhaps affecting by 1 or 2 the value of the second significant figure. However, the results did depend significantly on JMAX and on $I1/IMAX$, which is the size of the region near the origin in which the solution is smoothed - see ref. [1].

TABLE 2: The maximum value of du for which the evolution was stable for 500 iterations. AB stands for the Adams-Bashforth algorithm, and PC for the Predictor-Corrector algorithm.

JMAX	16		32		64	
$I1/(\text{IMAX-1})$	IMAX= 51	IMAX= 101	IMAX= 51	IMAX= 101	IMAX= 51	IMAX= 101
0.1 AB	0.0058	0.0053	0.0012	0.0011	0.00029	0.00029
PC	0.017		0.0034		0.00080	
0.2 AB	0.010	0.010	0.0021	0.0020	0.00051	0.00050
PC	0.020		0.0069		0.0016	
0.3 AB	0.011	0.0098	0.0030	0.0028	0.00070	0.00068
PC			0.010		0.0023	

In most entries in Table 2, the PC value of du is about 3 times the AB value. Since the PC algorithm requires twice as much computation as the AB algorithm, we would find that, over a specified time period $u = 0$ to $u = u_1$, the PC algorithm would require about $1/1.5$ as many computations as the AB algorithm; $i.e.$, it would be about 33% faster.

The functional dependence of du on $dy(= 2/(\text{JMAX} - 1))$ and on $(I1/(\text{IMAX} - 1))$ is estimated by means of least squares fitting, using a formula of the form

$$ du = a \left(\frac{I1}{\text{IMAX} - 1} \right)^{b} dy^{c} \tag{21} $$

The calculation is performed as described in section 3, and for the case IMAX = 51 we find

$$ AB: \qquad du = 1.2 \left(\frac{I1}{\text{IMAX} - 1} \right)^{0.56} dy^{2.0} \tag{22} $$

$$ PC: \qquad du = 0.95 \left(\frac{I1}{\text{IMAX} - 1} \right)^{0.42} dy^{1.6} \tag{23} $$

In both cases the values found for a and b are sensitive to the precise data points used. However, one conclusion that comes out clearly, however the calculations are performed, is that the power of dy in (22) and (23) is always of order 2.

The power of dy in (22) and (23) is a problem in that it can make the choice of du much smaller than would be made on normal accuracy requirements, causing a lack of efficiency in the evolution code. The behaviour can be understood by observing that, in terms of u and y, the equation being solved is essentially

$$\frac{\partial q}{\partial u} = \frac{k\partial^2 q}{\partial y^2} \tag{24}$$

This parabolic equation is being approximated by an explicit finite difference scheme, and the stability condition is well known

$$k^2 du < \frac{1}{2}dy^2 \tag{25}$$

For equation (25), the problem is solved by changing to an implicit finite difference scheme such as Crank-Nicolson, and it may well be that some form of implicit scheme will alleviate matters in the CIVP code. This reasoning is supported by the fact that stability is largely independent of dr, and that the structure of the equations means that the algorithm is essentially implicit in the r direction. This problem has also been discussed by Gomez and Winicour [7].

5 THE CIVP AT INFINITY

In both the Cauchy and CIVP formulations we need an outer boundary condition representing no incoming gravitational radiation. Because of the nonlinearities in Einstein's equations, the boundary condition cannot be exact and consequently backscattering of gravitational radiation is ignored [10]. In the CIVP we expect $q \sim f(u)$ for large r, so that in a numerical computation δr can become large. Note that δR cannot be made large in the Cauchy formulation: the gravitational field variables $\sim f(T-R)$, so that both δT and δR must be much smaller than the wavelength of gravitational radiation (R and T are Cauchy radial and time coordinates).

We therefore investigate the compactification of the r coordinate within the context of the CIVP. An obvious and simple way to do this is to make the coordinate transformation $r \to z$ where [4]

$$r = \frac{z}{1-z}, \quad z = \frac{r}{1+r}. \tag{26}$$

Making this transformation in the metric (2) leads to Einstein equations that are strongly singular at $z = 1$, in that they involve terms $(1-z)^{-n}$ with n as large as 4. It is therefore clear that for progress to be made the metric will have to be written in another form. Inspired by the form of the Bondi metric at future null infinity [3,11] and after some trial and error, we write the metric as

$$ds^2 = -hunr^2e^{2qn}du^2 - 2\frac{e^{bn}r^2}{z^2}dudz + 2hynr^2e^{2qn}dudy$$

$$+r^2\left(\frac{e^{2qn}}{F}dy^2 + Fe^{-2qn}d\phi^2\right). \tag{27}$$

The coordinates are u, z, y and ϕ with $r = z/(1-z)$ and, as before, $F = (1-y^2)$. The metric variables hun, bn, hyn and qn are functions of u, z and y. The relationship between this metric and the metric (2) of section 2 is

$$qn = q, \quad bn = \ln(hr^{\frac{1}{2}}), \quad hyn = \frac{hy}{r^2e^{2qn}}, \quad hun = \frac{hu}{r^2e^{2qn}}. \tag{28}$$

The appropriate Einstein equations for the metric (27) were calculated using the computer algebra system REDUCE. The resulting equations appear to have some singularities at $z = 1$ in the form of terms with $(1-z)^{-1}$. However, we will show that all the solutions of the differential equations are, in fact, regular at $z = 1$. In order to do this we first note the behaviour, at $x = 0$, of the singular equation

$$x\frac{dy}{dx} + (g_0 + xg_1(x))y + (f_0 + xf_1(x)) = 0 \tag{29}$$

where f_0, g_0 are constant and $f_1(x)$, $g_1(x)$ are regular at $x = 0$. The solution $y(x)$ is regular at $x = 0$ provided $g_0 < 0$; in this case the solution is

$$y = -\frac{f_0}{g_0} + Cx^{-g_0} + x \times \text{ (regular function of } x) \tag{30}$$

In order to use the result (30) for the solution of Einstein's equations, we make the substitution $(1 - z) = x$ and note that $d/dz \rightarrow -d/dx$.

$E_{11} = 0$ for the metric (28) is

$$\frac{db}{dz} = \frac{z(1-z)}{2}\left(\frac{dqn}{dz}\right)^2. \tag{31}$$

Numerical integration at $z = 1$ is straightforward; and further, a boundary condition on dqn/dz at $z = 1$ is not required. $E_{12} = 0$ is of the form

$$\frac{dp}{dx} - \frac{2p}{x}[1 + xg_1(x)] - \frac{4e^{2(bn-qn)}}{x}\frac{dbn}{dy}[1 + xf_1(x)] = 0 \tag{32}$$

where $p = dhyn/dx$. The result (30) applies with $g_0 = -2$. Thus $dhyn/dz$ is regular at $z = 1$ and is

$$\frac{dhyn}{dx} = 2e^{2(bn-qn)}\frac{dbn}{dy}. \tag{33}$$

The value of hyn at $z = 1$ can then be found by standard numericals methods. $E_{10} = 0$ is

$$\frac{dhun}{dx} - \frac{3hun}{x}[1 + xg_1(x)] - \frac{3hyn^2}{x}[F(y)1 + xf_1(x)] = 0. \qquad (34)$$

The result (30) applies with $g_0 = -3$. Thus hun is regular at $z = 1$ and is

$$hun = -F(y)hyn^2. \qquad (35)$$

The expression $EV = 0$ (from equation (8)) can be written

$$-x^2\frac{d}{dx}\left(\frac{1-x}{x}qn_{,u}\right) - qn_{,y}F(y)hyn - \frac{1}{2}hyn_{,y}F(y) + xf_0 + x^2f_1(x) = 0. \qquad (36)$$

Integrating

$$qn_{,u} = F(y)(qn_{,y}hyn + \frac{1}{2}hyn_{,y}) + \frac{x}{1-x}f_0\log_e x + xf_2(x). \qquad (37)$$

Thus, $qn_{,u}$ is continuous at $z = 1$ and is

$$qn_{,u} = F(y)(qn_{,y}hyn + \frac{1}{2}hyn_{,y}). \qquad (38)$$

There are two comments to make about the above results. Firstly, it has been shown explicitly that $qn_{,u}$ has the value given by equation (38) at $z = 1$ and that $qn_{,u}$ is continuous there. The condition of asymptotic flatness implies that qn is differentiable at $z = 1$ and does not have a dependence of the form $(1-z)\log_e(1-z)$. Thus $f_0 = 0$, and numerical values of qn are expected to vary smoothly near $z = 1$.

The second comment is that although the Einstein equations for the metric (27) are well-behaved at $z = 1$, in their present form they are not well-behaved at $z = 0$. However, there is no intrinsic singularity and the problem is simply one of using suitable metric variables. One possibility is to use the Einstein equations (5) to (8) with the metric (2) for a region $0 \le r \le r_+$. Let $z = z_+$ at $r = r_+$. Then at $z = z_+$, the metric coefficients qn, bn, hyn, hun are calculated from equations (28) and the Einstein equations for the metric (27) are then used to calculate the metric in the region $z = z_+$ to $z = 1$.

6 THE BOUNDARY BETWEEN CAUCHY AND CIVP

Here we examine the feasibility of evolving the gravitational field by means of Cauchy algorithms within some central region and CIVP algorithms in the external region to future null infinity. There are many aspects to this problem. Here we discuss two of the issues that have to be addressed: (1) the construction of the CIVP coordinates and metric from Cauchy data; and (2) using the results of CIVP calculations to obtain the outer boundary condition at $R = R_+$ for the Cauchy calculation.

The formulation of the Cauchy problem is well-known *e.g.* [12], and it will be useful to summarise here the notation that will be needed. The coordinates are $X^\alpha = (T, R, \Theta, \Phi)$. The metric is denoted by $H_{\alpha\beta}$: N is the lapse function, $N^i (i = 1, 2, 3)$ is the shift vector; A and B are components of the spacelike part of the metric defined by

$$A^2 dR^2 + R^2(B^{-2}d\Theta^2 + B^2\sin^2(\Theta)d\Phi^2) \tag{39}$$

K_+ is a component of the extrinsic curvature. The Cauchy calculation is carried out in a domain $R < R_+$, and it is assumed that matter is present only in a central region $0 \leq R < R_m$ with $R_m < R_+$.

6.1 Construction of the CIVP coordinates
The construction that will be described here is based upon the CIVP coordinates of sections 2 and 5, and coordinate lines are null geodesics starting from $R = 0$. It may be preferable to construct CIVP coordinates in another way, for example from a 2-surface $T = $ constant just inside $R = R_+$ [14]. Another possibility is to start with a Bondi sphere at $r = \infty$, $z = 1$ and to construct null geodesics going inwards and backwards in time. However, the practicalities of such approaches have yet to be explored.

The numerical implementations of the scheme described here suffers from a startup difficulty. In order to calculate the null geodesics from $R = 0$ to $R = R_+$, we need Cauchy data not only on the initial hypersurface $T = 0$, but in the hypersurfaces between $T = 0$ and $T = T_p$ (say). This could be obtained either by imposing an *ad hoc* boundary condition at $R = R_+$, or preferably by having Cauchy data that is static between $T = 0$ and $T = T_p$, with the onset of a collapse modelled by a change in the equations of state at $T = T_p$. The first step in calculating null geodesics is to evaluate the components of the metric connection $\Gamma^\alpha_{\beta\gamma}$ at the Cauchy grid points, by applying the central difference formula. Of course, we will need to know $\Gamma^\alpha_{\beta\gamma}$ at points not on the grid and this is done by interpolation of data at the surrounding grid points. The geodesics are calculated from the geodesic equation $(d^2x^\alpha/d\lambda^2 + \Gamma^\alpha_{\beta\gamma}(dx^\beta/d\lambda)(dx^\gamma/d\lambda))$ and with initial conditions

$$R = 0, \quad \Theta = \Theta_a, \quad T = T_a, \quad \frac{dR}{d\lambda} = 1, \quad \frac{d\Theta}{d\lambda} = 0, \quad \frac{dT}{d\lambda} = N^{-1}. \tag{40}$$

The values chosen for Θ_a and T_a depend upon the nature of the required CIVP grid. The initial direction of these geodesics is null, as required. The geodesics are used to define CIVP-type coordinates

$$u = \int_0^{T_a} N(T,0)dT, \quad \bar{r} = \lambda, \quad y = -\cos\Theta_a, \quad \phi = \Phi. \tag{41}$$

These are not quite the CIVP coordinates of section 2, because the radial coordinate $\bar{r}$ represents affine distance rather than luminosity distance; the appropriate adjustment will be made in due course. For the moment we set up a uniform grid in $R < R_+$ based on the coordinates (41). Then at any grid point (CIVP or Cauchy) we may calculate by finite differences

$$\frac{\partial X^a}{\partial \bar{x}^\beta} \quad \text{and} \quad \frac{\partial \bar{x}^a}{\partial X^\beta} \tag{42}$$

so that vectors and tensors may be transformed from one coordinate system to the other. In particular the metric tensor $\bar{g}_{\alpha\beta}$ may be found at all CIVP grid points. The transformation to the CIVP coordinates of section 2 can now be made. Let

$$r = \sqrt[4]{\bar{g}_{22}\bar{g}_{33}} \quad \text{and} \quad q = \frac{1}{4}\log_e(\bar{g}_{22}(1-y^2)^2/\bar{g}_{33}). \tag{43}$$

Then,

$$hu = \bar{g}_{00} \quad hr = \left(\frac{\partial\bar{r}}{\partial r}\right)\bar{g}_{01} \quad \text{and} \quad hy = \bar{g}_{02} \tag{44}$$

where $(\partial\bar{r}/\partial r)$ is estimated by finite differences. In this way the CIVP coordinates and metric of section 2 are found in the region $R < R_+$. If desired they can be transformed to the form used in section 5 by means of equations (26) and (28).

In the region $R > R_+$, we need as initial data q or qn everywhere on the null cone $u = 0$. The other metric functions are found in this null cone by solving the CIVP equations in the range $R > R_+$. (Of course, in general $R = R_+$ is not a surface of constant r, but that does not affect matters.) The initial conditions for the CIVP equations are the values of q, hr, hy and hu at or near $R = R_+$ found by transformation from Cauchy metric. The evolution equation is then used to find q or qn on the "next" null cone in $R \lesssim R_+$.

6.2 Outer boundary condition for Cauchy problem

The procedure proposed here is iterative. The first estimate for the values of variables at the outer boundary comes from a currently-used formulation of the condition of no incoming radiation. The values are then refined by one (or more) applications of the following procedure. The procedure will be described only for the outer grid-point, but it should be noted that it could be used to correct values at, say, $R = R_+$, $R_+ - \delta R$ and $R_+ - 2\delta R$ and that this could be advantageous for achieving a smooth transition from Cauchy to CIVP evolution.

Refering to Figure 1, the problem is to refine the Cauchy variables B and K_+ at Q. First knowing the Cauchy coordinates of Q we find the CIVP coordinates of Q. This is calculated by linear interpolation of the Cauchy and CIVP coordinates of the points P_i — the CIVP grid points surrounding Q. Next, also by linear interpolation of data at the points P_i, we find the CIVP metric $g_{\alpha\beta}$ at Q. We transform $g_{\alpha\beta}$ to $H_{\alpha\beta}$ at Q by

$$H_{\alpha\beta} = \frac{\partial x^\gamma}{\partial X^\alpha}\frac{\partial x^\delta}{\partial X^\beta}g_{\gamma\delta}. \tag{45}$$

In order to do this we need values for $\partial x^\gamma/\partial X^\alpha$ at Q; these are found by finite differencing, although for some values the end-point formula, rather than the central difference formula, will have to be used. The value of B is found explicitly and that of K_+ is obtained by solving one of the constraint equations at Q : this is straightforward as, numerically, the problem is one (non-linear) equation in one unknown.

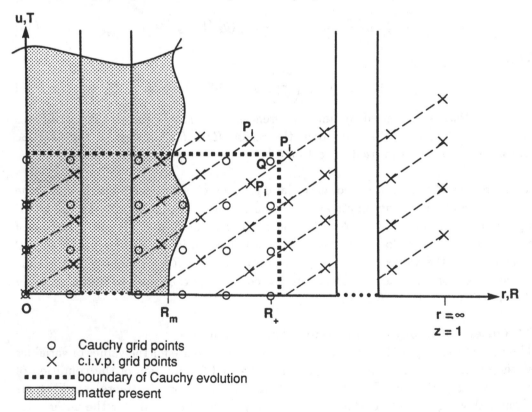

Figure 1. Schematic view of the interface between the Cauchy and CIVP coordinate grids. The points P_i are CIVP grid points and Q is on the outer boundary of the Cauchy grid.

ACKNOWLEDGEMENTS
Some of this work was carried out at the University of Southampton, and I thank
C.J.S. Clarke, R.A. d'Inverno and A.C.W. Garratt for their collaboration. I thank
the FRD for financial support.

REFERENCES

[1] Bishop, N. T., Clarke, C. J. S. and d'Inverno, R. A. (1990). *Class. Quant. Grav. Lett.*, **7**, L23–L27.

[2] Bishop, N. T., Clarke, C. J. S. and d'Inverno, R. A. (1990). *S.A. Jnl. Sci.*, **86**, 64–65.

[3] Isaacson, R. A., Welling, J. S. and Winicour, J. (1983). *J. Math. Phys.*, **24**, 1824–1834.

[4] Gomez, R., Isaacson, R. A., Welling, J. S. and Winicour J. (1986). In *Dynamical spacetimes and numerical relativity*, J.M. Centrella (ed.), Cambridge University Press, Cambridge.

[5] Hultquist, P. F. (1988). *Numerical methods*, Benjamin/Cummings, California, 246.

[6] Misner, C. W., Thorne, K. S. and Wheeler, J. A. (1973). *Gravitation*, Freeman, San Francisco, 608–611.

[7] Gomez, R. and Winicour, J. (1989). In *Frontiers in numerical relativity*, C.R. Evans, L.S. Finn and D.W. Hobill (eds.), Cambridge University Press, Cambridge.

[8] Bishop, N. T. (1991). In *Proceedings of the sixth Marcel Grossmann meeting on general relativity*, World Scientific Press, Singapore.

[9] Sachs, R. K. (1962). *Proc. Roy. Soc. A*, **270**, 103–126.

[10] Abrahams, A. M. (1989). In *Frontiers in numerical relativity*, C.R. Evans, L.S. Finn and D.W. Hobill (eds.), Cambridge University Press, Cambridge.

[11] Bondi, H., van der Burg, M. G. J. and Metzner, A. W. K. (1962). *Proc. Roy. Soc. A*, **269**, 21–52.

[12] Bardeen, J. M. and Piran, T. *Phys. Rep*, **96**, 205–250.

[13] Stewart, J. M. and Friedrich, H. (1982). *Proc. Roy. Soc. A*, **384**, 427–454.

[14] Clarke, C. J. S. (1991). Private communication.

THE CHARACTERISTIC INITIAL VALUE PROBLEM IN GENERAL RELATIVITY

J. M. Stewart

Department of Applied Mathematics and Theoretical Physics, University of Cambridge, Cambridge, UK

Abstract. The difficulties in solving the characteristic initial value problem in general relativity when matter fields are included are discussed. A scheme is proposed with new dependent variables, coordinates and tetrad which should alleviate some of the problems.

1 THE CHARACTERISTIC INITIAL VALUE PROBLEM

Almost all calculations in numerical relativity are based on the Arnowitt-Deser-Misner (1962) formalism in which spacetime is foliated by spacelike hypersurfaces. The Einstein equations are decomposed into elliptic constraint equations which are intrinsic to the slices, and hyperbolic evolution equations which govern the evolution from slice to slice. In principle the constraint equations have only to be solved on the initial slice, but even this requires considerable effort and a large computer.

Stewart and collaborators, Friedrich and Stewart (1982), Corkill and Stewart (1983), developed an alternative approach based on a fundamentally different principle. In this spacetime was foliated by null hypersurfaces, and originally two families of null hypersurfaces were used. A helpful analogy is to consider the Schwarzschild solution described by double null coordinates (u, v, θ, φ). Examination of that solution will reveal that such coordinates have many theoretical advantages. One can explore the spacetime through the regular event horizon right up to the singularity at $r = 0$. Further one can proceed out along surfaces $u = const.$ to future null infinity, the area inhabited by distant observers. Indeed one can bring infinity in to a finite point by the technique of *conformal compactification*, due originally to Penrose. (See e.g., Penrose and Rindler (1984).) One rescales the "physical" 4-metric g_{ab} to obtain an "unphysical" metric $\hat{g}_{ab}$ via

$$\hat{g}_{ab} = \Omega^2 g_{ab}.$$

Here Ω is a positive function which is $O(r^{-1})$ as $r \to \infty$ along $u = const.$ This ensures that the components of $\hat{g}$ are regular at future null infinity. The Ricci tensor transforms inhomogeneously so that a physical vacuum spacetime contains an artificial fluid after rescaling. Friedrich (1983) showed how to construct a system of equations which was regular at null infinity. Thus for the first time one could incorporate infinity into a finite grid with regular equations. A final theoretical advantage was that by reversing time one can do cosmology. The data for that subject is given on our past light cone.

Using null surfaces has a profound practical advantage — the elliptic constraints disappear! In fact this is not quite true. Once one decomposes the equations with respect to a tetrad/coordinate system there are still some elliptic equations, but they occur only within 2-surfaces. Provided one chooses initial data consistent with them (or solves them initially), no further elliptic equations have to be solved. The initial data are the matter and radiation fluxes transverse to the initial hypersurfaces, and may be chosen freely. This seems to have been realized within relativity first by Sachs (1962), but is actually a property of the characteristic initial value problem, which has been exploited in a series of papers by Friedrich and is reviewed in Stewart (1991).

This absence of constraints opens up the possibility of doing numerical relativity on workstations. For example data in the near zone of an axisymmetric radiating system can be evolved out to infinity on a $50 \times 50 \times 50$-grid in about 20 minutes using 2 Mbytes of memory on a Sun 3/60 machine, Stewart (1989).

There are however two grave disadvantages to this approach which have made it unpopular. The first is the presence of caustics. The null hypersurfaces are ruled by null geodesics or light rays. As is well-known gravity can bend light and so refocus initially diverging light rays. This caused many-valuedness and hence singularities in coordinate systems based on light rays. A detailed study of caustics in Minkowski spacetime was made in Friedrich and Stewart (1983), Stewart (1991) and algorithms to evolve them were described in Corkill and Stewart (1983) and Stewart (1986). The second is that although the vacuum and conformal vacuum problems appear straightforward it seems very difficult to include matter, e.g., a perfect fluid. In this paper we show how this can be done, by changing both the set of dependent variables and by changing the foliation used.

2 THE DEPENDENT VARIABLES

Although it is certainly possible to proceed without using a tetrad, e.g., Winicour (this volume), it is usually most convenient to introduce a Newman Penrose (NP) null tetrad, Newman and Penrose (1962), Penrose and Rindler (1984), Stewart (1991). One chooses a set of four null vectors, two real, two complex (and conjugates of each other). By choosing coordinates adapted to the tetrad, one can write the tetrad components in terms of six spacetime functions. This information is fully equivalent to the metric tensor. Unlike the $3 + 1$ approach where the gauge conditions, specification of the lapse and shift, are determined dynamically usually by imposing certain differential conditions, here the gauge is imposed algebraically and ab initio. A seventh scalar function is the conformal factor if compactification is being used. The connection is determined by the tetrad components which are complex scalar functions or "spin coefficients". The defining relations are turned into first order equations for the frame coefficients. Similarly the expressions for the curvature tensor components in terms of the connection are turned into first order equations for the spin coefficients. The chain is closed by utilizing the Bianchi identities as first order equations for the curvature components. Proceeding in this way one can build a first order system which is symmetric hyperbolic for the vacuum or conformal vacuum problem Friedrich (1983). Unfortunately when physical matter is present it does not seem to be possible to implement this scheme, except in very special cases, e.g., a massless scalar field.

One way out of this problem which seems to offer promise is to restrict the range of dependent variables. One can remove from this set the five complex scalars giving the Weyl tensor components, the Ψ_n in NP notation. The NP "field equations" giving the curvature components in terms of the connection components need modification. In the usual form, Penrose and Rindler (1984), Stewart (1991), there is precisely one equation containing Ψ_0, and one containing Ψ_4. These are discarded. There are two equations containing a term Ψ_1 and by subtracting these an equation without this term is found. The same applies to Ψ_3. Finally there are four equations containing Ψ_2 which can be reduced to three equations without this term. The five discarded equations can always be used later to generate the Ψ_n if desired. Next the Bianchi identities are discarded. The remaining Ricci tensor components in the "field equations" are replaced by energy-momentum tensor components, via Einstein's field equations. Finally the conservation law $T^{ab}{}_{;b}$, together with equations of state is used to generate evolution equations for the matter variables. For a perfect fluid (at least) and with a suitable choice of frame one has a usable evolution scheme.

3 THE CHOICE OF COORDINATES AND FRAME

In the original scheme with two families of null hypersurfaces the one corresponding to "ingoing" light rays invariably developed caustics as the evolution developed. This was an inconvenience rather than a serious problem in the vacuum case, for the generic form of caustics was known, Friedrich and Stewart (1983), and numerical techniques for handling them were developed, Corkill and Stewart (1983), Stewart (1986). Once matter is included the analytic form of caustics becomes much more complicated, and the original numerical techniques do not work as well. It seems worthwhile to move to a different coordinate system to try to minimise their effect.

The current system uses polar coordinates (r, θ, φ) to describe the spatial dependencies. The origin of coordinates $r = 0$ is taken to be a timelike geodesic with affine parameter u. At each point the future light cone is constructed, and this is labelled $u = const.$, giving the first coordinate. There is an S^2's worth of generators for each cone. We may label these by polar angles θ, φ, and then take r to be an affine parameter along each generator. Such a coordinate system is well-defined in the neighbourhood of the central worldline.

We may attach a NP null tetrad to the same construction. We choose the covector l_a to be normal to the cones $u = const.$ This means that the vector l^a is tangent to the generators and we may normalize it by setting $l = \partial/\partial r$. On each cone the hypersurfaces $r = const$, are topologically 2-spheres and it is convenient to choose the complex null vectors m^a and $\overline{m}^a$ to span the corresponding tangent space. This means that $m = P^A \partial/\partial x^A$, where A ranges over $2, 3$, and the P^A are complex functions. There would appear to be four (real) degrees of freedom involved here. However the normalization condition $m^a \overline{m}_a = -1$ reduces the number to three. The final real component of the tetrad cannot be aligned neatly to the geometry. We set $n = \partial/\partial u + Q \partial/\partial r + C^A \partial/\partial x^A$, where the only simplification results because $l_a n^a = 1$. This introduces three (real) degrees of freedom, and so the tetrad can be described by six real functions. Unlike other formalisms we use up the available freedom in the metric tensor algebraically and ab initio.

Consistent with the above construction and the weak equivalence principle, we may demand that the coordinate/tetrad system be asymptotically Minkowskian at the central worldline. This enforces the asymptotic conditions

$$Q \sim -\tfrac{1}{2}, \qquad P^A \sim P_o^A = 2^{-1/2}r^{-1}(1, \, i\operatorname{cosec}\theta), \qquad C^A = O(r), \qquad \text{as } r \to 0.$$

Thus although there is a coordinate singularity at $r = 0$ it is the "usual" polar coordinate singularity and is well-understood.

In particular there is a significant feature, implicit in Friedrich (1979), Ellis et al (1985), in that the singular NP quantities deviate from their flat space values only by (multiplicative) terms which are $O(r^2)$. Thus

$$P^A = P_o^A(1 + O(r^2)), \qquad \rho = -r^{-1}(1 + O(r^2)), \qquad \mu = -\tfrac{1}{2}r^{-1}(1 + O(r^2)).$$

This means that the r, θ and φ dependence of all the remaining regular quantities cannot be arbitrary. If they are decomposed into linear combinations of spherical harmonics then the r-dependence of the coefficients is constrained. See e.g., Bonazzola and Marck (1989), Nakamura and Oohara (1989). Further one can regularize the equations, i.e., construct near-regular (in a sense to be described) equations for regular quantities e.g., $\rho_1 = \rho + r^{-1}$.

4 EVOLUTION STRATEGY

The evolution of the geometry is determined by one complex scalar function σ, the shear of the cone generators. For all geometry variables except σ there are radial evolution equations. This means that they may be obtained by integrating an initial value problem along the generators, starting from Minkowskian values at the vertex. For σ we have a cone-to-cone evolution equation. We evolve the matter using the conservation equations $T^{ab}{}_{;b} = 0$ which may be decomposed into equations governing the cone-to-cone evolution of the fluid variables.

We need to specify σ and the matter variables freely, i.e., without constraints on the initial cone as data. Using these values we may integrate the radial equations for the rest of the geometry using Minkowskian data at the vertex. Once this is done we may compute the u-derivatives of σ and the matter variables and evolve them to the next cone. Using these values plus the Minkowskian data at the vertex we now integrate radially the equations for the remaining variables on the new cone. By repeating this process we gradually evolve the spacetime. Let us examine some of the features in more detail.

The evolution of σ can proceed either along lines of constant r, θ and φ or along the integral curves of n^a. The matter can be evolved either along one of these directions or along the flow lines (integral curves of the 4-velocity). It is not yet clear what the optimal choice is.

The radial evolution equations have to be regularized. Consider e.g.,

$$\frac{d\rho}{dr} = \rho^2 + \sigma\bar\sigma + \Phi_{00},$$

where the last term is a matter variable (part of the Ricci tensor). Setting $\rho = -r^{-1} + \rho_1$ we obtain

$$\rho_1 = \frac{1}{r^2} \int_0^r s^2(\rho_1{}^2 + \sigma\bar\sigma + \Phi_{00})\,ds,$$

an integral equation for ρ_1 which can be solved by iteration. It should be noted however that simple trapezoidal algorithms provide insufficient accuracy near $r = 0$.

If we set $r_c = r_b + h$, where h is the steplength then

$$\rho_{1c} = \frac{1}{r_c{}^2}\left[r_b{}^2\rho_{1b} + \int_{r_b}^{r_c} s^2 f(s)\,ds\right],$$

where f is an abbreviation for $\rho_1{}^2 + \sigma\bar{\sigma} + \Phi_{00}$. A suitable integration algorithm is

$$\rho_{1c} \approx \lambda^2\rho_{1b} + \frac{h}{12}\left[(1 + 2\lambda + 3\lambda^2)f_b + (3 + 2\lambda + \lambda^2)f_c\right],$$

where $\lambda = r_b/r_c$. For $f(s)$ is a linear function then this algorithm is exact for all values of r. For large values of r we have $\lambda \approx 1$ and it reduces to the trapezoidal rule

$$\rho_{1c} \approx \rho_{1b} + \tfrac{1}{2}h(f_b + f_c).$$

However this better-known simpler algorithm is wildly inaccurate for small r. This is in fact the most difficult radial evolution equation, for all the others are linear.

Considerable care is needed also in the computation of angular (θ, φ) derivatives. E.g., for small r

$$\sigma = \sigma_1(u)\,_2Y_{2m}(\theta, \varphi)r + O(r^2),$$

where $_2Y_{2m}(\theta, \varphi)$ is a spin-weighted spherical harmonic. However

$$\sigma_{,u} = [\Delta\,_2Y_{lm} + \ldots]r^0 + O(r),$$

where the coefficient multiplying r^0 vanishes analytically because of properties of spherical harmonics. The numerical algorithm must emulate this accurately or else the accuracy of the code decreases dramatically leading to a numerical instability. The current code is axisymmetric and computes θ-derivatives by a pseudo-spectral (FFT) technique.

Care is needed with the choice of data. Although the choice is arbitrary not all smooth data correspond to an initially regular spacetime.

Further progress will be reported elsewhere. Workstation facilities were provided by the SERC.

REFERENCES

Arnowitt, R., Deser, S. and Misner, C. W. (1962). In *Gravitation: an introduction to current research*, L. Witten (ed.), Wiley, New York.

Bonazzola, S. and Marck, J-A. (1989). In *Frontiers in numerical relativity*, C. R. Evans, L. S. Finn and D. W. Hobill (eds.), Cambridge University Press, Cambridge.

Corkill, R. W. and Stewart, J. M. (1983). *Proc. Roy. Soc. Lond.* A, **386**, 373–391.

Ellis, G. F. R., Nel, S. D., Maartens, R., Stoeger, W. R. and Whitman, A. P. (1985). *Phys. Rep.*, **124**, 315–417.

Friedrich, H. (1979). Ph.D. Thesis, Hamburg.

Friedrich, H. (1983). *Commun. Math. Phys.*, **100**, 525–543.

Friedrich, H. and Stewart, J. M. (1983). *Proc. Roy. Soc. Lond. A*, **385**, 345–371.

Nakamura, T. and Oohara, K. (1989). In *Frontiers in numerical relativity*, C. R. Evans, L. S. Finn and D. W. Hobill (eds.), Cambridge University Press, Cambridge.

Newman, E. T. and Penrose, R. (1962). *J. Math. Phys.*, **3**, 566–578.

Penrose, R. and Rindler, W. (1984). *Spinors and space-time 1: Two-spinor calculus and relativistic fields*, Cambridge University Press, Cambridge.

Sachs, R. K. (1962). *J. Math. Phys.*, **3**, 908–914.

Stewart, J. M. (1986). In *Astrophysical Radiation Hydrodynamics*, K-H. A. Winkler and M. L. Norman (eds.), 531–568, Reidel, Dordrecht.

Stewart, J. M. (1989). *Proc. Roy. Soc. Lond. A*, **424**, 211–222.

Stewart, J. M. (1991). *Advanced general relativity*, Cambridge University Press, Cambridge.

Stewart, J. M. and Friedrich, H. (1982). *Proc. Roy. Soc. Lond. A*, **384**, 427–454.

ALGEBRAIC APPROACH TO THE CHARACTERISTIC INITIAL VALUE PROBLEM IN GENERAL RELATIVITY

Jörg Frauendiener

Max Planck Institute for Astrophysics, Munich, Germany

Abstract. Penrose has described a method for computing a solution for the characteristic initial value problem for the spin-2 equation for the Weyl spinor. This method uses the spinorial properties in an essential way. From the symmetrized derivatives of the Weyl spinor which are known from the null datum on a cone one can compute all the derivatives by using the field equation and thus one is able to write down a power series expansion for a solution of the equation. A recursive algorithm for computing the higher terms in the power series is presented and the possibility of its implementation on a computer is discussed.

1 INTRODUCTION

Due to the nonlinear nature of general relativity it is very difficult to obtain exact solutions of the field equations that are in addition of at least some physical significance. Prominent examples are the Schwarzschild, Kerr and Friedmann solutions. Given a concrete physical problem it is more often than not rather hopeless to try to solve the equations using analytical techniques only. Therefore, in recent years, attention has turned towards the methods of numerical relativity where one can hope to obtain answers to concrete questions in a reasonable amount of time given enough powerful machines. However, it is still a formidable task to obtain a reliable code. There is first of all the inherent complexity of the field equations themselves when written out in full without the imposition of symmetries or other simplifying assumptions. In the "3+1" formulation one has the problem that the equations are separated into "constraints" and evolution equations. Therefore, one has to solve a set of elliptic equations to obtain the initial data for the subsequent evolution. An additional problem is to find an appropriate choice of coordinates in which the metric is expressed.

Once the metric is computed one needs to know what it means. One would like to determine some invariant properties of the space-time like its geodesic structure or the type of the gravitational field (the Weyl curvature) etc. But this means that one has to either evolve appropriate tetrad, connection and curvature components together with the metric components or obtain those quantities from the calculated metric functions by numerical differentiation.

In his landmark paper (Penrose 1960) Penrose applied spinor methods to general relativity and suggested a method for computing an analytic solution of Einstein's field equation in an essentially invariant way. The idea is to consider the second Bianchi identity as the basic field equation to be solved for the Weyl spinor. The (derivative of the) energy-momentum tensor acts as the source for the gravitational field via insertion of Einstein's equation $G_{ab} = 8\pi T_{ab}$ into the Bianchi identity. Penrose's method is algebraic in nature and none of the above mentioned problems appear in that approach. The curvature is computed in a coordinate independent way which makes it easy to determine its Petrov type etc. Only after the curvature is obtained from this procedure does one introduce coordinates in order to express the metric.

However, in this method as well, there are severe problems to overcome and at the moment it is not at all clear whether it is a practicable one. In this work we want to describe Penrose's approach and then reformulate it in such a way that it seems possible to implement it on a computer using one of the current algebraic manipulation programs like Maple, Reduce or Mathematica. In section 2 we will give a brief outline of Penrose's method. In section 3 we formulate the method using a formalism developed by Sparling (Sparling 1984) that is based on a well known correspondence between spinor fields on space-time and homogeneous functions on the spin bundle over space-time. We will give here only a formal description without too much mathematical rigour. An exact mathematical formulation will be given elsewhere.

2 DESCRIPTION OF PENROSE'S METHOD

The algebraic approach to the characteristic initial value problem has been formulated by Penrose in terms of "exact sets of spinor fields" on space-time. Let us briefly review what this means. Let (M, g_{ab}) be a 4-dimensional manifold with spin-structure. We will always be working in a local context, so there is no restriction imposed on the manifold in question. Let ϕ_A denote an irreducible spinor field $\phi_{(A...BC'...D')}$ on M. Now we pick an arbitrary point $Q \in M$ and a vector $x^a \in T_Q M$ in the tangent space at Q. Let $P = \exp(x^a)$ be the image of x^a under the exponential map (assuming that exp is defined at x^a) i.e., $P = \gamma(1)$ where γ is the geodesic starting at $\gamma(0) = Q$ with tangent vector $\dot{\gamma}(0) = x^a$. Then we can expand ϕ_A in a Taylor series around Q along γ:

$$\phi_A(P) = \sum_{n=0}^{\infty} \frac{1}{n!} (x^a \nabla_a)^n \phi_A|_Q$$

$$= \phi_A(Q) + x^a \nabla_a \phi_A(Q) + \frac{1}{2} x^a x^b \nabla_a \nabla_b \phi_A(Q) + \dots$$

If we are working in the analytic category then knowing all derivatives of ϕ_A at Q is equivalent to knowing ϕ_A along γ. If we are interested in C^k-fields, then the Taylor polynomial constructed from the first k derivatives of ϕ_A provides an approximation to ϕ_A in a neighbourhood of Q by Taylor's theorem.

Next, suppose that P lies on the (future) null cone C_Q of Q. Then we can write $x^a = \xi^A \xi^{A'}$ for some spinor ξ^A and we may propagate ξ^A along γ by the requirement that it be covariantly constant: $x^a \nabla_a \xi^A = 0$. The function $\phi : C_Q \to \mathbb{C}$, $\phi(P) = \xi^A \ldots \xi^B \xi^{C'} \ldots \xi^{D'} \phi_{A \ldots BC' \ldots D'}(Q)$, which we call the null datum for ϕ_A, has the Taylor expansion

$$\phi(P) = \phi(Q) + \xi^A \xi^{A'} \nabla_{AA'} \phi(Q) + \frac{1}{2} \xi^A \xi^{A'} \xi^B \xi^{B'} \nabla_{AA'} \nabla_{BB'} \phi(Q) + \ldots$$
$$= \phi(Q) + \xi^A \ldots \xi^B \xi^E \xi^{C'} \ldots \xi^{D'} \xi^{E'} \nabla_{EE'} \phi_{A \ldots BC' \ldots D'}(Q) + \ldots$$

Note, that in this expansion only the totally symmetric spinor derivatives of ϕ_A appear. By the same argument as above it is clear that knowing all (the first k) derivatives of ϕ at Q is equivalent with knowing (an approximation of) ϕ on C_Q.

So far we did not make any particular assumption about ϕ_A apart from its differentiability properties. It was Penrose's observation that it is the role of the field equations together with the available structure on a metric manifold to provide just enough information to bridge the gap between the totally symmetric and the unsymmetrized derivatives. He formalized this by introducing the notion of "exact sets of spinor fields" (Penrose 1980, Penrose and Rindler 1984):

Definition. Let F be a collection of irreducible spinor fields "in interaction", i.e., connected by a set of covariant differential relations, the "field equations". Then F is called *exact* iff

(i) at any point $Q \in M$ all totally symmetric covariant derivatives are algebraically independent and

(ii) at any point $Q \in M$ all covariant derivatives are algebraically determined by the totally symmetric derivatives.

For an exact set of spinor fields $F = \{\phi_A, \psi_B, \ldots\}$ the characteristic initial value problem is *formally* well posed. By (i) the null data $\{\phi, \psi, \ldots\}$ may be specified freely on C_Q (there are no constraints to satisfy) and by (ii) the fields are uniquely determined in the interior of the null cone. The higher derivatives can be computed from the totally symmetric derivatives using the differential relations by purely algebraic manipulations. The problem however is that the number of necessary manipulations grow very rapidly with the order of the derivative considered.

Most equations of mathematical physics can be put in the form of exact sets of interacting spinor fields. Examples (see Penrose and Rindler 1984) include the Maxwell equations on a curved background, any zero rest mass field on flat space and the Dirac equation. Most prominent is the case of vacuum general relativity which will be considered exclusively in the sequel but there are also examples of gravity coupled to sources like e.g., the Maxwell field or the Dirac equation. In the vacuum case F consists of the Weyl spinor Ψ_{ABCD} only. The differential relations are the vacuum Bianchi identities $\nabla^A_{A'}\Psi_{ABCD} = 0$ and the commutator relations generated by

$$\{\nabla^{P'}_A \nabla^{Q'}_B - \nabla^{Q'}_B \nabla^{P'}_A\}\alpha_C = \epsilon^{P'Q'}\Psi_{ABCD}\alpha^D$$

and its complex conjugate. It is shown in (Penrose 1980) that F is an exact set and the first terms of the expansion were already computed in (Penrose 1960). In this paper Penrose derived the general plane wave solution using this algebraic method.

3 A DIFFERENT FORMULATION OF PENROSE'S METHOD

3.1 Operators on the spin bundle

Although it is clear that the method outlined above for deriving expressions for the higher order terms in a Taylor expansion of the spinor fields in question works in principle and in special cases it seems to be quite tedious to be applied in full generality. The use of programs for algebraic manipulations suggests itself. However, the calculations to be performed are mainly index manipulations which are quite cumbersome to implement on machines (Thomas 1991). We therefore suggest an alternative formulation that does not use abstract indices. Instead, the whole structure is coded into algebraic relations between operators on the spin bundle. This reduces the calculations to pure string manipulations which are much easier to handle by machines. This approach is based on Sparling's "exterior calculus on the (co)spin bundle" (Sparling 1984).

Since we are working in a local context (that is, in a neighbourhood of a point $Q \in M$) we may consider the spin bundle $\mathbf{S}$ to be the collection of pairs $\{(Q, \pi^A) : Q \in M, \pi^A \in \mathbb{C}^2\}$. To motivate our approach we remember that in section 2. the function $\phi : C_Q \to \mathbb{C}$, $\phi(Q) = \xi^A \ldots \xi^B \xi^{C'} \ldots \xi^{D'} \phi_{A \ldots BC' \ldots D'}(Q)$ appeared as the null datum for the characteristic initial value problem for an exact set of spinor fields. ϕ can be extended in a natural way to a function (also denoted by ϕ) on the spin bundle defined by $\phi(Q, \pi^A) = \pi^A \ldots \pi^B \pi^{C'} \ldots \pi^{D'} \phi_{A \ldots BC' \ldots D'}(Q)$. With this definition we find that for each irreducible spinor field $\phi_{A \ldots BC' \ldots D'}(Q)$ on M there exists a function ϕ on $\mathbf{S}$ that is globally defined on each fiber of $\mathbf{S}$ and has the homogeneity property $\phi(Q, \lambda \pi^A) = \lambda^p \bar{\lambda}^{p'} \phi(Q, \pi^A)$, where p and p' are the numbers of unprimed and primed indices, respectively. It can be shown by application of Liouville's theorem that the converse is also true. Note, that for fixed $Q \in M$, $\phi(Q, \pi^A)$ is a monomial in the two complex variables (π^0, π^1). We call ϕ a (p, p')-function according to its homogeneities.

We now have to introduce several structures that we will need in the following:

(1) π^A and $\pi^{A'}$ will denote the coordinate along the fibers of $\mathbf{S}$ and its complex conjugate.

(2) ∂_A and $\partial_{A'}$ are the corresponding derivatives along the fibers: $\partial_A = \partial/\partial\pi^A$ and $\partial_{A'} = \partial/\partial\pi^{A'}$. Hence $\partial_A \pi^B = \delta_A^B$ and $\partial_{A'} \pi^{B'} = \delta_{A'}^{B'}$.

(3) $H = \pi^A \partial_A$ and $H' = \pi^{A'} \partial_{A'}$ are the Euler operators on $\mathbf{S}$: $H\phi = p\phi$, $H'\phi = p'\phi$ for a (p, p')-function ϕ.

(4) Combining $\nabla_{AA'}$, the covariant derivative on spinors with π^A, $\partial_A,\ldots$ we can construct four derivative operators on $\mathbf{S}$:

$$L = \pi^A \pi^{A'} \nabla_{AA'} \qquad\qquad M = \pi^A \partial^{A'} \nabla_{AA'}$$
$$N = \partial^A \partial^{A'} \nabla_{AA'} \qquad\qquad M' = \partial^A \pi^{A'} \nabla_{AA'}$$

To illustrate their meaning we consider a (p, p')-function ϕ which we may write as $\phi = \phi_{A\ldots BC'\ldots D'}\pi^A\ldots\pi^B\pi^{C'}\ldots\pi^{D'}$ with p unprimed and p' primed indices. Then $L\phi = \nabla_{EE'}\phi_{A\ldots BC'\ldots D'}\pi^A\ldots\pi^B\pi^{C'}\ldots\pi^{D'}\pi^E\pi^{E'}$ and we find that $L\phi$ corresponds to the totally symmetric part of $\nabla_{EE'}\phi_{A\ldots BC'\ldots D'}$. In a similar way, the remaining operators project onto the other irreducible parts of the spinor derivative.

(5) From the commutator of the covariant derivative we construct the curvature as usual: $\Box_{AB} = \nabla_{A'(A}\nabla_{B)}^{A'}$, $\Box_{A'B'} = \nabla_{A(A'}\nabla_{B')}^{A}$. Proceeding in the same way as above we define six curvature operators on $\mathbf{S}$:

$$S = \pi^A \pi^B \Box_{AB}, \qquad T = \pi^A \partial^B \Box_{AB}, \qquad U = \partial^A \partial^B \Box_{AB},$$
$$S' = \pi^{A'} \pi^{B'} \Box_{A'B'}, \qquad T' = \pi^{A'} \partial^{B'} \Box_{A'B'}, \qquad U' = \partial^{A'} \partial^{B'} \Box_{A'B'}.$$

(6) For a (p, p')-function ϕ we find the expression

$$S\phi = (-p\Psi_{ABC}{}^P \phi_{D\ldots PE'\ldots F'} - p'\Phi_{ABE'}{}^{P'} \phi_{C\ldots DF'\ldots P'})\pi^A\pi^B\pi^C\ldots\pi^D\pi^{E'}\ldots\pi^{F'}$$

and similar expressions for $T\phi$, $U\phi$, etc. The form of the right hand side suggests to introduce the following bilinear pairings between two functions ϕ_1, ϕ_2 with homogeneities (p_1, p_1'), (p_2, p_2') $(k > 0)$:

$$C_k(\phi_1, \phi_2) = \frac{1}{(p_1)_k(p_2)_k}\partial_{A_1}\ldots\partial_{A_k}\phi_1\partial^{A_1}\ldots\partial^{A_k}\phi_2$$

$$C_k'(\phi_1, \phi_2) = \frac{1}{(p_1)_k(p_2)_k}\partial_{A_1'}\ldots\partial_{A_k'}\phi_1\partial^{A_1'}\ldots\partial^{A_k'}\phi_2$$

Here we use the notation $(n)_k := n(n-1)\ldots(n-k+1)$ and we define $C_0 = C_0'$ to be the ordinary pointwise product of complex valued functions. Note, that C_k, C_k' are symmetric (skew) for k even (odd).

With these notations we can express the curvature operators in terms of the curvature spinors and the C_k's, e.g.:

$$S\phi = pC_1(\Psi, \phi) + p'C_1'(\Phi, \phi)$$
$$T\phi = p(p-1)C_2(\Psi, \phi) + p'(p'-1)C_2'(\Phi, \phi) + 2p^2\Lambda\phi, \text{ etc.}$$

Here Ψ, Φ and Λ are the functions on S corresponding to the Weyl, Ricci and the scalar curvature spinors.

As usual, second covariant derivatives are related to curvature and the d'Alembert operator $\Box$. This fact is expressed here in the form of commutation relations between the derivative operators:

$$[L, M] = -(H'+1)S \qquad\qquad [L, M'] = -(H+1)S'$$
$$[N, M] = (H+1)U' \qquad\qquad [N, M'] = (H'+1)U$$
$$[L, N] = -(H+1)T' - (H'+1)T - 1/2(H+H'+2)\Box$$
$$[M, M'] = -(H+1)T' + (H'+1)T - 1/2(H-H')\Box$$

and the additional relation

$$LN - MM' = -(H'+1)T + 1/2H(H'+1)\Box.$$

Now we are in a position to reformulate the notion of an exact set: let F be a collection of homogeneous (globally defined) functions on S subject to differential relations involving only the operators L, N, M and M'. Then F is called *exact* iff for all $\phi \in F$

(i) all the powers $L^k\phi$ are algebraically independent and
(ii) the results of applying all possible strings of operators L, N, M and M' to ϕ are completely determined by those powers $L^k\phi$.

3.2 Application to vacuum general relativity

In this case the Einstein equation yields $\Phi = 0$ and $\Lambda = 0$ and the basic field is the Weyl curvature Ψ. The curvature operators acting on a (p, p')-function are then simply

$$S\phi = pC_1(\Psi, \phi)$$
$$T\phi = p(p-1)C_2(\Psi, \phi)$$
$$U\phi = p(p-1)(p-2)C_3(\Psi, \phi)$$

together with the complex conjugate expressions. The field equation is the vacuum Bianchi identity on Ψ which in the present formalism is expressed as

$$M'\Psi = 0.$$

In addition, on homogeneity grounds, we have the equations

$$M\Psi = 0, \qquad N\Psi = 0.$$

These equations are the first step to show that the knowledge of all $L^k\Psi$ at a point Q is enough to determine all the derivatives of Ψ at Q ($L\Psi$ is the only possible first derivative). In fact, we have the following

Proposition. The outcome of applying an arbitrary string of derivative operators to Ψ can be expressed as a sum of terms containing only the powers $L^k\Psi$ their complex conjugates and the bilinear pairings C_k and C'_k with $0 \le k \le 2$.

The proof is by induction on the length of the string and inspection of the commutation relations, the form of the curvature operators and the "generalized Leibniz rule"

$$OC_k(\phi_1, \phi_2) = C_k(O\phi_1, \phi_2) + C_k(\phi_1, O\phi_2)$$
$$+ \alpha_1 C_{k-1}(\tilde{O}\phi_1, \phi_2) + \alpha_2 C_{k-1}(\phi_1, \tilde{O}\phi_2),$$

where O, $\tilde{O}$ symbolize any of the operators L, N, M and M' and where α_1 and α_2 are rational numbers which together with $\tilde{O}$ depend on O and k. Let s be an arbitrary string of derivative operators and let n be its length. For $n = 0, 1$ the statement is true. Now consider the function $(Os)\Psi = O(s\Psi)$. By the induction hypothesis $s\Psi$ is a sum of terms of the stated form. By linearity we may concentrate on one of those. This term will consist of a factor and a bilinear pairing C_k whose arguments are again C_k or a power $L^k\Psi$. When applying O to such a term we use the generalized Leibniz rule to obtain terms with O or a different O applied to one of the arguments. Continuing in this way we descend the tree built up from the C's until O hits a power L^k. If $O = L$ we are done. Otherwise we commute O with each L in L^k producing a curvature term at each step which in turn is again a bilinear pairing between Ψ or $\bar\Psi$ and some power. Finally O will hit Ψ and this will produce zero by the field equation and the homogeneity equations. One thing to note is that when commuting L and N the wave operator appears. But using the additional relation given above $\square$ can again be expressed in terms of derivative and curvature operators. Since for $n = 2$ no C_k, C'_k with $k > 2$ appears and since the generalized Leibniz rule involves only k and $k - 1$ no higher k than $k = 2$ appears in the expressions. This proves the statement.

It is clear that this proof provides a recursive algorithm to produce all $s\Psi$, s an arbitrary string of derivative operators in terms of the $L^k\Psi$.

3.3 Calculation of the Taylor series
As before we fix a point $Q \in M$ and choose an arbitrary vector $x^{AA'} \in T_Q M$. In the Taylor series there appear the powers of the directional derivative $x^a \nabla_a$. It is easy to

show that the following equation holds

$$(H+1)(H'+1)\nabla_{AA'} = \partial_A\partial_{A'}L + \pi_A\pi_{A'}L - \partial_A\pi_{A'}M - \pi_A\partial_{A'}M'$$

This motivates the introduction of four operators on the fiber of **S** over Q:

$$x^0 = x^{AA'}\partial_A\partial_{A'}, x^1 = -x^{AA'}\partial_A\pi_{A'},$$
$$x^2 = -x^{AA'}\pi_A\partial_{A'}, x^3 = x^{AA'}\pi_A\pi_{A'}.$$

Then $x^a\nabla_a \propto x^0L + x^1M + x^2M' + x^3N$. Note, that these operators are derivative and multiplication operators acting on monomials in two complex variables. Now an additional complication appears: when we expand $(x^a\nabla_a)^l$ we get terms of the form $(xOxO\ldots xO)$ but we need terms of the form $x\ldots xO\ldots O$. Therefore, we need to express terms of the form Ox by terms of the form xO. This is a straightforward calculation which yields e.g.

$$(H+1)(H'+1)Lx^0 = HH'x^0L - Hx^1M - H'x^2M' + x^3N$$
$$(H+1)Lx^1 = Hx^1L + x^3M'$$
$$(H'+1)Lx^2 = H'x^2L + x^3M$$
$$Lx^3 = x^3L$$

and similar relations for the other operators. This concludes the procedure to compute the kth term in the Taylor expansion.

To summarize: given an integer k

1. compute $(x^a\nabla_a)^k$ as a sum of terms of the form $x\ldots xO\ldots O$ using the formulae above,
2. for each operator string obtained in step 1. compute $O\ldots O\Psi$ using the recursive algorithm outlined above.
3. To each of the results obtained in step 2. apply the corresponding $x\ldots x$ and add them together.

Note, that this procedure has to be done only once. One may consider the result to be a list of operations that have to be performed on the initial data. In a concrete application one would give the null datum Ψ on C_Q in terms of its derivatives at the vertex Q. This is a sequence of monomials $\{m_k\}_{0\le k\le\infty}$ in $(\pi^A, \pi^{A'})$ of degree $(k+4, k)$. The operations that have to be performed according to the list are then the usual operations that can be performed on polynomials like multiplication and differentiation and finally addition.

CONCLUSION

We have presented here an outline of a procedure to compute algebraically the solution of the characteristic initial value problem for the Weyl spinor in vacuum general relativity. The hope is that one can implement this procedure on a computer to actually perform all the necessary manipulations. Why would this be a sensible thing to do? The first answer to this question is that this approach is an interesting alternative to the methods of numerical relativity and that it is always advantageous to have two different methods for the same problem. Depending on the order of the derivatives that can be handled one calculates an approximation to the actual solution of the problem. In numerical relativity the same thing happens, one determines an approximation to a solution, but by a totally different method. Therefore, comparison of the results of the two methods should give a hint about their reliability.

Another point is that the usual numerical approach to the characteristic initial value problem has to take the appearance of caustics into account (Corkill and Stewart 1983). This seems to be no problem here, because as mentioned above, one can obtain the entire space-time for a plane wave without any difficulty although there do appear caustics and crossing regions in the exponential map. Penrose has suggested (Penrose 1985) initial data where the null cone is reconverging. This means that trapped surfaces appear and that space-time must contain a singularity. One could try to use this method to examine the structure of space-time as the singularity is approached.

How much of this programme can be realized is a question of how many terms can be handled by a computer and this remains to be seen.

REFERENCES

Corkill, R. W. and Stewart, J. M. (1983). *Proc. Roy. Soc. Lond. A*, **386**, 373.

Penrose, R. (1960). *Ann. Phys.*, **10**, 171.

Penrose, R. (1980). *Gen. Rel. Grav.*, **12**, 225.

Penrose, R. (1985). In *Galaxies, axisymmetric systems and relativity*, M. A. H. MacCallum (ed.), Cambridge University Press, Cambridge.

Penrose, R. and Rindler, W. (1984). *Spinors and space-time*, Volume 1, Cambridge University Press, Cambridge.

Sparling, G. A. J. (1984). "Twistor theory and the characterization of Fefferman's conformal structures", University of Pittsburgh, preprint.

Thomas, V. "The initial value problem in general relativity by power series", *Twistor Newsletter*, **9**, 8.

ON HYPERBOLOIDAL HYPERSURFACES

Helmut Friedrich

Max Planck Institute for Astrophysics, Munich, Germany

Abstract. Initial value problems involving hyperboloidal hypersurfaces are pointed out. Characteristic properties of hyperboloidal initial data and rigorous results concerning the construction of smooth hyperboloidal initial data are discussed.

1 INTRODUCTION

In this article I shall discuss some properties of "hyperboloidal hypersurfaces". These occur naturally in a number of interesting initial value problems. I became first interested in them in the context of abstract existence proofs for solutions of Einstein's field equations which fall off in null directions in such a way that they admit the construction of a smooth conformal boundary at null infinity (Friedrich (1983)). But it appears to me that hyperboloidal hypersurfaces should also be of interest, in particular if questions concerning gravitational radiation are concerned, in various numerical studies.

Let us consider solutions to Einstein's field equations with vanishing cosmological constant and possibly massive sources of spatially compact support and long range fields like Maxwell fields. We call a space-like hypersurface in such a space-time "hyperboloidal" if it extends to infinity in such a way that it ends on null infinity. We assume that the hypersurface remains space-like in the limit when it "touches null infinity". The standard examples of such hypersurfaces are the space-like unit hyperbolas in Minkowski space, which motivate the name hyperboloidal. In the standard picture of Minkowski space it is seen that these hypersurfaces are asymptotic to certain null cones. This indicates that they intersect the null cones at null infinity but it may be remarked here that the conformal structure is such that, in the smooth conformal extension of Minkowski space, the hyperboloidal hypersurfaces intersect the null cones as well as null infinity transversally.

We shall denote by "hyperboloidal data" initial data for Einstein's field equations which are implied on a hyperboloidal hypersurface. Space-times determined from null cone data and from hyperboloidal data are often quite similar. As an example we consider the evolution of a space-time from data on a null cone which are such that the null cone remains smooth away from the vertex and has smooth intersection with null infinity. If in the resulting space-time we choose any hyperboloidal hypersurface which intersects null infinity in the same 2-surface as the null cone, we find that the space-time determined from the null cone data coincides with the space-time constructed from the hyperboloidal data if we consider the past as well as the future evolution of the hyperboloidal data.

The numerical evolution of space-time from null cone data appears to be technically easier than from hyperboloidal data, since in the second case one has to deal with the constraint equations on space-like hypersurfaces in a similar way as in the standard Cauchy problem. The initial data on smooth null hypersurfaces may be specified freely and the problem of the constraint equations appears to be absent, since the interior equations implied by Einstein's equations on null hypersurfaces degenerate essentially into a hierarchy of ordinary differential equations along the null generators. This advantage, however, is paid for by the problem of caustics. The occurrence of this type of "singularity" is an intrinsic property of null hypersurfaces and cannot be avoided unlike the appearance of coordinate singularities in other initial value problems which can be removed by prescribing different coordinate conditions.

In the theoretical investigations of space-times evolving from hyperboloidal data it has been assumed, so far, that the data show a certain smoothness at infinity (after a suitable conformal compactification) such that they determine a space-time with a smooth conformal structure at null infinity. Under this condition the analysis is performed most effectively in terms of the regular conformal field equations (cf. Friedrich (1991), Friedrich (1992) and the references given there). For numerical evolutions of such data these equations have the draw-back that one has to deal with a somewhat lengthy list of dependent unknowns. However, these equations have the advantage that they allow the numerical evolution of space-times from hyperboloidal data up to null infinity in a finite grid without the introduction of an artificial cut-off. The calculation may be performed in a slicing of the space-time by hyperboloidal hypersurfaces and for detailed calculations of various asymptotic quantities like radiation fields etc., it may be useful to extend the calculation slightly beyond null infinity by suitably extending the initial data (cf. Friedrich (1991)).

Beside the type of initial value problem indicated above there are other interesting initial boundary value problems involving hyperboloidal hypersurfaces. To study the effect of gravitational radiation falling in from infinity onto a system of massive sources, one could prescribe data on a hyperboloidal hypersurface S which intersects past null infinity and the radiation field on the part of past null infinity which is in the future of S. If one is interested in studies of solutions of Einstein's field equations with cosmological constant $\Lambda < 0$ (our signature is $-,+,+,+$) it appears natural to study the initial boundary value problem, where data are prescribed on a space-like hypersurface extending to space-like infinity — these are again hyperboloidal data

— and on the conformal boundary at infinity which in the present case is time-like. Nothing much has been done on this problem yet theoretically or numerically but it appears manageable since with respect to the regular conformal field equations this is again a finite initial boundary value problem.

In such calculations it is of course of interest to provide hyperboloidal initial data which show the right smoothness behaviour at infinity. The discussion of this behaviour and an outline of a result relevant for this question will be the the main topic of this article. In the next section I shall outline the basic properties of hyperboloidal hypersurfaces in space-times with vanishing or positive cosmological constant which allow a smooth conformal structure at null infinity. Under this assumption the hyperboloidal data show a specific fall-off behaviour at infinity. After that I shall describe a result obtained by Lars Anderson, Piotr Chruściel, and myself (Andersson et al (1992)) which shows that one can easily construct a large class of hyperboloidal initial data with the desired fall-off behaviour.

2 PROPERTIES OF HYPERBOLOIDAL DATA

To simplify the discussion, let $(\tilde{M}, \tilde{g})$ denote a solution of Einstein's vacuum field equations $Ric(\tilde{g}) = 0$ which has a smooth conformal structure at future null infinity. Let $\tilde{S}$ be a space-like hypersurface in $\tilde{M}$ which extends to future null infinity in such a way that it touches null infinity in a space-like 2-surface ∂S and such that $S = \tilde{S} \cup \partial S$ is a smooth space-like hypersurface with boundary in any smooth conformal extension of $(\tilde{M}, \tilde{g})$. The physical metric $\tilde{g}$ implies an interior metric $\tilde{h}$ on $\tilde{S}$. From our assumptions follows that there are a smooth Riemannian metric h and a smooth function Ω on S with the following properties. The function Ω is a defining function of the boundary of S in the sense that $\Omega > 0$ on $\tilde{S}$ while $\Omega = 0$ and $d\Omega \neq 0$ on ∂S. Furthermore it satisfies together with h the relation

$$\tilde{h} = \Omega^{-2} h \quad \text{on} \quad \tilde{S}. \tag{2.1}$$

The fact that the Riemannian manifold $(\tilde{S}, \tilde{h})$ can be derived in such a way from a Riemannian space (S, h) with boundary by means of a defining function of the boundary is the characterizing property of a hyperboloidal hypersurface. It is easy to see that the space $(\tilde{S}, \tilde{h})$ is complete. Using the transformation law for curvature tensors under conformal rescalings one finds that along any curve in $\tilde{S}$ which approaches ∂S the curvature tensor of $\tilde{h}$ in any orthonormal frame with respect to $\tilde{h}$ takes in the limit the form of a curvature tensor of a space of negative constant curvature. Thus it is seen that the geometry of hyperboloidal data is quite distinct from that of asymptotically euclidean standard Cauchy data.

Let $\tilde{\chi}$ denote the second fundamental form induced on $\tilde{S}$. Then $\tilde{h}$ and $\tilde{\chi}$, which together satisfy the vacuum constraints, constitute the hyperboloidal initial data on

$\tilde{S}$. We shall need to discuss the behaviour of various conformal fields derived from $\tilde{h}$, $\tilde{\chi}$, and Ω. To simplify the following discussion and also because the results I shall present later on make use of the assumption below, we shall assume that our data are such that

$$\tilde{\chi} = \frac{1}{3}\text{tr}(\tilde{\chi})\tilde{h}. \qquad (2.2)$$

This condition is an analogue of the condition of time symmetry which is considered in the case of the standard Cauchy problem for Einstein's equations. It may be noticed here that the vanishing of the trace-free part of the second fundamental form of $\tilde{S}$ is a property which is invariant under conformal rescalings of $\tilde{g}$. The condition (2.2) implies that the magnetic part of the conformal Weyl tensor on $\tilde{S}$ vanishes.

It follows now from the momentum constraint that $\text{tr}(\tilde{\chi}) = c = const.$ The fact that $\tilde{S}$ ends on null infinity implies that $c \neq 0$. By a conformal rescaling of the metric $\tilde{h}$ with a positive constant conformal factor the Hamiltonian constraint can written as the condition

$$R(\tilde{h}) = -6 \qquad (2.3)$$

on the Ricci-scalar of $\tilde{h}$.

Under the condition (2.2) the intial data for the regular conformal field equations on S are given by the tensor fields

$$h, \quad \Omega, \quad t = \frac{1}{3}D_a D^a \Omega \qquad (2.4)$$

$$s_{ab} = -\Omega^{-1}(D_a D_b \Omega - \frac{1}{3}h_{ab}D_c D^c \Omega) \qquad (2.5)$$

$$d_{ab} = \Omega^{-1}(R_{ab}(h) - \frac{1}{3}h_{ab}R(h) - s_{ab}). \qquad (2.6)$$

Here the covariant derivative D is defined with respect to the metric h. The tensor field s_{ab} is essentially the orthogonal projection of the unphysical Ricci tensor into S while $C_{ab} = \Omega d_{ab}$ represents the electric part of the conformal Weyl tensor on S. Since it has been shown by R. Penrose, that the conformal Weyl tensor must vanish on null infinity where null infinity is smooth and admits spherical sections, it may be assumed that the tensor fields s_{ab}, d_{ab} extend smoothly to all of S. We call hyperboloidal data "smooth on S" if the initial data for the regular conformal field equations extend smoothly to S. We shall see below in which sense this smoothness requirement for hyperboloidal data imply particular fall-off conditions at null infinity.

A more general discussion of the initial data on S for the regular conformal field equations, which covers the case where condition (2.2) is not required, can be found in Friedrich (1988). It remains the question to what extent our discussion changes if there is a negative cosmological constant Λ in the field equations. It is clear that a relation of the type (2.1) must hold again for the interior metric on a space-like hypersurface which extends to the conformal boundary at infinity, assuming that the latter has a smooth structure. If again the condition (2.2) is assumed on our hypersurface, the Hamiltonion constraint will also take the form of equation (2.3) where now the negative constant on the right hand side is given by an expression involving the trace of the second fundamental form and the cosmological constant. After a conformal rescaling of h by a suitable constant conformal factor we obtain the Hamiltonian constraint exactly in the form (2.3). The hyperboloidal initial data for the conformal field equations are again given by the tensor fields (2.4), (2.5) and (2.6).

3 CONSTRUCTION OF SMOOTH HYPERBOLOIDAL DATA

If instead of deriving properties of hyperboloidal data we want to construct them from "simple" data we can proceed as follows. We choose any 3-dimensional, orientable, compact Riemannian space (S, h) with boundary ∂S, pick on S any smooth defining function ω of the boundary and set $\Omega = \phi^{-2}\omega$ where the function ϕ is to be determined such that it is positive on S and that the metric $\tilde{h} = \Omega^{-2}h$ satisfies (2.3). This condition can be read as an elliptic equation for ϕ which, if expressed with respect to h and ω, takes the form

$$8\left(\omega^2 \Delta_h \phi + \omega < grad_h\omega, d\phi >\right) + R(\omega^{-2}h)\phi = -6\phi^5. \tag{3.1}$$

Here for the Laplace operator has been used the convention which makes it a positive operator. It is important to note that the Ricci scalar $R(\omega^{-2}h)$ extends to a smooth function on S. The most conspicuous property of equation (3.1) is the fact that it degenerates on the boundary ∂S because of the factor ω^2 in front of the principal part. Of course one could get rid of this problem by rewriting the equation completely in terms of the metric $\omega^{-2}h$, but then one will have difficulties in analysing, in fact even in talking about "the smoothness of ϕ at ∂S", which will be of prime importance for us.

After the discussion in the previous section we are confronted with the following questions:

(i) What can be said about the existence and uniqueness of positive solutions to equation (3.1)? Is there any freedom to prescribe data for ϕ on ∂S?

(ii) If the solutions to (3.1) existed, what could be said about the smoothness of ϕ? The smoothness of ϕ on $\tilde{S}$ would follow from standard elliptic theory. However, because of the degeneracy of equation (3.1) on ∂S the question of the smoothness of ϕ near ∂S is quite delicate and does not follow from standard theory. In fact, it is not clear a priori that the requirement that ϕ extend smoothly to all of S should not lead to restrictions on the choice of the spaces (S, h).

(iii) If ϕ could be shown to exist and to be smooth on S, what could be said about the smooth extensibility of the tensor fields (2.4), (2.5) and (2.6) to ∂S? Could this be ensured by choosing our "free data" (S, h) in a suitable way?

It will be seen in the following that under the assumption (2.2) complete and rigorous answers can be given to the questions above. The results, which I shall give only in outline here, have been obtained in the article (Andersson et al (1992)) to which I refer for details and proofs.

It is well known that equation (3.1) — a special case of the Lichnerowicz equation which is called in mathematical circles nowadays the Yamabe equation — possesses a certain conformal covariance due to which our results will not depend on the chosen defining function ω or on any conformal rescaling of the given metric h. This conformal behaviour of the equation can be exploited to bring it into a form which is technically useful and which exhibits at the same time the essential obstruction to the smoothness if its solution.

Lemma 3.1 *The metric h can be conformally rescaled by a smooth positive conformal factor on S and the defining function ω can be chosen in such a way that $|grad_h\omega|_h = 1$ and $R(\omega^{-2}h) = -6 + \bar{R}\omega^3$ near ∂S, where $\bar{R}$ is a smooth function on S.*

The first point of importance here is the fact, that both conditions can be satisfied at the same time. The function ω can now be taken as the ingoing coordinate of a Gauss coordinate system based on ∂S which allows us to analyse the behaviour of the solutions to (3.1) in a convenient way. If we denote by k the first and by λ the second fundamental form induced by the metric h on ∂S, we find in the present scaling of h that $tr_k(\lambda) = 0$. The vanishing of the trace free part of λ on ∂S is a conformally invariant property and thus cannot be affected by a rescaling.

There arises the question here why we did not choose ω immediately such that $R(\omega^{-2}h) = -6$. If we consider this equation as an ansatz for ω and try to determine a formal expansion for ω in terms of its derivatives on ∂S, we find, that one cannot solve for the third order derivative. In fact, the condition $\bar{R}|_{\partial S} = 0$ is invariant under smooth conformal rescalings of h and thus signals a first source of trouble.

Our first question is answered by

Theorem 3.2 *There exists a unique solution ϕ of equation (3.1) on S which is positive on S. In the gauge considered in lemma 3.1 it is of the form $\phi = 1 + \psi$ with $\psi \in C^0(S) \cap C^\infty(\tilde{S})$ and such that $\omega^{-p}\psi$ is bounded on $\tilde{S}$ for any real $p < 3$.*

The existence and the behaviour of ϕ near ∂S described in the last statement follow from the construction of explicit upper and lower solutions. The uniqueness part follows from certain a priori estimates and a repeated use of the maximum principle. What may be noted here is the fact, that requiring the solution to be positive on the compact manifold S removes the freedom to describe boundary data for ϕ.

Technically the most difficult question is concerned with the smoothness of the solution near ∂S. We find

Theorem 3.3 *Let ϕ be the solution considered in theorem 3.2, given in the gauge considered in lemma 3.1. Then*

1. *The solution ϕ has an asymptotic expansion*

$$\phi = 1 + \sum_i^\infty \sum_j^{N_i} \phi_{ij}(log\omega)^j \omega^i \tag{3.2}$$

   with certain non-negative integers N_i and smooth functions ϕ_{ij} on S which satisfy $< grad_h\omega, d\phi_{ij} >= 0$ near ∂S.
2. *The solution ϕ is smooth on S if and only if the conformal class of the metric h is such that the condition*
$$\bar{R}|_{\partial S} = 0 \tag{3.3}$$
   or, equivalently, the condition

$$C(h) \equiv \delta^A \delta^B \lambda^*_{AB} + R^{AB}(h)\lambda^*_{AB} + \frac{1}{2}tr_h(\lambda)\lambda^{*AB}\lambda^*_{AB} = 0 \quad on \quad \partial S \tag{3.4}$$

   is satisfied. Here λ^*_{AB} denotes the trace free part of the second fundamental form and δ_A the interior covariant derivative on ∂S while $R_{AB}(h)$ denotes the orthogonal projection of the Ricci tensor of h into ∂S.

There can be given examples of data (S, h), where the smoothness conditions in the second part are violated such that the solution ϕ has logarithmic terms. The quantity $C(h)$ transforms like a conformal density of weight -3 under conformal rescalings of the metric h. Thus the second smoothness criterion is independent of the choice of conformal gauge. It is remarkable here that the smoothness depends only on the behaviour of h in the immediate neighbourhood of ∂S.

Our third question is settled by

Theorem 3.4 *Let ϕ be the solution to (3.1) considered in theorem 3.2. Then $\Omega \equiv \phi^{-2}\omega$ and the tensor fields (2.5) and (2.6) calculated on $\tilde{S}$ from Ω and h extend to smooth tensor fields on S if and only if the conformal class of h is such that the trace free part of the second fundamental form of ∂S satisfies*

$$\lambda^*_{AB}|_{\partial S} = 0 \qquad (3.5)$$

or if the equivalent condition

$$C_{ab} = \Omega d_{ab} \longrightarrow 0 \quad at \quad \partial S \qquad (3.6)$$

is satisfied.

The second condition is saying that the smoothness of the hyperboloidal data derived from (S, h) is equivalent to the vanishing of the conformal Weyl tensor in the limit to ∂S, which in the space-time evolving from these data is the limit to null infinity on $\tilde{S}$. This is one way of relating the smoothness of the hyperboloidal data to fall-off conditions for physical fields. Condition (3.5) gives the specific fall-off requirement directly in terms of our "freely" specified data (S, h). It shows that smooth hyperboloidal data are easily constructed on any oriented 3-manifold S.

If one studies the meaning of condition (3.5) in the 4-dimensional space-time evolving from the data one finds that it is just the expression for the well known fact that the family of null generators on a smooth null infinity is shear free. It is quite remarkable that even if the Hamiltonian constraint is taken into account, which after all is a global requirement on the hyperboloidal hypersurface, one finds in the end no condition on the data which has not been found by the "local" analysis of the field equations near null infinity. In the proofs of theorems 3.3 and 3.4 only the analysis of various fields and operators in a neighbourhood of ∂S is required. Thus it may be expected that similar results may be obtained if massive sources of compact support on $\tilde{S}$ are taken into account. Finally it may be worthwhile to point out, that besides the orientability no further topological restriction on the manifold S have been imposed.

4 CONCLUSIONS

I haven't made an effort to hide the fact, that my personal interest in the questions considered above was motivated essentially by theoretical problems concerning the existence and the properties of solutions to Einstein's equations. But after all one would like to obtain precise quantitative information on the space-times under study. I think numerical investigations of the initial value problems which I indicated above, should lead to interesting results concerning questions of gravitational radiation. In fact, I think that in careful studies of the effects of gravitational radiation falling in from far distances to interact with a system of massive objects, the use of hyperboloidal data can hardly be avoided. If one embarks on such a calculation it is of course convenient to know in advance, what kind of smoothness (respectively fall-off, if one prefers to think in terms of the physical picture) need to be imposed on the data respectively can be expected of the solution which one wants to construct. Theorem 3.4 gives precise information how to construct smooth hyperboloidal data. We know then from Friedrich (1991) that the evolution of these data will preserve the smoothness of the asymptotic fields at least for some time. It is of course also possible to consider the evolution of hyperboloidal data, which do not show the smoothness at infinity which has been studied here. But then it may be much harder to extract the information on the asymptotic behaviour of the fields which one would like to find. A more general analysis of hyperboloidal data, where the condition (2.2) is dropped and which also involves weaker smoothness properties will be given in Andersson and Chruściel.

REFERENCES

Andersson, L., Chruściel, P. T. and Friedrich, H. (1992). "On the Regularity of Solutions to the Yamabe Problem and the Existence of Smooth Hyperboloidal Initial Data for Einstein's Field Equations", *Comm. Math. Phys.*, to appear.

Andersson, L. and Chruściel, P. T., in preparation.

Friedrich, H. (1983). *Comm. Math. Phys.*, **91**, 445–472.

Friedrich, H. (1988). *Comm. Math. Phys.*, **119**, 51–73.

Friedrich, H. (1991). *J. Differential Geometry*, **34**, 275–345.

Friedrich, H. (1992). In *Recent Advances in General Relativity*, A. Janis and J. Porter (eds.), Birkhäuser, Boston.

THE INITIAL VALUE PROBLEM ON NULL CONES

J. A. Vickers

Faculty of Mathematical Studies, University of Southampton, Southampton, UK

Abstract. The characteristic initial value problem is reviewed and a number of possible schemes for implementing it are discussed. Particular attention is given to choosing variables and choosing a minimal set of equations in the Newman–Penrose formalism. A particular scheme which is based on null cones and involves giving the free gravitational data in terms of Ψ_0 is presented. The question of regularity at the vertex is briefly discussed and asymptotic expansions for the spin coefficients near the vertex are given.

1 INTRODUCTION

In studying problems in which gravitational radiation plays an important rôle, a description of the geometry which is adapted to the wavefronts of the radiation is obviously useful. Thus both the Bondi formalism and Newman–Penrose formalism have proved very helpful in understanding gravitational radiation at null infinity J^+. From the point of view of an initial value problem this suggests that rather than specifying data on a spacelike surface one should specify data on a null surface and look instead at the characteristic initial value problem (CIVP).

There are a number of technical advantages that one gets from looking at the CIVP. The first of these is that the variables one uses are precisely those one needs to calculate the physically important quantities such as the amount of gravitational radiation, the Bondi momentum and so on. The second advantage is that the elliptic constraints which play such an important rôle in the spacelike case are effectively eliminated and one can freely specify the appropriate null data. Finally one has the property, pointed out by Penrose (see Penrose and Rindler (1984)) amongst others, that on a characteristic initial hypersurface one need specify half as many real numbers at each point as data, as on a spacelike hypersurface, so that the free data is given by one complex field, or two real fields.

In some pionering numerical work Stewart (see e.g. Corkill and Stewart (1983)) performed a number of calculations in which he applied the 2+2 formalism of d'Inverno, Smallwood and Stachel (d'Inverno and Smallwood (1980) d'Inverno and Stachel (1978)), where the spacetime is foliated into two families of null hypersurfaces. This numerical work also employed theoretical work of Friedrich (1981a), which showed how to use a tetrad description to formulate the problem as a symmetric hyperbolic system of differential equations. A rather different approach to the characteristic initial value problem was taken by Winicour and co-workers (see e.g. Gomez *et al* (1986)), who based their numerical calculations on characteristic surfaces given by the null cones of a point on some timelike geodesic, and employed the coordinate based Bondi formalism. A completely different method of looking at the characteristic initial value problem has also been given by Penrose (see Penrose and Rindler (1984)) who uses the concept of an exact set of fields to determine the evolution of the gravitational field in terms of symmetrised derivatives of the Weyl curvature (see Frauendiener (1992) this volume), although so far this method has not been used numerically.

2 POSSIBLE FORMALISMS FOR THE CIVP

Although the CIVP seem to be well suited to problems involving gravitational radiation, it is apparent that there are many different formalisms that can be used. In this section we consider further some of the choices available. The first choice involves the type of null hypersurface that is used. The two main choices are a 2+2 description or a null cone description. In the 2+2 approach the vacuum field equations are very simple, but it is harder to include matter. One is also likely to encounter problems with caustics, although techniques exist for evolving through these. Caustics are also a potential problem when one uses null cones as the characteristic surfaces (although this is not always obvious in the Bondi formalism), but for systems in which rotational effects are not too large one can choose the vertex of the null cone in such a way as to avoid the formation of caustics. A second problem with the latter approach concerns regularity of the solution at the vertex of the cone, although this is now much better understood (Rendall 1990) and appropriate conditions on the data which ensure regularity are now known. However the 2+2 and null cone foliations are not the only possible choices. One possibility is to use a null hypersuface in the neighbourhood of J^+ and join this on to an interior region where one had a spacelike hypersuface (see Bishop (1992) this volume). Another alternative would be to use a spacelike, but asymptotically null hypersuface (the hyperboloidal situation examined by Friedrich (1992) this volume), although this method does not give a CIVP (see figure 1).

The second main decision is whether to use a coordinate or tetrad based description. If one chooses the former one is quickly led to using Bondi coordinates based on null cones. This formalism has the advantage that the equations have a simple hierarchical structure and are simple to solve (see Winicour (1983)). However in my view despite this substantial advantage it also has a number of disadvantages. In the Bondi formalism the free gravitational data is given by the 2-metric on the $r = $ constant, $u = $ constant surfaces, where r is a luminosity parameter. This is

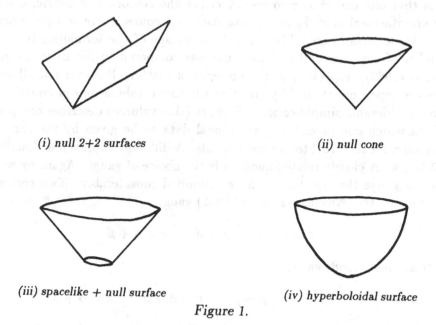

(i) null 2+2 surfaces   (ii) null cone

(iii) spacelike + null surface   (iv) hyperboloidal surface

Figure 1.

perfectly correct mathematically and makes the equations easy to solve, but from a physical point of view it is natural to regard the gravitational field Ψ_0 as the free data — see especially Penrose for an elaboration of this viewpoint (Penrose and Rindler (1984)). A second problem is that it is not at all straightforward to conformally compactify in this formalism, so one has to specify data all the way out to infinity. Since one cannot do this numerically, one tends to impose some arbitrary truncation of the data at a suitably large value of r. A third point is that the Bondi formalism does not apply to null cones with caustics, since the Newman–Penrose (NP) quantity ρ is given in terms of the Bondi quantity β by $\rho = -e^{-2\beta}/r$. Thus although the initial data might be free of caustics, the evolution equations break down once caustics start to form. Although caustics do cause problems in the tetrad description, it is at least possible to describe them in the NP formalism, and there do exist techniques for evolving through caustics.

If one chooses a tetrad formalism such as that of Newman and Penrose (1962) (or its generalisation the Geroch–Held–Penrose (1973) (GHP) formalism), one is not compelled to use null cones as the characteristic surface but can use some of the other choices considered above. The equations also have the advantage that it is fairly straightforward to incorporate a conformal compactification. By adding in the conformal factor as an additional variable satisfying appropriate equations Friedrich

(1981b) was able to construct a system of equations regular at J^+. The final advantage is that one can choose to specify either the conformal 2-metric, the shear σ or the gravitational field Ψ_0 as the free data. Of course there is a price one has to pay for these advantages. The main disadvantage of the formalism is the very large number of variables and equations one has compared to the Bondi formalism (Bondi *et al* (1962)), especially in the presence of matter. However not all the NP equations are independent, and by choosing suitable combinations of equations one can make considerable simplifications. Stewart (this volume) describes one possible approach in which one takes the gravitational data to be given by the shear, and then eliminates all the Weyl tensor components. A different approach is described in section 3 below. A closely related question is the choice of gauge. Again by working in a suitable gauge the equations can be simplified considerably. Two choices are commonly made; the 'Newman–Penrose'(1962) gauge

$$\kappa = \pi = \epsilon = 0 \qquad \rho = \bar{\rho} \qquad \tau = \bar{\alpha} + \beta$$

and the 'Hawking' (1968) gauge

$$\kappa = \epsilon = 0 \qquad \rho = \bar{\rho} \qquad \mu = \bar{\mu} \qquad \tau = \bar{\alpha} + \beta \qquad \pi = \alpha + \bar{\beta} \ .$$

Both these choices are adapted to using an affine parameter r, but one could presumably use a luminosity parameter in the NP formalism, in which case other gauge choices might be more suitable.

The final choice I wish to consider concerns whether or not to compactify. In the NP formalism it is simple to conformally compactify. This can be done either by specifying the conformal factor Ω as a given function of r or alternatively by including the conformal factor as one of the dynamic variables. It is far less easy to compactify in the Bondi formalism, but one possibility involves choosing a new compactified radial coordinate and then showing that the equations when written in terms of this new variable are regular at J^+. To summarise one can make the following choices:

(i) choice of hypersurface
(ii) coordinates or tetrad
(iii) choice of NP gauge
(iv) choice of NP variables
(v) choice of NP equations
(vi) evolve through caustics or choose (restricted) caustic free data
(vii) compactify; yes or no; dynamically or a priori ?

All these questions have a bearing on how simple the resulting equations are to deal with numerically. Furthermore it seems likely that the above choices are relevant to the ease of producing algorithms which are numerically stable, and for which good error bounds may be obtained.

3 CHOICE OF NEWMAN–PENROSE VARIABLES AND EQUATIONS

In this section we consider the way in which the structure of the Newman–Penrose equations may be simplified by making suitable choices of gauge, variables and equations. In the Newman–Penrose formalism one really has three sets of equations; the commutator equations, the Ricci identity, and the Bianchi identity. By applying the commutator equations to the coordinates x^a one obtains differential equations for the tetrad components in terms of the spin coefficients, the Ricci identity gives differential equations for the spin coefficients in terms of the curvature, the contracted Bianchi identity gives differential equations for the Ricci curvature and hence for the matter variables by Einstein's equations, whilst the evolution of the Weyl curvature is given by the remaining equations of the Bianchi identity. In simplifying the equations the strategy will be to choose the characteristic surfaces to be given by the null cones of some timelike geodesic (with affine parameter u), and to take Ψ_0 as the free data for the gravitational field, and the matter variables which determine Φ_{ab} as the free data for the matter. We will then obtain all the other variables on the null cone by using only radial integration along the affinely parameterised null geodesic generators of the cone (and regularity at the vertex) and will avoid the use of angular differential equations. The gravitational and matter free data is then evolved in the u direction using the Bianchi identity.

We begin by briefly discussing the matter variables for a perfect fluid. The energy-momentum tensor is
$$T_{ab} = (\rho + p)v_a v_b - p g_{ab}$$
where we assume we also have some equation of state $p = p(\rho)$ and
$$g^{ab}v_a v_b = 1 \quad .$$
Thus we can find T_{ab} in terms of ρ and v_a. In the NP setting the appropriate variables are ρ together with the tetrad components of v_a
$$a = \ell^a v_a, \quad b = m^a v_a, \quad c = n^a v_a \quad .$$
These are not independent but satisfy $2ac - 2\bar{b}b = 1$ since v_a is a unit vector. One can easily write down all the components of Φ_{ab} in terms of these variables, but it turns out that for a perfect fluid a more convenient set of independent variables is Φ_{00}, Φ_{01} and $\Phi_{11} + 3\Lambda$, which also determine the other components of Φ_{ab} and Λ in this case.

We next illustrate the way in which the information in the angular differential equations is encoded in the radial equations on the null cone. This can be done in either the NP gauge or the Hawking gauge. If one uses the GHP formalism it is easier to use the Hawking gauge but if one uses the standard NP formalism the NP gauge turns out to be more convenient. We will describe the NP case here and give the details of the GHP case elsewhere. An example of the sort of equation that we want to eliminate is the angular equation for ρ and σ

$$\delta\rho - \bar{\delta}\sigma = \tau(\rho - \bar{\rho}) + \kappa(\mu - \bar{\mu}) + \rho(\bar{\alpha} + \beta) - \sigma(3\alpha - \bar{\beta}) - \Psi_1 + \Phi_{01}$$

which in the NP gauge becomes

$$\delta\rho - \bar{\delta}\sigma = \rho(\bar{\alpha} + \beta) - \sigma(3\alpha - \bar{\beta}) - \Psi_1 + \Phi_{01} \tag{1}$$

However if one uses the commutator equation

$$D\delta - \delta D = (\bar{\alpha} + \beta)D - \sigma\bar{\delta} - \rho\delta \quad, \tag{2}$$

the Ricci equations

$$D\rho = \rho^2 + \sigma\bar{\sigma} + \Phi_{00} \tag{3}$$

$$D\sigma = 2\rho\sigma + \Psi_0 \tag{4}$$

$$D\alpha = \alpha\rho + \beta\bar{\sigma} + \Phi_{10} \tag{5}$$

$$D\beta = \beta\rho + \alpha\sigma + \Psi_1 \tag{6}$$

and the Bianchi equation

$$D\Psi_1 - D\Phi_{01} = \bar{\delta}\Psi_0 - \delta\Phi_{00} + 2(\bar{\alpha} + \beta)\Phi_{00} + 4\rho\Psi_1 - 4\alpha\Psi_0 - 2\rho\Phi_{01} - 2\sigma\Phi_{01} \tag{7}$$

one can show that equations (2–7) imply that

$$D\{\delta\rho - \bar{\delta}\sigma - \rho(\bar{\alpha} + \beta) + \sigma(3\alpha - \bar{\beta}) + \Psi_1 - \Phi_{01}\} = 0 \quad. \tag{8}$$

On the other hand regularity at the vertex ensures that (1) holds initially and hence by (8), equation (1) holds everywhere on the null cone. In a similar way one can show that equation (1) is preserved by evolution in the u direction. Proceeding in this fashion one may obtain a set of equations which are equivalent to the NP equations and which allow one to obtain the values of all the variables on the null cone by radial integration, and for which one has evolution equations for the free data off the null cone. Of course we still have available the other NP equations, and from a numerical point of view this could prove valuable, as the extent to which equation (1) failed to hold numerically would provide a good way of monitoring the errors in the system.

We now consider our proposed set of variables and equations. We start by listing the variables we will use in the NP gauge.

(i) Tetrad components:

$$D = \ell^a \nabla_a = \frac{\partial}{\partial r}$$

$$\delta = m^a \nabla_a = \omega \frac{\partial}{\partial r} + \xi^A \frac{\partial}{\partial x^A} \qquad A = 2, 3$$

$$\Delta = n^a \nabla_a = \frac{\partial}{\partial u} + U \frac{\partial}{\partial r} + X^A \frac{\partial}{\partial x^A} \qquad A = 2, 3$$

(ii) Spin Coefficients:

$$\rho, \quad \sigma, \quad \alpha, \cdot \ \beta, \quad \gamma, \quad \lambda, \quad \mu, \quad \nu$$

(iii) Curvature Components:

$$\Psi_0, \quad \Psi_1, \quad \Psi_2, \quad \Psi_3, \quad (\Psi_4 \text{ is not needed})$$

$$\Phi_{ab} \quad (a, b = 0, 1, 2) \quad \Lambda$$

(for a perfect fluid these are given by algebraic equations in Φ_{00}, Φ_{01} and $\Phi_{11} + 3\Lambda$)

We now give the hypersurface equations

Radial Equations

Given Ψ_0 and Φ_{00} we solve

$$D\rho = \rho^2 + \sigma\bar{\sigma} + \Phi_{00}$$

$$D\sigma = 2\rho\sigma + \Psi_0$$

for ρ and σ. The remaining equations form a hierarchical system of 6 linear equations each of the form

$$D\mathbf{x} = A\mathbf{x}$$

for the remaining variables. Here the matrix A has coefficients which depend on the variables already solved for earlier on in the hierarchy.

Evolution Equations

Here one obtains 4 linear equations for the evolution with respect to u of the free gravitational data Ψ_0 and the three pieces of the free matter data.

We remark that the only non-linear equations to be solved are the first pair of equations for ρ and σ. However as pointed out by Newman and Penrose (1962) this pair of equations may be written as a matrix Riccati equation. Let

$$P = \begin{pmatrix} \rho & \bar{\sigma} \\ \sigma & \rho \end{pmatrix} \qquad Q_0 = \begin{pmatrix} \Phi_{00} & \bar{\Psi}_0 \\ \Psi_0 & \Phi_{00} \end{pmatrix}$$

Then

$$DP = P^2 + Q_0 \quad .$$ (9)

If we now let $P = -M^{-1}DM$, then we obtain the equivalent equation

$$D^2M + MQ_0 = 0 \quad ,$$ (10)

a linear second order matrix equation for M. Thus one can completely avoid solving non-linear differential equations if desired. It is also worth remarking that it is this equation which tells one when there are caustics; these occur when M is not invertible. Thus the condition that the null cone be free of caustics for given initial data is just the condition that there is an invertible solution to equation (10). Of course it may be hard to tell if such a solution exists for general data, but in special cases one can employ comparison theorems to show that caustics do not occur.

4 REGULARITY AT THE VERTEX

In this final section we briefly discuss regularity of the solutions to our system of equations at the vertex. To do this we go into greater detail about how the coordinates and tetrad are chosen. Let u be an affine parameter along the timelike geodesic γ which forms the origin of the null cones, then we extend u to to a coordinate on the whole manifold M by letting the null cones be given by $u = $ constant. We now let $\ell_a = \nabla_a u$ and let r be an affine parameter along the null geodesic generators of the future null cone. Let T^a be the tangent to the central geodesic $T^a = \dot\gamma^a$, then we choose r so that the vertex is given by $r = 0$ and at the vertex we have $g(T, \frac{\partial}{\partial r}) = 1$. This fixes r uniquely and ensures that $\ell = \frac{\partial}{\partial r}$. We may decompose ℓ^a at the vertex into $\ell^a = T^a + X^a$ where X^a is an outward normal to the unit tangent sphere S_0 lying in the tangent space at the vertex and orthogonal to T^a. We then define $n^a = \frac{1}{2}(T^a - X^a)$ at the vertex and extend it to the rest of the null cone by parallel propagation along the null generators of the cone. The coordinates x^A ($A = 2, 3$) may be taken to be any convenient pair of coordinates which label the null geodesics, but a useful choice which we will employ here is to use polar coordinates (θ, ϕ) on S_0 to label the geodesics. We now complete ℓ^a, n^a to a null tetrad $\{\ell^a, n^a, m^a, \bar{m}^a\}$ on γ with m^a parallel along γ and again extend this to a tetrad on the whole space-time by parallel propagation along the null generators of the cone. By choosing our coordinates and frame in this way we ensure that the $r = $ constant, $u = $ constant 2-surfaces are asymptotically metric 2-spheres as $r \to 0$. This together with our choice of tetrad means that the asymptotic behaviour of the tetrad components and spin coefficients is given to leading order in r by the values in Minkowski space in (u, r, θ, ϕ) coordinates and the obvious frame, i.e. the leading order behaviour is

$$\omega = 0$$
$$\xi^2 = 1/r\sqrt{2} \quad \xi^3 = i\sin\theta/r\sqrt{2}$$
$$U = -\tfrac{1}{2}$$
$$X^2 = 0 \quad X^3 = 0$$
$$\rho = -1/r$$
$$\mu = -1/2r$$
$$\alpha = \cot\theta/r\sqrt{2}$$
$$\beta = -\cot\theta/r\sqrt{2}$$

If one has a spacetime and chooses such a tetrad and coordinate system, then the Weyl, Ricci, and scalar curvature have finite directional dependent limits at the vertex of a null cone. The directional dependence of the limits is encoded in the spin weights of the quatities concerned. Thus Ψ_0 is a spin weight two quantity and since the $r = $ constant, $u = $ constant 2-surfaces are asymptotically 2-spheres, this means that the limiting value $\mathring{\Psi}_0(\theta,\phi)$ may be expressed in terms of the usual spin weight two spherical harmonics. Conversely the solution will be regular at the vertex if we ensure that the curvature has the correct directional dependence. Because of the structure of the NP equations we only require the correct directional dependence of the free data, together with some differentiability requirements. Thus we need $\mathring{\Psi}_0$ to have spin weight two, $\mathring{\Phi}_{01}$ to have spin weight one, and both $\mathring{\Phi}_{00}$ and $\mathring{\Phi}_{11} + 3\mathring{\Lambda}$ to have spin weight zero to ensure regularity at the vertex.

In the same way that it is possible to calculate the asymptotic expansions of the spin coefficients and the tetrad components in terms of $1/r$ at J^+ (Newman and Unti (1962)) it is possible to calculate asymptotic expansions in terms of r at the vertex. The significant point to note when one does this is that the deviation from flat space occurs at one higher order in r than one might naively expect. Thus for example

$$\rho = -1/r + \tfrac{1}{3}r\mathring{\Phi}_{00} + O(r^2)$$
$$\sigma = \tfrac{1}{3}r\mathring{\Psi}_0 + O(r^2)$$

and there are no $O(1)$ terms as one might expect. Because of the sensitivity of the equations to the behaviour near the vertex it might well be worthwhile explicitly enforcing these asymptotic conditions in the numerical code.

REFERENCES

Bishop, N. (1992). This volume.

Bondi, H., van der Burg, M. G. J. and Metzner, A. W. K. (1962). *Proc. R. Soc. A*, **269**, 21-52.

Corkill, R. W. and Stewart, J. M. (1983). *Proc. R. Soc. A*, **386**, 373-391.

Frauendiener, J. (1992). This volume.

Friedrich, H. (1981a). *Proc. R. Soc. A*, **378**, 401-421.

Friedrich, H. (1981b). *Proc. R. Soc. A*, **375**, 169-184.

Friedrich, H. (1992). This volume.

Geroch, R., Held, A. and Penrose, R. (1973). *J. Math. Phys.*, **14**, 874-881.

Gomez, R., Isaacson, R. A., Welling, J. S. and Winicour, J. (1986). In *Dynamical spacetimes and numerical relativity*, J. M. Centrella (ed.), Cambridge University Press, Cambridge.

d'Inverno, R. A. and Smallwood, J. (1980), *Phys. Rev. D.*, **22**, 1233-1247.

d'Inverno, R. A. and Stachel, J. (1978). *J. Math. Phys.*, **19**, 2447-2460.

Newman, E. T. and Penrose, R. (1962). *J. Math. Phys.*, **3**, 566-578.

Newman, E. T. and Unti, T. W. J. (1962). *J. Math. Phys.*, **3**, 891-901.

Penrose, R. and Rindler, W. (1984). *Spinors and space-time*, Cambridge University Press, Cambridge.

Rendall, A. D. (1990). *Proc. R. Soc. A*, **427**, 221-239.

Winicour, J. (1983). *J. Math. Phys.*, **24**, 1193-1198.

INTRODUCTION TO DUAL-NULL DYNAMICS

S. A. Hayward

Max Planck Institute for Astrophysics, Munich, Germany

Abstract. Much of physics concerns temporal dynamics, which describes a spatial world (or Cauchy surface) evolving in time. In Relativity, the causal structure suggests that null dynamics is more relevant. This article sketches Lagrangian and Hamiltonian formalisms for dual-null dynamics, which describes the evolution of initial data prescribed on two intersecting null surfaces. The application to the Einstein gravitational field yields variables with recognisable geometrical meaning, initial data which divide naturally into gravitational and coordinate parts, and evolution equations which are covariant on the intersection surface and free of constraints.

1 INTRODUCTION

The ADM or "3+1" formalism [1,2] is a natural approach to the Cauchy problem in General Relativity, and has been used widely both analytically and numerically. By comparison, null (or characteristic) evolution problems are more appropriate to the study of problems involving radiation, whether gravitational or otherwise, since radiation propagates in null directions. Null surfaces also have a central place in the causal structure of General Relativity which spatial surfaces do not.

A distinction should be drawn between the null-cone problem discussed elsewhere in this volume, in which the initial surface is a null cone, and the dual-null problem, in which there are two intersecting null initial surfaces. The latter problem was originally described by Sachs [3], with existence and uniqueness proofs being given by Müller zum Hagen and Seifert [4], Friedrich [5] and Rendall [6], and a general "2+2" formalism being developed by d'Inverno, Smallwood and Stachel [7–9]. The dual-null case is easier to treat, since in the null-cone case there is an apex problem which necessitates additional conditions, and a choice of coordinates adapted to an axis on which the apex evolves.

The aim is to write the the Einstein equations in a geometric form adapted to two null 3-surfaces S^+ and S^- intersecting in a spatial 2-surface S. In particular it is desired that the variables be tensors on S with recognisable geometrical meaning, such as the first, second and normal fundamental forms of S, and that the equations should be covariant on S, with the evolution of S being described by the Lie derivatives L_u and L_v along two evolution vectors u and v which commute: $L_u v = 0$. The use of the Lie derivative is geometrically natural, and commutativity of the evolution vectors is necessary to ensure that the surfaces generated from S along u and v are integrable. This retains the maximal amount of explicit geometrical information appropriate to the dual-null evolution problem. In my view, this compares well with the commonly used spin-coefficient formalisms, in which the variables are scalar and complex, and somewhat distant from the geometry.

The motivation is to provide a clear formalism which describes the propagation and self-interaction of gravitational radiation in general. Of particular interest is the occurrence of trapped surfaces, caustics and singularities due to gravitational effects, and the asymptotics of gravitational scattering.

The approach adopted, by analogy with the ADM approach to the Cauchy problem, is to develop a general Lagrangian-Hamiltonian formalism with two evolution directions instead of one.

2 TEMPORAL DYNAMICS

It is appropriate to begin with a brief review of the standard Lagrangian and Hamiltonian formalisms for temporal dynamics. The initial surface is assumed to be a compact orientable manifold S, of dimension n, where spacetime has dimension $n+1$. The space of smooth functions on S is denoted by FS, and the smooth n-forms (or scalar densities) by F_dS. The configuration fields are assumed to be the smooth sections $q \in CQ$ of a vector bundle Q over S, called the configuration bundle. The velocity bundle is the tangent bundle TQ over TS, with velocity fields $(q, \dot{q}) \in CTQ$. Evolution occurs in an open interval T, with evolution parameter (time-coordinate) $\tau \in T$ such that $\dot{q} = dq/d\tau$.

Dynamics are determined from a Lagrangian (density) $L : CTQ \to F_dS$ by the principle of stationary action $\delta \int_T \int_S L \, d\tau = 0$, from which follows the Euler-Lagrange equation

$$\frac{d}{d\tau}\left(\frac{\delta L}{\delta \dot{q}}\right) = \frac{\delta L}{\delta q}.$$

The second-order Euler-Lagrange equation can be expressed as first-order Hamilton equations by transforming to the momentum bundle T_d^*Q over T_d^*S, whose fibres consist of the linear maps from the corresponding fibre of TQ to F_dS. The momentum fields $(q, \bar{q}) \in CT_d^*Q$ are given by the Lagrange transformation

$$\Lambda : CTQ \to CT_d^*Q, \qquad \bar{q} = \frac{\delta L}{\delta \dot{q}}.$$

The Lagrangian determines a Hamiltonian $H : CT_d^*Q \to F_dS$ by $H = \Lambda_*^{-1}[\bar{q}(\dot{q}) - L]$

if Λ is invertible, and the Lagrange equation is equivalent to the Hamilton equations

$$\frac{dq}{d\tau} = \frac{\delta H}{\delta \bar{q}}, \qquad \frac{d\bar{q}}{d\tau} = -\frac{\delta H}{\delta q}.$$

The initial data are then q and $\bar{q}$ on S. If Λ is not invertible, there are constraints instead of evolution equations.

3 DUAL-NULL DYNAMICS

Proceeding by analogy with temporal dynamics, consider a compact orientable manifold S of dimension n, where spacetime has dimension $n + 2$, and a configuration bundle Q over S with fields $q \in CQ$. The velocity bundle is the Whitney sum $(TQ)^2 = TQ \oplus TQ$ over $(TS)^2$, with velocity fields $(q, q^+, q^-) \in C(TQ)^2$. Evolution occurs in open intervals U and V, with evolution parameters $\xi \in U$ and $\eta \in V$ such that $q^+ = \partial q/\partial \xi$ and $q^- = \partial q/\partial \eta$. The initial surfaces are $S^+ = S \times U$ and $S^- = S \times V$.

Dynamics are determined from a Lagrangian $L : C(TQ)^2 \to F_d S$ by the principle of stationary action $\delta \int_V \int_U L \, d\xi \, d\eta$, which leads to the Euler-Lagrange equation

$$\frac{\partial}{\partial \xi} \left(\frac{\delta L}{\delta q^+} \right) + \frac{\partial}{\partial \eta} \left(\frac{\delta L}{\delta q^-} \right) = \frac{\delta L}{\delta q}.$$

The momentum bundle is $(T_d^* Q)^2$ over $(T_d^* S)^2$, with momentum fields $(q, \bar{q}, \hat{q}) \in C(T_d^* Q)^2$ given by the Lagrange transformation

$$\Lambda : C(TQ)^2 \to C(T_d^* Q)^2, \qquad \bar{q} = \frac{\delta L}{\delta q^+}, \qquad \hat{q} = \frac{\delta L}{\delta q^-}.$$

The Hamiltonian $H : C(T_d^* Q)^2 \to F_d S$ is given by $H = \Lambda_*^{-1}[\bar{q}(q^+) + \hat{q}(q^-) - L]$, and the Hamilton equations follow as

$$\frac{\partial q}{\partial \xi} = \frac{\delta H}{\delta \bar{q}}, \qquad \frac{\partial q}{\partial \eta} = \frac{\delta H}{\delta \hat{q}}, \qquad \frac{\partial \bar{q}}{\partial \xi} + \frac{\partial \hat{q}}{\partial \eta} = -\frac{\delta H}{\delta q}.$$

There is one important difference to the standard temporal Hamiltonian theory, which is that the Hamilton equations by themselves do not give the full first-order field equations, but need to be supplemented with the integrability condition

$$\frac{\partial^2 q}{\partial \xi \partial \eta} = \frac{\partial^2 q}{\partial \eta \partial \xi}.$$

An illustrative example follows.

4 THE KLEIN-GORDON FIELD

In a flat spacetime, the metric can be written in the dual-null form

$$g = \begin{pmatrix} 0 & -1 & 0 \\ -1 & 0 & 0 \\ 0 & 0 & \delta \end{pmatrix},$$

where δ is the two-dimensional Kronecker delta. In such a basis, the coordinate derivative is decomposed as $\nabla = (\partial/\partial\xi, \partial/\partial\eta; D)$, where D is the two-dimensional coordinate derivative. Thus $\nabla\phi = (\phi^+, \phi^-; D\phi)$ and the dual-null Lagrangian is

$$L(\phi, \phi^+, \phi^-) = \phi^+\phi^- - \tfrac{1}{2}(D\phi)^2 - \tfrac{1}{2}m^2\phi^2.$$

The Lagrange equations become

$$2\frac{\partial^2\phi}{\partial\xi\partial\eta} = D^2\phi - m^2\phi,$$

which is the dual-null form of the Klein-Gordon equation.

The Lagrange transformation to momentum fields $(\phi, \bar{\phi}, \hat{\phi})$ gives simply $\bar{\phi} = \phi^-$ and $\hat{\phi} = \phi^+$, and the dual-null Hamiltonian is

$$H = \bar{\phi}\hat{\phi} + \tfrac{1}{2}(D\phi)^2 + \tfrac{1}{2}m^2\phi^2,$$

with Hamilton equations

$$\frac{\partial\phi}{\partial\xi} = \hat{\phi}, \qquad \frac{\partial\phi}{\partial\eta} = \bar{\phi}, \qquad \frac{\partial\bar{\phi}}{\partial\xi} + \frac{\partial\hat{\phi}}{\partial\eta} = D^2\phi - m^2\phi.$$

In addition, applying the integrability condition

$$\frac{\partial^2\phi}{\partial\xi\partial\eta} = \frac{\partial^2\phi}{\partial\eta\partial\xi}$$

to the Hamilton equations gives

$$\frac{\partial\bar{\phi}}{\partial\xi} = \frac{\partial\hat{\phi}}{\partial\eta},$$

which gives a total of four first-order equations determining the evolution of $\bar{\phi}$ in the ξ-direction, $\hat{\phi}$ in the η-direction, and ϕ in both directions. Thus the initial data are ϕ on S, $\bar{\phi}$ on S^- and $\hat{\phi}$ on S^+.

5 INTEGRABILITY CONDITIONS

The nature of the integrability condition in the above case may be generalised as follows. Defining the second variations

$$A = \left(\frac{\delta}{\delta \bar{q}} \otimes \frac{\delta}{\delta \bar{q}} \right) H, \qquad B = \left(\frac{\delta}{\delta \bar{q}} \otimes \frac{\delta}{\delta \hat{q}} \right) H, \qquad C = \left(\frac{\delta}{\delta \hat{q}} \otimes \frac{\delta}{\delta \hat{q}} \right) H,$$

it can be shown [10] that the integrability condition is solvable simultaneously with the Hamilton equations for $\partial \bar{q}/\partial \xi$ and $\partial \hat{q}/\partial \eta$ in terms of the momentum fields $(q, \bar{q}, \hat{q})$ if and only if

$$A = C = 0, \qquad B \text{ invertible.}$$

In this standard case the initial data are q on S, $\bar{q}$ on S^- and $\hat{q}$ on S^+.

The condition $A = C = 0$ is referred to as the dual-null condition, since it is the condition for S^+ and S^- to be null in spacetime, for various examples. If it is not satisfied, the equations do not take a form in which the evolution derivatives $(\partial/\partial \xi, \partial/\partial \eta)$ of the fields are expressed uniquely in terms of the momenta $(q, \bar{q}, \hat{q})$. Invertibility of B is less crucial, with non-invertibility leading to constraint equations instead of evolution equations, and consequently a different structure for the initial data.

6 KINETIC-POTENTIAL SYSTEMS

The status of the dual-null condition is clarified somewhat in the case of kinetic-potential systems, where the Lagrangian is the sum of a kinetic term quadratic in the velocities and a potential term independent of the velocities. To define quadratics requires a metric h^* on the configuration bundle, mapping CQ to CQ_d^* by $q^* = h^*(q)$, with the inverse h_* mapping CQ_d^* to CQ by $p_* = h_*(p)$, where the density dual Q_d^* has fibres consisting of the linear maps from the corresponding fibre of Q to $F_d S$. For tensor fields, the metric h on S induces a natural configuration metric h^*. Using this structure, a symmetric kinetic-potential Lagrangian is one of the form

$$L(q, q^+, q^-) = \tfrac{1}{2} c (q^+)^* (q^+) - b (q^+)^* (q^-) + \tfrac{1}{2} a (q^-)^* (q^-) - V,$$

where a, b, c and V depend only on q. The Euler-Lagrange equation takes the form

$$c \frac{\partial^2 q}{\partial \xi^2} - 2b \frac{\partial^2 q}{\partial \xi \partial \eta} + a \frac{\partial^2 q}{\partial \eta^2} = f \left(q, \frac{\partial q}{\partial \xi}, \frac{\partial q}{\partial \eta} \right),$$

which is a quasi-linear second-order partial differential equation. For the hyperbolic case $b^2 > ac$, the linear transformation freedom in (ξ, η) can be used to find characteristic coordinates such that

$$a = c = 0,$$

which is the dual-null condition for such systems. In this case the equation takes the canonical form of the wave equation:

$$\frac{\partial^2 q}{\partial \xi \partial \eta} = -\frac{1}{2b} f.$$

This agrees with the general definition of the dual-null condition, since

$$A = -\frac{ah_*}{b^2 - ac}, \qquad B = -\frac{bh_*}{b^2 - ac}, \qquad C = -\frac{ch_*}{b^2 - ac}.$$

7 NUMERICAL MODELS

A numerical model for a dual-null system may be constructed by analogy to the familiar grid model for the Cauchy problem. The neighbourhood $S \times U \times V$ of S is modelled by a grid for S and a Cartesian grid for $U \times V$. The fields $(q, \bar{q}, \hat{q})$ are replaced by their values on the grid, and the equations are modelled by finite difference equations. In the standard dual-null case, where the initial data are q on S, $\bar{q}$ on S^- and $\hat{q}$ on S^+, there are three integration routines to perform: integrate q and $\bar{q}$ in the ξ-direction up S^+; integrate q and $\hat{q}$ in the η-direction up S^-; and integrate q and $\bar{q}$ in the ξ-direction and q and $\hat{q}$ in the η-direction into the interior.

This gives two estimates q_1 and q_2 of q at each point, which provides improved accuracy simply by taking the average, and also provides a well-defined error measure

$$\frac{\| (q_1 - q_2, \bar{q}, \hat{q}) \|}{\| (q_1 + q_2, \bar{q}, \hat{q}) \|},$$

which gives an accuracy check. There is no analogue of this for the Cauchy problem. The above norm may be taken to be either the kinetic norm

$$\| (q, \bar{q}, \hat{q}) \|_k = \sqrt{\int_S [q^*(q) + \bar{q}_*(\bar{q}) + \hat{q}_*(\hat{q})]}$$

defined using the configuration metric, or the dynamic (energy) norm

$$\| (q, \bar{q}, \hat{q}) \|_d = \sqrt{\int_S H(q, \bar{q}, \hat{q})}$$

if $\int_S H \geq 0$, which has the status of a positive-energy conjecture for the field considered.

One other important advantage of a dual-null approach over the Cauchy approach is that in the latter, radiation propagates only approximately at light-speed, due to numerical error, whilst in the former the correct null propagation in the normal directions is forced by the structure. Thus a dual-null (or null-cone) numerical model may be expected to be more accurate for radiation problems.

8 THE MAXWELL ELECTROMAGNETIC FIELD

The Maxwell field provides a natural example of the use of dual-null dynamics. Using the flat metric of §4, the electromagnetic field tensor may be decomposed as

$$F = \begin{pmatrix} 0 & \bar{\psi} & \hat{a} \\ -\bar{\psi} & 0 & \bar{a} \\ -\hat{a} & -\bar{a} & b \end{pmatrix},$$

where $\bar{\psi}$ is a scalar, $\bar{a}$ and $\hat{a}$ are covectors and b is a 2-form. The sourceless Maxwell equations are found to be [10]

$$\frac{\partial b}{\partial \xi} = 2D \wedge \hat{a}, \qquad \frac{\partial \bar{a}}{\partial \xi} = \tfrac{1}{2}(-D\bar{\psi} - D \cdot b), \qquad \frac{\partial \bar{\psi}}{\partial \xi} = D \cdot \hat{a},$$

$$\frac{\partial b}{\partial \eta} = 2D \wedge \bar{a}, \qquad \frac{\partial \hat{a}}{\partial \eta} = \tfrac{1}{2}(D\bar{\psi} - D \cdot b), \qquad \frac{\partial \bar{\psi}}{\partial \eta} = -D \cdot \bar{a}.$$

Thus the initial data are b and $\bar{\psi}$ on S, $\bar{a}$ on S^-, and $\hat{a}$ on S^+. Note that there are no constraints, as compared with the two constraints of the temporal form of the Maxwell equations. The reduction of the constraint problem is a general feature of dual-null dynamics, and has numerous advantages both analytically and numerically, and potentially for quantum theory.

Interpreting the variables, b is the magnetic field normal to S, $\bar{\psi}$ is the normal electric field, and $\bar{a}$ and $\hat{a}$ are the polarisation vectors of the electromagnetic radiation propagating normal to S in either direction.

9 THE EINSTEIN GRAVITATIONAL FIELD

In General Relativity, the metric itself is a variable, and the formalism is consequently more complex. The starting point is the assumption of a spacetime (M, g) which admits an embedding $\phi : S \times U \times V \to M$, where U and V are open intervals, S is a compact orientable 2-manifold, and the induced metric $h = \phi_*^{-1}|_S g$ is spatial. Taking parameters $\xi \in U$ and $\eta \in V$, the evolution vectors are defined by $u = \partial/\partial\xi$ and $v = \partial/\partial\eta$, and therefore commute: $L_u v = 0$. Taking a basis $(u, v; e)$ for M, where e is a basis for S, the metric decomposes as

$$g = \begin{pmatrix} r \cdot r - ae^{-m} & r \cdot s - e^{-m} & r \\ r \cdot s - e^{-m} & s \cdot s - ce^{-m} & s \\ r & s & h \end{pmatrix},$$

where r and s are shift vectors and a and c are lapse functions describing the embedding of S in $S^+ = S \times U$ and $S^- = S \times V$ respectively, and m is a function.

The dual-null condition is $a = c = 0$, which is the condition for S^+ and S^- to be null surfaces. The configuration fields are $q = (h; r, s; m, a, c)$, and the velocity fields (q, q^+, q^-) are given by Lie derivatives in spacetime: $q^+ = L_u q$, $q^- = L_v q$.

A dual-null Lagrangian $L(q, q^+, q^-)$ may then be calculated from the Einstein-Hilbert Lagrangian. The expressions for the Lagrangian and the corresponding Lagrange equations are rather lengthy, even when expressed in an index-free notation, and will be given separately [11]. However, the structure in the dual-null case may be given as follows: the equations for h and m are wave equations

$$\frac{\partial^2 h}{\partial \xi \partial \eta} = f_h, \qquad \frac{\partial^2 m}{\partial \xi \partial \eta} = f_m,$$

where the fs are functions of $(q, \partial q/\partial \xi, \partial q/\partial \eta)$, expressed covariantly on S; the equations for r and s can be written in the form

$$\frac{\partial^2 s}{\partial \xi \partial \eta} = f_r, \qquad \frac{\partial^2 s}{\partial \xi^2} = f_s,$$

by taking the coordinate choice $r = 0$; and the equations for a and c can be written in the form

$$\frac{\partial^2 \Omega}{\partial \eta^2} = f_a, \qquad \frac{\partial^2 \Omega}{\partial \xi^2} = f_c,$$

where $\Omega = \sqrt{\det h}$ is the conformal factor of h. Note that there are no constraints, in contrast to the "3+1" version [1,2]. The initial data on S^+ are then m and the conformal metric $k = \Omega^{-1} h$, representing respectively the coordinate and gravitational data; the initial data on S^- are k, m and s; and the initial data on S are Ω, $\partial \Omega/\partial \xi$, $\partial \Omega/\partial \eta$ and $\partial s/\partial \xi$. Rendall [6] has proved local existence and uniqueness by first showing the existence of harmonic coordinates fixing m on S^+ and m, s on S^-. Encoding the gravitational initial data in the conformal metric was first suggested for the dual-null problem by Sachs [3], and in general by d'Inverno and Stachel [7]. Note also that when considering only a single null surface S^-, the only relevant equation is that for $\partial^2 \Omega/\partial \eta^2$, which is an ordinary differential equation, i.e. it does not involve derivatives in S. Breakdown of this equation indicates caustics.

The corresponding dual-null Hamiltonian system may be constructed in terms of the independent momenta, which are $\bar{h}$, $\hat{h}$, $\bar{m}$, $\hat{m}$ and $\bar{s}$ [11]. It is to be noted that whilst the extrinsic curvature of a 2-surface S can be described by two second fundamental forms and a normal fundamental form, a foliation of 2-surfaces requires two extra functions, which by coordinate choice may be set to zero on S, but not globally. These extra functions and the second fundamental forms are encoded in $\bar{h}$, $\hat{h}$, $\bar{m}$ and $\hat{m}$, and $\bar{s}$ is essentially the normal fundamental form. The Hamilton equations are then equations for

$$\frac{\partial}{\partial \xi}(h, \bar{h}, m, \bar{m}, \hat{m}, \bar{s}, s), \qquad \frac{\partial}{\partial \eta}(h, \hat{h}, m, \bar{m}, \hat{m}, \bar{s})$$

as functions of the momentum fields $(h, \bar{h}, \hat{h}, m, \bar{m}, \hat{m}, s, \bar{s})$. Again, the absence of constraints is noteworthy. The initial data are $\hat{h}$ on S^+, $(\bar{h}, s)$ on S^-, and $(h, m, \bar{m}, \hat{m}, \bar{s})$ on S. Interpreting the variables, the traceless parts of $\bar{h}$ and $\hat{h}$ encode the gravitational radiation normal to S^- and S^+ respectively, and $\bar{s}$ encodes the gravitational radiation tangent to S. The coordinate information is encoded in m, s and the traces of $\bar{h}$ and $\hat{h}$, and $\bar{m}$ and $\hat{m}$ are the expansions of S in S^- and S^+ respectively. Incidentally, the numbers of variables and equations agree with the GHP formalism [12], but not the over-parametrised NP formalism [13].

Thus the vacuum Einstein equations are obtained in a first-order form in terms of geometrical quantities with physical meaning, expressed covariantly on S. Furthermore, a dual-null decomposition of the Weyl tensor yields variables directly representing the gravitational radiation normal to S^+ and S^- and tangent to S. The corresponding dual-null form of the Bianchi identities describes the interaction of the radiation, and yet again has no constraints, unlike the "3+1" electric-magnetic form [14]. Thus is obtained a convenient basis for the study of gravitational interactions.

10 COMMENTS
Numerical methods have an important role in General Relativity which remains to be fully exploited. The main unresolved questions in the theory concern gravitational radiation, its propagation, self-interaction and interaction with matter, and under what circumstances this produces features such as trapped surfaces, causal loops, caustics and singularities. Such exotic features of the theory are known to occur in exact solutions and in symmetric classes of spacetimes, but their ubiquity is still unclear. On the one hand, there are theorems which prove generic incompleteness under various assumptions [15], but more recently generic sets of complete spacetimes have been shown to exist [14,16]. There is scope for further analytical progress on such questions, but the possibility of obtaining relevant results numerically is obvious, and appealing in terms of comparative speed and ease. In either approach, null evolution problems are likely to be at least as relevant to an understanding of General Relativity as the Cauchy problem, and a geometrically lucid formalism should prove useful.

Acknowledgements
It is a pleasure to thank Chris Clarke for various constructive comments, and Ray d'Inverno for originally interesting me in "2+2". This research was supported initially by the Science and Engineering Research Council, and latterly by the European Science Exchange Programme.

REFERENCES

[1] Arnowitt, R., Deser, S. and Misner, C. W. (1962). In *Gravitation: an Introduction to Current Research*, L. Witten (ed.), Wiley, New York.

[2] Fischer, A. E. and Marsden, J. E. (1979). In *General Relativity, an Einstein Centenary Survey*, S. W. Hawking and W. Israel (eds.), Cambridge University Press, Cambridge.

[3] Sachs, R. K. (1962). *J. Math. Phys.*, **3**, 908.

[4] Müller zum Hagen, H. and Seifert, H. J. (1977). *Gen. Rel. Grav.*, **8**, 259.

[5] Friedrich, H. (1981). *Proc. R. Soc. Lond. A*, **378**, 401.

[6] Rendall, A. D. (1990). *Proc. R. Soc. Lond. A*, **427**, 221.

[7] d'Inverno, R. A. and Stachel, J. (1978). *J. Math. Phys.*, **19**, 2447.

[8] d'Inverno, R. A. and Smallwood, J. (1980). *Phys. Rev. D*, **22**, 1233.

[9] Smallwood, J. (1983). *J. Math. Phys.*, **24**, 599.

[10] Hayward, S. A. (1992). *Dual-Null Dynamics*, preprint.

[11] Hayward, S. A. (1992). *Dual-Null Dynamics of the Einstein field*, in preparation.

[12] Geroch, R., Held, A. and Penrose, R. (1973). *J. Math. Phys.*, **14**, 874.

[13] Newman, E. T. and Penrose, R. (1962). *J. Math. Phys.*, **3**, 566.

[14] Christodoulou, D. and Klainerman, S. (1989). "The Global Non-linear Stability of the Minkowski Space", preprint.

[15] Hawking, S. W. and Ellis, G. F. R. (1973). *The Large Scale Structure of Space-Time*, Cambridge University Press, Cambridge.

[16] Friedrich, H. (1991). *J. Diff. Geom.*, **34**, 275.

ON COLLIDING PLANE WAVE SPACE-TIMES

J. B. Griffiths

Department of Mathematical Sciences, University of Technology, Loughborough, UK

Abstract. General features of the colliding plane wave problem are described with emphasis on those aspects that are relevant to its treatment by numerical techniques.

1 INTRODUCTION

Because of the non-linearity of Einstein's equations it follows that, in general relativity, even electromagnetic waves must interact with each other through their gravitational fields. It is clearly important to investigate the character of such an interaction, and the most simple situation in which to do so is in the interaction following the collision of two plane waves in a flat background.

2 COLLIDING PLANE WAVES

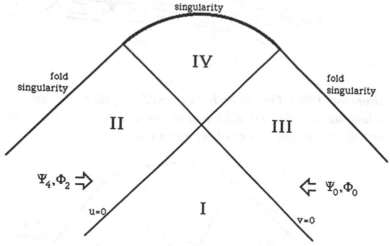

Figure 1. The structure of colliding plane wave solutions. Region I is taken to be flat, regions II and III contain the approaching plane waves, and region IV is the interaction region following the collision. A curvature singularity usually develops in the interaction region, but this may sometimes be replaced by an unstable quasiregular singularity.

Colliding plane wave space-times can conveniently be described in terms of four regions as illustrated in figure 1 (for a general review see Griffiths 1991). It is normally convenient to adopt two null coordinates u and v such that the various regions are bounded by the wave fronts $u = 0$ and $v = 0$. The most significant feature of these solutions is the development of a singularity to the future of the collision. This is generically a curvature singularity, but large classes of solutions exist in which this is replaced by an unstable Killing–Cauchy horizon.

It is also important to remember that caustics necessarily form in the regions behind the plane waves. This is illustrated in figure 2, and partly explains why topological singularities necessarily occur in regions II and III. As pointed out by Penrose (1965), a plane wave space-time contains no global Cauchy hypersurface. It follows that the general structure of colliding plane wave space-times is as illustrated in figure 3.

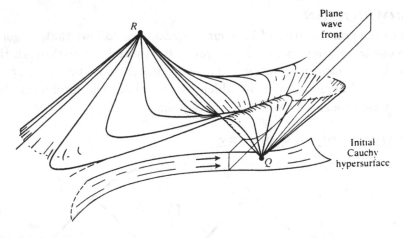

Figure 2. (Penrose, 1965) The future light cone of the point Q is distorted as it passes through a plane wave and is again focused to another vertex R which may be a point or a line (one spatial dimension has been suppressed).

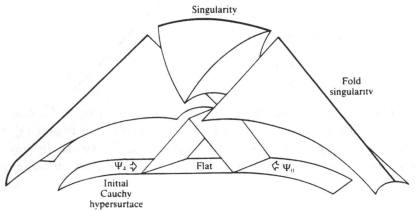

Figure 3. The general structure of colliding plane wave solutions related to some initial hypersurface on which initial Cauchy data may be set.

From figure 3 it can be seen that it is not appropriate to attempt to formulate this problem using Cauchy data. Clearly it is best formulated as a characteristic initial value problem with initial data set on the two null hypersurfaces $u = 0$ and $v = 0$, in which positive values of u and v are taken only as far as the caustic.

This approach contrasts that in which explicit analytic solutions are usually obtained. It is usual practice to first obtain a particular class of solutions in the interaction region and only subsequently to determine the initial conditions that give rise to it.

The more basic problem is to set up initial conditions in regions I, II and III and then to determine the character and effects of the interaction between the waves that occurs in region IV. Analytic methods for dealing with this problem are being developed by Hauser and Ernst (1989a,b, 1990, 1991). Methods for dealing with the characteristic initial value problem numerically have been developed by Stewart and Friedrich (1982) and applied to colliding plane wave problems by Corkill and Stewart (1983). Their approach appropriately uses a (2+2) splitting of space-time. They have also indicated how to integrate the evolution equations right up to the singularity.

The approaching plane wave in region II can be described by the Kerr–Schild line element

$$ds^2 = 2dudr + h_{ij}(u)x^i x^j du^2 - \delta_{ij}dx^i dx^j$$

where $i, j = 2, 3$. It is convenient to transform this to the Rosen form which involves two null coordinates u and v

$$ds^2 = 2dudv - g_{ij}(u)dx^i dx^j.$$

The same form can then be used in region III with u and v interchanged. Without specifying the matter field in the interaction region, it can be shown that the line element there takes the general form

$$ds^2 = 2e^{-M}dudv - g_{ij}\left(X^i du + Y^i dv - dx^i\right)\left(X^j du + Y^j dv - dx^j\right)$$

where M, g_{ij}, X^i and Y^i are functions of both u and v. For many physically interesting situations, such as colliding gravitational waves, electromagnetic waves, scalar waves or combinations of these, the Ricci tensor components Φ_{10} and Φ_{21} both remain zero throughout the interaction region. Under this condition, it is possible (Szekeres 1972, Griffiths 1991) to transform both X^i and Y^i to zero simultaneously, so that the metric in this region is the product of two 2-spaces and the two space-like Killing vectors are surface-forming.

Consider initially colliding gravitational plane waves. When the polarization of the approaching waves is constant and aligned, the main equation is linear and general classes of solutions can easily be obtained. In the more general case, the main equation is identical to the Ernst equation. All of the solution-generating techniques for this equation that have been developed in the context of stationary axisymmetric space-times may be carried over to this situation unchanged. The boundary conditions for this situation, however, are very different. It can be shown that the two wave components Ψ_0 and Ψ_4 are both modified in the interaction, and that the component Ψ_2 also necessarily appears. A curvature singularity almost always develops, but large classes of solutions without curvature singularities also occur.

It would be of interest if numerical procedures could reproduce any of the exact solutions with Killing–Cauchy horizons rather than curvature singularities. However, as pointed out at the meeting by John Stewart, in view of the instability of these solutions there is very little chance of their being obtained numerically.

For colliding plane electromagnetic waves, it can be shown (Bell and Szekeres 1974) that gravitational waves with components Ψ_0 and Ψ_4 (and usually also Ψ_2) are always generated by the collision, even when the initial regions are conformally flat. Again, a curvature singularity almost always develops, but large classes of solutions with Killing–Cauchy horizons also occur.

For colliding neutrino fields (Griffiths 1992), the line element in the interaction region takes the general form described above with X^i and Y^i not both zero. In this region all components of the Ricci tensor, and the Weyl tensor components Ψ_1, Ψ_2 and Ψ_3, necessarily become non-zero. In this sense, colliding neutrino fields may be considered to generate longitudinal gravitational waves. A curvature singularity always develops in this case.

REFERENCES

Bell, P. and Szekeres, P. (1974). *Gen. Rel. Grav.*, **5**, 275–86.

Corkill, R. W. and Stewart, J. M. (1983). *Proc. Roy. Soc. A*, **386**, 373–91.

Griffiths, J. B. (1991). *Colliding plane waves in general relativity*, Oxford University Press.

Griffiths, J. B. (1992). *Class. Quantum Grav.*, **9**, 207–15.

Hauser, I. and Ernst, F. J. (1989a). *J. Math. Phys.*, **30**, 872–87.

Hauser, I. and Ernst, F. J. (1989b). *J. Math. Phys.*, **30**, 2322–36.

Hauser, I. and Ernst, F. J. (1990). *J. Math. Phys.*, **31**, 871–81.

Hauser, I. and Ernst, F. J. (1991). *J. Math. Phys.*, **32**, 198–209.

Penrose, R. (1965). *Rev. Mod. Phys.*, **37**, 215–20.

Stewart, J. M. and Friedrich, H. (1982). *Proc. Roy. Soc. A*, **384**, 427–52.

Szekeres, P. (1972). *J. Math. Phys.*, **13**, 286–94.

BOUNDARY CONDITIONS FOR THE MOMENTUM CONSTRAINT

Niall Ó Murchadha

Physics Department, University College, Cork, Ireland

Abstract. The momentum constraint of general relativity can be transformed into an elliptic equation. This means that the "far-field" part of the extrinsic curvature is dominated by the harmonic functions of the relevant second-order elliptic operator. This, in turn, allows us to write down a multipole expansion for the extrinsic curvature. Hence we can identify the dominant terms in the extrinsic curvature far away from the sources. One application of this identification is that it permits us to tailor the boundary conditions on a finite region so as to pick out the dominant multipole moments and thus obtain a more accurate numerical solution.

1 INTRODUCTION

Initial data for the gravitational field consist of four objects $(g_{ab}, K^{ab}, \rho, \vec{j})$; where g_{ab} is a Riemannian three-metric which gives the geometry of a spacelike three-slice through a pseudo-Riemannian four-manifold; K^{ab} is a symmetric three-tensor which is the extrinsic curvature of the three-slice embedded in the four-manifold; and ρ and $\vec{j}$ represent the energy density and current density of whatever source-fields we have coexisting with the gravitational field.

These quantities cannot be given arbitrarily; they must be chosen so as to satisfy the initial value constraints

$$^{(3)}R - K^{ab}K_{ab} + (trK)^2 = 16\pi\rho, \tag{1}$$

and

$$\nabla_b[K^{ab} - (trK)g^{ab}] = -8\pi j^a. \tag{2}$$

$^{(3)}R$ is the scalar curvature of g_{ab}, equation (1) is known as the hamiltonian constraint and equation (2) is known as the momentum constraint.

In this article I will focus my attention on the extrinsic curvature especially on its asymptotic behaviour as determined by the momentum constraint. The standard method of analysing this equation is to recognise that it is a generalization of one of the Maxwell constraints $\nabla_a D^a = \rho$. The Maxwell constraint is solved by splitting $\vec{D}$ into a transverse part $\vec{D}_T$, which satisfies $\nabla_a D_T^a = 0$ and the gradient of a scalar $(\nabla_a \phi)$.

$$D^a = D_T^a + \nabla^a \phi. \tag{3}$$

When this is substituted into the constraint, it reduces to $\nabla^2 \phi = \rho$, a well-behaved elliptic equation which allows us to determine ϕ if we are given ρ. Further, we recognise that the independent part of $\vec{D}$ is the transverse part.

The next step is to restrict the choice of the extrinsic curvature K^{ab} so as to select a preferred slice. The standard condition that is imposed on K^{ab} is to demand that it be tracefree, the so-called 'maximal' slicing condition. This choice has the advantage that it decouples the momentum from the hamiltonian constraint . When this condition is imposed, the momentum constraint, equation (2), simplifies to

$$\nabla_a K^{ab} = -8\pi j^b \; ; \quad \mathrm{tr} K = g_{ab} K^{ab} = 0 \tag{4}$$

It turns out that there exists a decomposition of tracefree symmetric tensors (York, 1973) which is exactly suited to equation (4). Let us define

$$(LW)^{ij} \equiv \nabla^i W^j + \nabla^j W^i - \frac{2}{3} g^{ij} \nabla_a W^a. \tag{5}$$

This is the tracefree (or conformal) Killing form of a vector $\vec{W}$. The relevant decomposition is that any symmetric tracefree tensor K^{ab} can be written as

$$K^{ab} = K^{ab}_{TT} - (LW)^{ab}, \tag{6}$$

where K^{ab}_{TT} is the transverse, tracefree part of K^{ab}, that is it satisfies $g_{ab} K^{ab}_{TT} = \nabla_a K^{ab}_{TT} = 0$. When (6) is substituted into equation (4), it becomes

$$\nabla_a (LW)^{ab} = 8\pi j^b. \tag{7}$$

The operator on the left-hand-side of equation (7), $\nabla \cdot L$, is regular and strongly elliptic. If the current-density of the source $(\vec{j})$ has compact support, and if the transverse-tracefree part of K^{ab} falls off quickly, then the far-field of the extrinsic curvature is dominated by the $(LW)^{ab}$ part, which satisfies $\nabla_a (LW)^{ab} = 0$. Thus, as the metric approaches flat-space, we know that the leading part of the extrinsic curvature is a flat-space harmonic function of the operator $\nabla \cdot L$. This means that we need to identify and classify the vectors $\vec{W}$ which satisfy

$$\nabla_a (LW)^{ab} = \nabla_a (\nabla^a W^b + \nabla^b W^a - \frac{2}{3} \nabla_c W^c g^{ab})$$

$$= \nabla^2 W^b + \frac{1}{3} \nabla^b (\nabla_c W^c) = 0. \tag{8}$$

Equation (8) is only approximately correct. We have neglected a term $R^{ab} W_b$; but in the far-field this will fall off more rapidly than the other terms and so can be neglected. However, to simplify the discussion and to avoid confusion I will assume from now on that the manifold is flat. Thus we can regard eqn.(8) as an exact formula and not an approximation.

2 MULTIPOLE MOMENTS

The easiest way to identify the solutions of equation (8) is to perform the standard vector decomposition (3) on the vector $\vec{W}$, i.e., $W^c = W^c_T + \nabla^c \theta$. When this is substituted into equation (8) we get

$$\nabla^2 (W^b_T + \frac{4}{3} \nabla^b \theta) = 0. \tag{9}$$

This means that we can write

$$W^b_T + \frac{4}{3} \nabla^b \theta = P^b, \tag{10}$$

where each component of the vector $\vec{P}$ is a harmonic function of the laplacian, $\nabla^2 P^b = 0$. Given $\vec{P}$ we can find the solution vector $\vec{W}$. Taking the divergence of equation (10) gives

$$\frac{4}{3} \nabla^2 \theta = \nabla_b P^b. \tag{11}$$

Equation (11) can be solved to find θ. Subtracting $\frac{1}{3} \nabla^b \theta$ from both sides of equation (10) gives $\vec{W}$. In other words

$$W^b = P^b - \frac{1}{4} \nabla^b [\nabla^{-2} (\nabla_c P^c)] \tag{12}$$

solves equation (8) for any $\vec{P}$ satisfying $\nabla^2 \vec{P} = 0$.

Given the possible choices of $\vec{P}$, we can list out the associated $\vec{W}$'s as given by (12). They share many properties of the harmonic functions of the laplacian. In particular, we get two families of harmonic functions of $\nabla \cdot L$; in one group each member blows up at infinity but is regular in the interior, and in the other group each member goes to zero at infinity but is singular at the origin. Further, each harmonic function in one group has a corresponding member in the other group. For example, there are three vectors which are constant at infinity and three vectors which fall off like $1/r$. At the next order, there are nine vectors that diverge at infinity like r, and nine vectors that fall off like $1/r^2$. This structure continues order by order.

Since equation (8) is linear, we can add or subtract solutions at will. This means that any identification of a 'fundamental' set must contain some degree of arbitrariness. Nevertheless, we can pick a 'natural' set if we impose an orthogonality relationship on the harmonic functions. Consider the harmonic functions that blow up at infinity like

r^N, call them $\vec{\lambda}_{(A)}$, and the harmonic functions that fall off at infinity like $1/r^{N+1}$, call them $\vec{\xi}_{(B)}$, where A and B are labels. Each group will have exactly the same number of members. The natural orthogonality relationship that one should impose on these sets is

$$\oint_{\infty} [\lambda^c_{(A)}(L\xi_{(B)})_{cd} - \xi^c_{(B)}(L\lambda_{(A)})_{cd}]dS^d = 8\pi\delta_{AB}. \tag{13}$$

To justify this condition, let us consider equation (7) $\nabla_a(LW)^{ab} = 8\pi j^b$. Assume that $\vec{j}$ has compact support and seek a solution $\vec{W}$ that goes to zero at infinity. This means that we can write (outside the support of the matter)

$$\vec{W} = \sum \alpha_A \vec{\xi}_{(A)}. \tag{14}$$

One interprets the α_A's as the multipole moments, the multipliers of the harmonic functions $\vec{\xi}_{(A)}$. Choose one of the diverging harmonic functions $\vec{\lambda}_{(C)}$ and consider

$$\oint_{\infty} [\lambda^e_{(C)}(LW)_{ef} - W^e(L\lambda_{(C)})_{ef}]dS^f. \tag{15}$$

Substitute expansion (14) into expression (15) and use (13) to get

$$\oint_{\infty} [\lambda^e_{(C)}(LW)_{ef} - W^e(L\lambda_{(C)})_{ef}]dS^f = 8\pi\alpha_C. \tag{16}$$

Equation (16) can be turned into a volume integral to give

$$8\pi\alpha_C = \int [\lambda^e_{(C)}(\nabla \cdot LW)_e - W^e(\nabla \cdot L\lambda_{(C)})_e]d^3x. \tag{17}$$

The extra terms in (17), $(\nabla\lambda_{(C)}) \cdot (LW) - (\nabla W) \cdot (L\lambda_{(C)})$, cancel because LW is symmetric and tracefree and so 'sees' only the symmetric, tracefree part of $(\nabla\lambda_{(C)})$, i.e., we can replace $(\nabla\lambda_{(C)})$ with $L\lambda_{(C)}$ and similarly with ∇W.

We can simplify formula (17) enormously. We recognise $(\nabla \cdot L\lambda_{(C)}) \equiv 0$, from the definition of $\vec{\lambda}_{(C)}$ and that we can replace $(\nabla \cdot LW)_e$ with $8\pi j_e$ from (7). This reduces equation (17) to

$$\alpha_C = \int \lambda^e_{(C)} j_e d^3x. \tag{18}$$

The procedure to find a "natural" set of harmonic functions should now be clear. We should start with eqn.(18) and work backwards. First, one should find set of $\vec{\lambda}_{(C)}$'s satisfying equation (8). Permute and combine them until one finds a set for which the integrals in equation (18) have some desirable physical interpretation. In turn, find the vectors $\vec{\xi}_{(B)}$ which satisfy the orthogonality conditions (13). Finally, write the far-field $\vec{W}$ as in (14) as a sum of these $\vec{\xi}$'s. Then equation (18) gives a physical meaning to the multipole moments, the α_C's. I will give concrete examples of this procedure in Sections 3 and 4.

This technique works remarkably well even in curved space. The whole procedure, right down to equation (18) which gives the multipole moments as integrals over the sources, carries over exactly, the only change is that one has to take the proper volume integral, not the coordinate volume integral. This is somewhat too facile, however. Neither formula (8) nor formula (12) for the harmonic functions hold true. They can now only be regarded as approximations, valid near infinity. This is not very important with regard to the harmonic functions that decay at infinity, the $\vec{\xi}$'s, because they are not globally defined anyway. Where it does matter is with regard to the $\vec{\lambda}$'s. If we cannot compute them globally we cannot perform the integrals in (18). Thus, even if we were given the current density $\vec{j}$ and the metric analytically, we might not be able to explicitly compute the multipole moments.

The key issue is that, even in curved space, the far-field behaviour of the extrinsic curvature is dominated by the same harmonic functions. We can continue to ascribe the same physical meaning to the multipoles, the coefficients of the harmonic functions. We do not care what is happening in the interior.

3 LINEAR MOMENTUM

The leading order harmonic functions of $\nabla \cdot L$ start at $O(1/r)$. There are three such vectors. The associated, 'diverging', harmonic functions in this case are constant vectors at infinity. The 'natural' choice of $\vec{\lambda}_{(C)}$ is easy . In flat space we have the three translational Killing vectors which must be harmonic functions of $\nabla \cdot L$. With these $\vec{\lambda}_{(C)}$'s we can write down the associated set of $\vec{\xi}_{(C)}$'s. The complete array is

C	$\vec{\lambda}_{(C)}$	$\vec{\xi}_{(C)}$	
1	$(1,0,0)$	$(-1/4)(7/r + x^2/r^3, xy/r^3, xz/r^3)$	(19.1)
2	$(0,1,0)$	$(-1/4)(xy/r^3, 7/r + y^2/r^3, yz/r^3)$	(19.2)
3	$(0,0,1)$	$(-1/4)(xz/r^3, yz/r^3, 7/r + z^2/r^3)$	(19.3)

From equation (18) we now can interpret the multipole moments as being just the components of the total linear momentum of the source. This identification of the $\vec{\xi}_{(C)}$'s remains correct even if the metric is non-trivial. The only change is that now the coefficients of the $\vec{\xi}_{(C)}$'s, the α_C's, have to be identified as the total linear momentum of the entire field, the gravitational waves as well as the sources. This means that the asymptotic behaviour of the extrinsic curvature (K_{ij}) is dominated by

$$\frac{-2\alpha_1}{3} \begin{pmatrix} x/r^3 + x^3/r^5 & y/r^3 + x^2y/r^5 & z/r^3 + x^2z/r^5 \\ & -x/r^3 + xy^2/r^5 & xyz/r^5 \\ & & -x/r^3 + xz^2/r^5 \end{pmatrix}, \tag{20}$$

and two other similar objects . The identification of the α_C's as the components of the linear momentum is totally in agreement with the standard definition of the

ADM linear momentum (Regge and Teitelboim, 1974) as

$$P_A = \oint_\infty \lambda^e_{(A)} \pi_{ef} dS^f \tag{21}$$

where $\lambda^e_{(A)}$ is one of the translation Killing vectors, and π^{ef} is the momentum density conjugate to the metric g_{ef}. It is related to K^{ij} by $\pi^{ef} = \sqrt{g}([\text{Tr}K]g^{ef} - K^{ef})$.

4 ANGULAR MOMENTUM

The next harmonic functions are those which fall off like $1/r^2$ and the associated ones which grow like r. There are nine of these, and a complete set is

C	$\vec{\lambda}_{(C)}$	$\vec{\xi}_{(C)}$	
1	$(y,-x,0)$	$(y/r^3, -x/r^3, 0)$	(22.1)
2	$(-z,0,x)$	$(-z/r^3, 0, x/r^3)$	(22.2)
3	$(0,z,-y)$	$(0, z/r^3, -y/r^3)$	(22.3)
4	$(0,0,2z)$	$1/8(x/r^3 - 3xz^2/r^5, y/r^3 - 3yz^2/r^5, -5z/r^3 - 3z^3/r^5)$	(22.4)
5	$(2x,0,0)$	$1/8(-5x/r^3 - 3x^3/r^5, y/r^3 - 3x^2y/r^5, z/r^3 - 3x^2z/r^5)$	(22.5)
6	$(0,2y,0)$	$1/8(x/r^3 - 3xy^2/r^5, -5y/r^3 - 3y^3/r^5, z/r^3 - 3y^2z/r^5)$	(22.6)
7	$(y,x,0)$	$(y/r^3 + x^2y/r^5, x/r^3 + xy^2/r^5, xyz/r^5)$	(22.7)
8	$(0,z,y)$	$(xyz/r^5, z/r^3 + y^2z/r^5, y/r^3 + yz^2/r^5)$	(22.8)
9	$(z,0,x)$	$(z/r^3 + x^2z/r^5, xyz/r^5, x/r^3 + xz^2/r^5)$	(22.9)

One property of the total linear momentum is that it equals the rate-of-change of the dipole moment. Consider, for example, the x-component of the total momentum

$$P_x = \int \lambda^a_{(1)} j_a d^3x, \tag{23}$$

where $\vec{\lambda}_{(1)} = (1,0,0)$. Now it is obvious that $(1,0,0) = \vec{\nabla}x$. We can substitute this into (23) and integrate by parts to get

$$P_x = -\int x\nabla_a j^a d^3x. \tag{24}$$

The continuity equation,

$$d\rho/dt + \nabla_a j^a = 0, \tag{25}$$

now gives,

$$P_x = \int x\frac{d}{dt}[\rho]d^3x = \frac{d}{dt}[\int x\rho\, d^3x] = \frac{d}{dt}[D_x], \tag{26}$$

where $D_x = \int x\rho\, d^3x$ is the x-component of the dipole moment.

This explains the independent multipoles at order $1/r^2$. The first three in (22) are the total angular momentum. The three $\vec{\lambda}_{(A)}$'s, with A = 1, 2 and 3, are the rotational Killing vectors and the three associated multipole moments give exactly the components of the total, conserved (ADM) angular momentum, defined just as in (21), except that one substitutes a rotational for a translational Killing vector. The remaining six multipole moments represent the time derivatives of the second moments of the sources (or of the total solution, in the non-flat case). For example, using (18) we get

$$\alpha_4 = \int \lambda^e_{(4)} j_e d^3x. \tag{27}$$

We have $\vec{\nabla}z^2 = (0,0,2z) = \vec{\lambda}_{(4)}$. Using this, and the continuity equation (25), in (27) gives us

$$\alpha_4 = \frac{d}{dt}[\int z^2\rho\, d^3x]. \tag{28}$$

Similarly we get

$$\alpha_5 = \frac{d}{dt}[\int x^2\rho\, d^3x], \qquad \alpha_6 = \frac{d}{dt}[\int y^2\rho\, d^3x],$$

$$\alpha_7 = \frac{d}{dt}[\int xy\rho\, d^3x], \qquad \alpha_8 = \frac{d}{dt}[\int yz\rho\, d^3x], \tag{29}$$

$$\alpha_9 = \frac{d}{dt}[\int xz\rho\, d^3x].$$

Thus we can immediately identify the combinations that represent the first time derivatives of the total quadrupole moments of the solution, i.e.,

$$\frac{d}{dt}[Q_{zz}] = \alpha_5 + \alpha_6 - 2\alpha_4, \tag{30}$$

$$\frac{d}{dt}[Q_{xy}] = \alpha_7, \qquad \text{and so on.} \tag{31}$$

This collection only picks up five of the six second moments. The missing one deals with the so-called 'moment of inertia about the origin',

$$B = \int r^2\rho\, d^3x. \tag{32}$$

It is

$$\alpha_4 + \alpha_5 + \alpha_6 = \frac{d}{dt}[B]. \tag{33}$$

The harmonic function that corresponds to this moment is $\vec{\xi}_{(4)} + \vec{\xi}_{(5)} + \vec{\xi}_{(4)}$. Let me call this

$$\vec{\xi}_{(B)} = -3/4(x/r^3, y/r^3, z/r^3). \tag{34}$$

5 'ROBIN' BOUNDARY CONDITIONS

In essentially all numerical formulations of general relativity the data are specified on a finite grid. Routinely, the finite grid is chosen to represent a finite volume of space. One obvious question concerns the boundary conditions to impose on the data. The boundary conditions for the metric are well understood. The natural choice for the geometry is to specify that the metric is asymptotically conformally flat. In the earliest calculations, the conformal factor was chosen to equal unity on the boundary of the grid. However, it was realised by York and Piran (York and Piran, 1982) that a mixed boundary condition, one that was tailor-made to pick up the leading harmonic function in the conformal factor, would allow a more accurate numerical code with relatively little extra work.

York and Piran also proposed the use of such mixed boundary conditions for the extrinsic curvature. They suggested

$$(LW)^{ab}n_a(\delta_b^c - 1/2n^c n_b) + \frac{6}{7R}W^a(\delta_a^c - 1/8n^c n_a) = 0, \qquad (35)$$

where $\vec{n} = (x/r, y/r.z/r)$, as a boundary condition for equation (7) on a boundary sphere of radius R. This condition is chosen so as to be compatible with the linear momentum harmonic functions (19).

However, many numerical problems are posed in such a way that the total linear momentum is zero. For such data, boundary condition (35) offers little advantage. For such problems it would be natural to choose a boundary condition which reflected the 'angular momentum' harmonic functions, array (22). Evans (Evans, 1984) has shown that if the the only multipole moments that were non-zero were the true angular momentum ones [C = 1,2,3 of array (22)], i.e., if all the second-moment terms [C = 4 . . . 9] vanished, one could use

$$(LW)^{ab}n_a + \frac{3}{R}W^a(\delta_a^b + n^b n_a) = 0. \qquad (36)$$

Objections can be made to this proposal. First, only a very restricted set of initial configurations would have constant second moments. Second, this 'constant second-moment' condition will not be preserved by the evolution. Thus, boundary condition (36) can only be used on a single slice.

I wish to suggest a resolution. Rather than give a complete solution to the problem, let me focus on a special case. Consider data which is axisymmetric and reflection symmetric, with zero linear and angular momentum. These conditions will be preserved by the Einstein evolution equations. A range of interesting problems fall into this class, from the head-on collision of two identical stars to the evolution of an axisymmetric gravitational wave. With these symmetry assumptions, the only harmonic functions that can be non-zero are $\vec{\xi}_{(4)}$ from (22), assuming the z-axis is the symmetry axis, and $\vec{\xi}_{(B)}$, as defined by (34). For the moment, let me assume that $\vec{W}$ is of the form

$$
\begin{aligned}
\vec{W} = &\frac{\alpha_4}{8}(x/r^3 - 3xz^2/r^5, y/r^3 - 3yz^2/r^5, -5z/r^3 - 3z^3/r^5) \\
&- \frac{3\alpha_B}{4}(x/r^3, y/r^3, z/r^3),
\end{aligned}
\tag{37}
$$

where α_4 and α_B are constants. It is a straight-forward calculation to show that

$$
(LW)^{ab}n_a = 3(\alpha_4 z^2/r^5 + \alpha_B/r^3)n^b,
\tag{38}
$$

$$
W^a n_a = \frac{r^2 - 3z^2}{8r^4}\alpha_4 - \frac{3}{4r^2}\alpha_B.
\tag{39}
$$

We also have

$$
\alpha_4 = \frac{5}{2\pi}\oint(1 - \frac{3z^2}{r^2})(W^a n_a)dS,
\tag{40}
$$

$$
\alpha_B = -\frac{1}{3\pi}\oint W^a n_a dS,
\tag{41}
$$

where we take the surface integral on any sphere of constant radius.

I claim that condition (38), where α_4 and α_B are defined via (40) and (41), form good boundary conditions for the equation $\nabla_b(LW)^{ab} = 8\pi j^a$. It forms an elliptic boundary value system (Hörmander 1969, 1985) and thus forms a Fredholm system. The next step is to evaluate the index, the difference between the dimension of the kernel and cokernel, of this system. The Fredholm nature of the operator and the index depend only on the ellipticity of the equation itself and the highest derivative term in (38). Thus, for the time being, we can replace the right-hand-side of (38) with zero. This would be the natural analogue of Neumann boundary conditions for eqn.(7). It is straightforward to show that the only harmonic functions (the kernel) of the operator are the conformal Killing vectors. Further, we require that the source, (5), must be orthogonal to the same conformal Killing vectors, and that this is the only restriction. Thus the index is zero.

Hence the system we are considering (with the nontrivial right-hand-side to (38)) satisfies the Fredholm alternative and so uniqueness implies existence. In other words, we need only prove that the equation

$$
\nabla_b(LW)^{ab} = 0,
\tag{42}
$$

when combined with the boundary condition on a sphere of constant radius R

$$(LW)^{ab}n_b = 3(\alpha_4 z^2/r^5 + \alpha_B/r^3)n^a, \tag{43}$$

$$\alpha_4 = \frac{5}{2\pi} \oint (1 - \frac{3z^2}{r^2})(W^a n_a)dS, \tag{44}$$

$$\alpha_B = -\frac{1}{3\pi} \oint W^a n_a dS, \tag{45}$$

implies $\vec{W} = 0$ to show that a solution exists to the inhomogeneous equation.

Let us multiply equation (42) by $\vec{W}$ and integrate over the sphere to give

$$0 = \int W_a \nabla_b (LW)^{ab} d^3 x$$
$$= -\frac{1}{2} \int (LW)^2 d^3 x + \oint W_a (LW)^{ab} n_b dS. \tag{46}$$

Substitution of (43) into the surface integral gives

$$\oint W_a (LW)^{ab} n_b dS = \oint 3(\alpha_4 z^2/r^5 + \alpha_B/r^3) W_a n^a dS. \tag{47}$$

Write $W_a n^a$ as in equation (39) but include a residual term Φ, i.e.,

$$W^a n_a = \frac{r^2 - 3z^2}{8r^4}\alpha_4 - \frac{3}{4r^2}\alpha_B + \Phi. \tag{48}$$

All we know about Φ, on using (44), (45) and (46), is that

$$\oint (1 - \frac{3z^2}{r^2})\Phi \, dS = 0, \tag{49}$$

$$\oint \Phi \, dS = 0. \tag{50}$$

If we substitute (48) back into (46), (49) and (50) give us that the integrals involving Φ must vanish , leaving us with

$$\oint 3(\frac{z^2}{r^5}\alpha_4 + \frac{1}{r^3}\alpha_B)W_a n^a dS = \oint \frac{3}{8}(\frac{z^2}{r^5}\alpha_4 + \frac{1}{r^3}\alpha_B)(\frac{r^2 - 3z^2}{r^4}\alpha_4 - \frac{6}{r^2}\alpha_B)dS. \tag{51}$$

The integration in (51) is straightforward to give

$$\oint W_a (LW)^{ab} n_b dS = -\frac{\pi}{R^3}[\frac{3}{20}\alpha_4^2 + (\frac{1}{2}\alpha_4 + 3\alpha_B)^2]. \tag{52}$$

This is clearly negative, and so therefore equation (46) implys that $\vec{W}$ is a conformal Killing vector and that both α_B and α_4 must vanish. These are incompatible (remember we assume axisymmetry and reflection symmetry). Thus we get the desired result that $\vec{W} = 0$ is the only solution to the homogeneous equation and so equation (43), (44), (45) define well-posed boundary conditions for the equation $\nabla_b (LW)^{ab} = 8\pi j^a$.

I suggest that

$$(LW)^{ab} n_b = 3(\alpha_4 z^2 / r^5 + \ldots + \alpha_7 xy / r^5 + \ldots) n^a, \qquad (53)$$

with a suitable definition of the α's (as in (44)) forms a good specification of the boundary data in the situation where the linear and angular momentum vanish, but no symmetry assumptions are made. The angular momentum terms do not fit directly into this scheme but a direct addition of the necessary terms may well work.

Acknowledgements
I would like to thank Helmut Friedrich for pointing me towards the Hörmander books. These have allowed me to convert the results in this article from, at best, persuasive arguments to something much stronger. I would also like to thank the British Council for a grant towards my travelling expenses to attend the conference. Much of the work described here was done in collaboration with James York many years ago.

REFERENCES
Evans, C. R. (1984). "A method for numerical relativity: Simulation of axisymmetric gravitational collapse and gravitational radiation generation." The University of Texas at Austin, Ph.D. dissertation.

Hörmander, L. (1969). *Linear Partial Differential Operators*. Springer-Verlag, Berlin, Chapter X.

Hörmander, L. (1985). *The Analysis of Linear Partial Differential Operators III*. Springer-Verlag, Berlin, Chapter 20.

Regge, T. and Teitelboim, C. (1974). *Ann. Phys. (N.Y.)*, **88**, 286-318.

York, J. W. (1973). *J.Math.Phys.*, **14**, 456-464.

York, J. W. and Piran, T. (1982). In *Spacetime and Geometry* R. A. Matzner and L. C. Shepley (eds.), University of Texas Press, Austin, 145-176.

ON THE CHOICE OF MATTER MODEL IN GENERAL RELATIVITY

A. D. Rendall

Max Planck Institute for Astrophysics, Munich, Germany

Abstract. Criteria are presented for choosing a matter model in analytical or numerical investigations of the Einstein equations. Two types of matter, the perfect fluid and the collisionless gas, are treated in some detail. It is discussed how the former has a tendency to develop singularities which have little to do with gravitation (matter-generated singularities) whereas the latter does not seem to suffer from this problem. The question of how the concept of a matter-generated singularity could be defined rigorously is considered briefly.

1 INTRODUCTION

In any investigation of the Einstein equations it is necessary to make some assumptions about the energy-momentum tensor. One possibility is simply to require that some energy conditions be satisfied. (In that case it might be more appropriate to say that the object of study is the 'Einstein inequalities'.) Despite the fact that this is sufficient to obtain important results including the singularity theorems and the positive mass theorem, it is very likely that there are significant results concerning the qualitative behaviour of solutions of the Einstein equations which require more specific assumptions. In any case the choice of a definite matter model is indispensable for numerical calculations and for analytic work based on the use of a well-posed initial value problem. The particular kind of matter chosen will of course depend on the problem being studied. If a concrete physical situation, such as a supernova explosion, is to be modelled then it makes sense to try to build in as much information about the microphysics as possible. If, on the other hand, one is interested in questions of principle in general relativity the priorities will be different. In this paper the aim is to discuss the best choice of matter model for studying those aspects of the behaviour of matter in gravitational fields which depend essentially on general relativity and very little on the detailed properties of the matter.

What has just been said may sound paradoxical for if certain phenomena are relatively insensitive to the choice of matter model this seems to imply that it is irrelevant which choice is made. One way of seeing that this need not be so is to realise that there could be 'pathological' matter models with physically unreasonable properties and that any universal behaviour of matter in gravitational fields will only be seen if these are avoided. Another important point is that, even if a certain kind of qualitative behaviour occurs for a wide range of types of matter, the difficulty of proving this rigorously may vary widely. Since the problem of proving theorems about situations such as gravitational collapse is a very hard one, it is necessary to take advantage of any simplifications which are available. Similar comments apply to numerical work.

What criteria are to be used when choosing a matter model for the purpose just outlined? Consider the following possibility. The dynamics of the matter leads, independently of the presence of gravitation, to a singularity and the resulting blow-up of the energy-momentum tensor (or its derivatives) causes the spacetime geometry to become singular as well. Here the 'singularity' could be one beyond which the solution cannot be meaningfully extended or it could merely be a point where the solution no longer has the minimal regularity required in order to be covered by existing theory. In the second case we fail to realise that there is no essential singularity because of technical problems. When this possibility, which I will refer to as a 'matter-generated singularity' is realised it can have two unfortunate consequences. The first is that an apparent spacetime singularity occurs which has no real physical significance. Secondly a matter-generated singularity may halt a numerical integration or an analytical investigation before the truly interesting gravitational singularities can be seen. An example of this will be given in section 2.

In this paper I will mainly discuss two types of matter model, namely the perfect fluid and the collisionless gas. Evidence will be presented that serious difficulties with matter-generated singularities can arise in the case of a fluid and that the collisionless gas offers a promising possibility of avoiding these. Dust, which is a common singular limit of these two models will also play a significant role. The question of how the concept of a matter-generated singularity, which has been introduced on an intuitive level above, can be made precise will be considered.

2 THE PERFECT FLUID

First some results of A. Titze (1989) will be described which provide a good example of how numerical and analytical investigations can complement each other. His starting point is a paper of Yodzis, Seifert and Müller zum Hagen (1974) where they demonstrated the development of naked singularities from regular initial data in certain circumstances. They treated a spherically symmetric perfect fluid with equation of state $p = f(\rho)$ where f is bounded. The singularities are not at the centre and so are not obviously an artefact of the symmetry. Titze first checked that he could, with his numerical scheme, reproduce the singularities found by these authors, starting from regular initial data. He then evolved the same initial data numerically using different equations of state. He found that for all unbounded equations of state he tried no singularity formed. This demonstrates concretely how the qualitative nature of gravitational collapse can depend strongly on the matter model chosen. He also obtained other results which are relevant to the question of matter-generated singularities. Namely, he evolved the same initial data with the equations for a Newtonian self-gravitating perfect fluid. He found that the development of singularities depended in that case on the equation of state in a manner very similar to the relativistic case. He then went further and switched off the gravitational field. Once again the behaviour was qualitatively similar. All this suggests that the singularities found by Yodzis *et. al.* are purely hydrodynamical in nature. They have little to do with gravitation and little to do with (even special) relativity.

Next I will present an example where a matter-generated singularity in a fluid hides a gravitational one. This is for a Newtonian self-gravitating fluid and I have unfortunately not been able to extend it to the general relativistic case. The equations to be solved are

$$\left.\begin{array}{c} \partial_t\rho + \partial_i(\rho v^i) = 0 \\ \rho(\partial_t v^i + v^j\partial_j v^i) + \delta^{ij}\partial_j p = \rho\delta^{ij}\partial_j U \\ \Delta U = -4\pi\rho \end{array}\right\} \tag{1}$$

where ρ is the density, p the pressure, v^i the velocity and U the Newtonian potential. The equation of state is polytropic, $p = K\rho^{\frac{n+1}{n}}$, with $1 \leq n \leq 3$ and initial data is chosen with the following properties.

(i) $\rho(0, x^i)$ is spherically symmetric, smooth, non-negative and has compact support
(ii) $\rho^{1/2n}(0, x^i)$ is smooth
(iii) $v^j(0, x^i)$ is smooth and spherically symmetric and vanishes where $\rho = 0$
(iv) $\int \rho(0, x^i)v^j(0, x^i)x_j \geq 0$
(v)   the energy is positive
The existence of a solution of (1) with the given data locally in time is far from obvious but is guaranteed by a theorem of Makino (1986). The technical condition (ii) required to ensure the applicability of this theorem. The local existence result has been extended to the general relativistic case in Rendall (1992a) and the obstacle to generalising the example on the formation of singularities to general relativity lies elsewhere. It can be shown that under the hypotheses (i)-(v) a local in time solution exists and the boundary of the support of ρ in this solution is freely falling. Thus if the solution existed long enough it would collapse to a single point in a predictable time $t_c = \pi\sqrt{R(0)^3/8M}$ where $R(t)$ is the radius of the support of the density at time t and M is the total mass of the fluid. In fact it does not exist that long, as is shown

by the following argument. Let

$$J = \frac{1}{2}\int \rho|x|^2,$$

$$E = \int(\frac{1}{2}\rho|v|^2 + np - \frac{1}{2}|\nabla U|^2)$$

The energy E is time independent. If the solution exists up to time t_c then it is clear that $J \to 0$ as $t \to t_c$ since $J \le \frac{2\pi}{3}MR^3(t)$. On the other hand, computing the second time derivative of J and doing some integration by parts gives $d^2J/dt^2 \ge E$. Integrating this inequality twice with respect to t gives

$$J(t) \ge J(0) + \dot{J}(0)t + Et^2. \tag{2}$$

Condition (iv) is equivalent to $\dot{J}(0) \ge 0$ and so (2) implies that $J(t) \ge J(0)$ and J cannot tend to zero. It follows that the solution cannot exist up to $t = t_c$. A matter singularity occurs before an interesting gravitational phenomenon (collapse to a point) has time to occur. It should be pointed out that no information is available about the nature of the singularity i.e. about which quantities become infinite there, since the proof of its existence was by contradiction. It seems that the only reasonable possibility of obtaining more precise information at the present time would be to do a numerical investigation. In particular it is not certain that this singularity is independent of the presence of gravitation. There is, however, indirect evidence that this is the case. The argument above is adapted from one used by Makino, Ukai & Kawashima (1986) to show the development of singularities in a Newtonian fluid body without gravitation and a similar argument also works in the special relativistic case (see Rendall 1992b). The fact that closely analogous arguments allow one to prove the existence of singularities in these various cases suggests strongly (although it does not prove) that the mechanism of singularity formation is similar for self-gravitating and non-self-gravitating fluids.

Lest the reader should think that worrying about the existence of a solution of (1) is a mere technicality, I would like to present an example with some couterintuitive properties. This works equally well in the Newtonian and general relativistic cases. Consider first spherically symmetric solutions of (1) in the case of an equation of state of the form $p = K\rho$ (which is not covered by Makino's theorem). For such an equation of state the second equation of (1) takes the form

$$\delta^{ij}\partial_i\rho = K^{-1}\rho(\delta^{ij}\partial_i U - \partial_t v^i - v^j\partial_j v^i). \tag{3}$$

I claim that there exists no C^1 spherically symmetric solution corresponding to any compactly supported initial data. For such a solution satisfies as a consequence of (3) an inequality of the form

$$\rho(r) \le C \int_{r_0}^r \rho(s)ds. \tag{4}$$

If ρ vanishes at some point $r = r_0$ then applying Gronwall's lemma (see e.g. Hartman 1982) to (4) shows that ρ vanishes identically. A more or less identical argument applies in the general relativistic case. The hypothesis that the solution is C^1 is more than is strictly required. In fact all that is needed is that ρ be continuous and that the Lagrangian derivative $\partial_t v^i + v^j \partial_j v^i$ of the velocity be bounded. This kind of equation of state occurs in the work of Ori & Piran (1990) on the formation of naked singularities in spherically symmetric fluid solutions of the Einstein equations. They first study self-similar solutions which cannot be asymptotically flat. The requirement of self-similarity fixes the form of the equation of state. They then cut off the density at a finite radius in order to obtain something which is asymptotically flat. The remarks just made show that in order to be able to do this consistently it is necessary either to alter the equation of state at low densities or to allow unbounded accelerations for the flowlines of the fluid. It is not clear whether unbounded accelerations for arbitrarily small amounts of fluid should be regarded as physically acceptable or not.

After these special examples I will say something about the formation of singularities in fluids in general. The *assumption* is made that for the issues considered gravitation and relativity make little difference to what happens, as in the calculations of Titze above. Of course the question of the *justification* of this assumption needs to be investigated but as no information is available at present no more can be said here. So what is known about the formation of singularities in classical hydrodynamics? Some of what follows is based on theorems while other statements are based on intuition which has been gained from numerical experiments. Consider first initial data on $\mathbb{R}^3$ for the Euler equations of the form $(\rho_0 + \epsilon \rho_1, \epsilon v_1)$. Here ρ_0 is a positive constant while ρ_1 and v_1 are smooth and compactly supported. If ϵ is small this is a small perturbation of an equilibrium solution. A theorem of Sideris (1985) will now be quoted, without however discussing the detailed technical conditions required. It says that for generic (ρ_1, v_1) the solution of the Euler equations corresponding to these initial data do not exist globally in time, no matter how small ϵ is. In fact the time of existence is bounded above by $C \exp(1/\epsilon^2)$ for some constant C. The solution exists for a very long time for small ϵ but not for ever: singularities are always formed. If we ask what the nature of these singularities is then the information available in the form of theorems is very limited (see however Chemin 1990). Nevertheless the following intuitive picture exists (Majda 1986, Sideris 1991). In an irrotational flow the singularities formed will be shock waves. In that case Sideris (1991) has obtained a lower bound of the form $C \exp(1/\epsilon)$ for the life span of the solution. In a general flow it is expected that much sooner than that (in a time of order $1/\epsilon$) singularities due to concentration of vorticity will occur. If we extrapolate this to the general relativistic case then the expectation is that singularities will almost always be formed in a fluid, even under mild conditions. In a violent situation such as gravitational collapse the likelihood is even greater. In our present state of knowledge such singularities are fatal for analytical investigations because we do not know how to continue the solution beyond them or even if it makes sense to do so. For classical hydrodynamics some limited results do exist (Majda 1984) but in general relativity there is nothing available up to now. It may be that the situation is much better for numerical calculations but I do not feel qualified to comment on that.

3 THE COLLISIONLESS GAS

A matter model which has a clean bill of health in the Newtonian case is the collisionless gas described by the Vlasov equation. The Newtonian equations are

$$\left. \begin{array}{c} \dfrac{\partial f}{\partial t} + v^a \dfrac{\partial f}{\partial x^a} - \delta^{ab} \dfrac{\partial U}{\partial x^a} \dfrac{\partial f}{\partial v^b} = 0 \\[2mm] \rho(t,x) = \displaystyle\int f(t,x,v)dv \\[2mm] \Delta U = -4\pi\rho \end{array} \right\} \tag{5}$$

Here the function $f(t, x^a, v^a)$ describes the density of particles in the (extended) phase space which in this case can be identified with $\mathbb{R}^7$. These particles are supposed to move without any interactions except that resulting from the gravitational field which they generate collectively. If C^1 initial data $f_0(x,v)$ with compact support are prescribed then there exists a corresponding global C^1 solution of (5). This was proved recently by Pfaffelmoser (1992) (see also Lions & Perthame 1991 and Schaeffer 1991). It is interesting to compare this with dust which results if the C^1 function f_0 is replaced by a distribution of the form

$$f_0(x,v) = \rho(x)\delta(v - \bar{v}(x)). \tag{6}$$

For self-gravitating dust it is easy to write down initial data which lead to a singularity in finite time. The theorem of Pfaffelmoser tells us that if the δ-function in (6) is smoothed out a little to give a sharply peaked function then no singularity occurs at all. Two superficially similar matter models lead to quite different results and dust comes under suspicion of being a pathological matter model.

This state of affairs is of relevance to the work of Shapiro and Teukolsky (1991) concerning the formation of naked singularities. Consider first the earlier work of these authors where they solved the Vlasov-Poisson system (5) numerically (Shapiro & Teukolsky 1987). There they took analytic solutions for dust and reproduced them numerically. The agreement between the numerical and analytic solutions is very good, even up to the point where a singularity is formed. This shows that the code can calculate dust collapse very accurately. However the global existence theorems just quoted show that the situation looks completely different if the initial data have a small but non-trivial velocity dispersion. It would be interesting to know if the numerical code could also reproduce this fact. The general relativistic calculations start with initial data analogous to the Newtonian ones; in particular they are initial data for dust. The obvious question which then arises is whether the resulting naked singularities might also be an artefact of dust and whether these singularities might not 'dissolve' if the data were smoothed out a little in phase space. Unfortunately there are no existing results on the initial value problem for the Vlasov-Einstein system which approach in generality those available in the Vlasov-Poisson case. Recently G. Rein and the author have begun a systematic investigation of the initial value problem for the Vlasov-Einstein system. The results obtained up to now all tend to support the idea that the singularities of Shapiro and Teukolsky are due to the use of dust as a matter model but are still a long way from proving this definitively. The theorems which have been proved up to now all concern the spherically symmetric case. One says that sufficiently small initial data (with 'small' defined in an appropriate way) develop into global singularity-free (i.e. geodesically complete) spacetimes (Rein & Rendall 1992a). Since it is known that, in the case of spherically symmetric dust solutions, naked singularities can develop for arbitrarily small initial data (Christodoulou 1984), this provides a confirmation of the idea that the introduction of velocity dispersion can get rid of naked singularities, at least in some situations. Another result is that under appropriate circumstances spherically symmetric solutions of the Vlasov-Einstein system converge uniformly to solutions of the Vlasov-Poisson system if the speed of light is allowed to tend to infinity (Rein & Rendall 1992b). This can be looked on as evidence that there is some justification on a mathematical level for using results in Newtonian theory to try to predict what happens in the relativistic case.

4 MATTER-GENERATED SINGULARITIES

How might the notion of a matter-generated singularity introduced informally above be given a precise mathematical definition? One possibility which works entirely within the framework of general relativity is to try to use the behaviour of the Weyl tensor to diagnose such singularities. If the energy-momentum tensor becomes infinite somewhere in some sense then the Ricci tensor is also forced to become infinite there. The same is not true of the Weyl tensor. Thus one might try to associate the notion of a matter-generated singularity with situations where the Ricci tensor becomes singular but the Weyl tensor remains regular. Since however the contracted Bianchi identities show that derivatives of the Ricci tensor act as a source for the Weyl tensor, it is not clear that matter-generated singularities do not themselves generate singularities in the Weyl tensor. It would be interesting to examine various examples, such as the shell-crossing singularities in dust solutions, from this point of view. Note that the singularities found by Shapiro and Teukolsky extend into the vacuum region and so must involve blow-up of the Weyl tensor. If, as suggested above, these really are matter-generated singularities then the Weyl tensor criterion would fail. At the other extreme, the homogeneous and isotropic cosmological models are conformally flat and nevertheless contain singularities which are truly gravitational in nature and do not depend on details of the matter model used.

Another way of trying to make the notion of a matter-generated singularity precise is to use the limits where the speed of light tends to infinity or the gravitational constant tends to zero. The first of these limits (the Newtonian limit of general relativity) is a singular one from the point of view of the theory of differential equations and therefore very difficult to control theoretically. It would therefore seem more practical to pass to the special relativistic limit and to see if the singularities persist there. The limit $c \to \infty$ is then regular and so it would be possible to pass to the case of non-relativistic matter without gravitation in a second step. It appears that this kind of program has up to now never been carried out in practice in the framework of analytical studies. The fact that something analogous works so nicely numerically in the work of Titze indicates that it would be well worth attempting an investigation of this kind.

Acknowledgements: My understanding of this subject has been furthered by discussions with several participants of the conference, especially John Stewart, and I would like to thank them for sharing their insights.

REFERENCES

Chemin, J-Y. (1990). *Commun. Math. Phys.*, **133**, 323-329.

Christodoulou, D. (1984). *Commun. Math. Phys.*, **93**, 171-195.

Hartman, P. (1982). *Ordinary differential equations*, Birkhäuser, Boston.

Lions, P-L. and Perthame, B. (1991). *Invent. Math.*, **105**, 415-430.

Majda, A. (1984). *Compressible fluid flow and systems of conservation laws in several space variables*, Springer, New York.

Majda, A. (1986). *Commun. Pure Appl. Math.*, **39**, S187-S220.

Makino, T. (1986). In *Patterns and Waves*, T. Nishida, M. Mimura and H. Fujii (eds.), North Holland, Amsterdam.

Makino, T., Ukai, S. and Kawashima, S. (1986). *Japan J. Appl. Math.*, **3**, 249-257.

Ori, A. and Piran, T. (1990). *Phys. Rev. D*, **42**, 1068-1090.

Pfaffelmoser, K. (1992). *J. Diff. Eq.*, **95**, 281-303.

Rein, G. and Rendall, A. D. (1992a). "Global existence of solutions of the spherically symmetric Vlasov-Einstein system with small initial data", preprint.

Rein, G. and Rendall, A. D. (1992b). "The Newtonian limit of the spherically symmetric Vlasov-Einstein system", preprint.

Rendall, A. D. (1992a). "The initial value problem for a class of general relativistic fluid bodies", *J. Math. Phys.*, to appear.

Rendall, A. D. (1992b). "The initial value problem for self-gravitating fluid bodies", *Proc. 10th International Conference on Mathematical Physics, Leipzig*, to appear.

Schaeffer, J. (1991). *Commun. Partial Diff. Eq.*, **16**, 1313-1336.

Shapiro, S. and Teukolsky, S. (1987). *Astrophys. J.*, **318**, 542-567.

Shapiro, S. and Teukolsky, S. (1991). *Phys. Rev. Lett.*, **66**, 994-997.

Sideris, T. (1985). *Commun. Math. Phys.*, **101**, 475-485.

Sideris, T. (1991). *Indiana Math. J.*, **40**, 535-550.

Titze, A. (1989). PhD Thesis, University of Hamburg.

Yodzis, P. and Seifert, H-J., Müller zum Hagen, H. (1974). *Commun. Math. Phys.*, **37**, 29-40.

A MATHEMATICAL APPROACH TO NUMERICAL RELATIVITY

J. W. Barrett

Department of Applied Mathematics and Theoretical Physics, University of Cambridge, Cambridge, UK

Abstract. This article contains some proposals for the construction of an algorithm for the evolution of initial data in general relativity which will apply to generic initial values. One of the main issues is to allow a dynamic refinement of the discretisation which will be local and vary according to local values of the initial data. I outline some of the main problems which will have to be addressed in any implementation of the general scheme. There are also some suggestions for a construction of a smooth solution of the Einstein equations which is near to the discrete evolution.

1 INTRODUCTION

At the present time, computer codes for general relativity are written specifically for particular problems such as stellar collapse or coalescing binary systems. In the longer run relativists are interested in using the computer as a mathematical tool to investigate the properties of solutions which seem inaccessible by analytic means, or to formulate hypotheses which may then be attacked analytically. This requires the construction of an algorithm which applies to generic initial data and which also has a sufficiently solid framework which allows analytic investigation of the error of the approximation.

The approach I would like to suggest is based on triangulations. One of the problems of numerical relativity is that the degree of discretisation that is required to approximate given data well is dependent on that data. However one cannot predict — in advance — how this will evolve as the data evolves with time. For that reason, I stipulate the algorithm should allow the use of arbitrary triangulations of space and allow the triangulation to change with time in as flexible a manner as possible. The algorithm should be allowed to make changes to the discretisation of space which are different at different points in space, according to the local values of the data.

A recent advance in the theory of triangulations is helpful at this stage. A local move on a triangulation is the replacement of a small part of it by a different triangulation of the same manifold. Alexander (1930) produced a list of moves which will change any triangulation of a manifold into any other. Unfortunately, whilst Alexander's moves are local, the list has an infinite number of moves of different types. Reducing this to a finite list of moves was an unsolved problem until recently. The solution is rather simple, and has an elegant interpretation. The change of an n-dimensional triangulation is a result of its 'time evolution' given by deforming it as a hypersurface in an $n + 1$-dimensional manifold. This will be described next.

2 MOVES ON TRIANGULATIONS

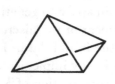

order 4 ⇕ order 1

order 3 ⇕ order 2

Figure 1.

Let M be an n-dimensional manifold. A triangulation of M is a decomposition of M into a union of a finite number of n-simplexes, $M = \bigcup\{\sigma\}$, such that the intersection $\sigma_1 \cap \sigma_2$ is either a common face or empty. (I omit some technicalities, e.g., M should be PL and compact. See Rourke and Sanderson (1982) for the definitions, and the papers referred to for a proper statement of the theorem). An example is the boundary of an $n + 1$-simplex, which is the union of $n + 1$ n-simplexes. Topologically, this is the n-sphere, S^n. Consider a partition of these simplexes into two (non-empty) subsets, K_1 and K_2. Then K_1 and K_2 are topologically n-dimensional disks, $K_1 \cup K_2 = S^n$ and $K_1 \cap K_2 = S^{n-1}$. For the case $n = 3$, I have drawn the two possible partitions in figure 1. On the left, there are two complexes with vertical arrows between them. This is the pair of complexes of the first partition, the remaining two on the right forming the second partition.

Now let T_1 and T_2 be triangulations of M. The triangulation T_2 is said to be obtained from T_1 by the application of an elementary move if it is obtained by removing a disk

K_1 and replacing it with a disk K_2. This is written $T_1 \rightarrow T_2$. In 3 dimensions, there are four elementary moves. The elementary move of order 1 replaces four tetrahedra with one. The elementary move of order 2 replaces three tetrahedra with two. The move of order 3 replaces two with three, and the move of order 4 replaces one with four. These are carried out according to figure 1. These moves have been used in practice in computer simulations, for example as in (Agistein and Migdal 1991) in statistical physics.

The following theorem was proved by Pachner (1991). Gross and Varsted (1991), and Bruce Westbury and I (1992) proved this independently for the case $n \leq 3$, and have a simple version of the argument.

Theorem. Let T and T' be triangulations of M. Then $T \rightarrow T_1 \rightarrow T_2 \ldots \rightarrow T_k \rightarrow T'$ for some finite k.

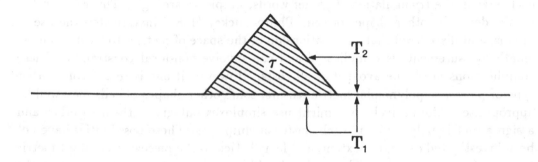

Figure 2.

The result is that the elementary moves are sufficient to generate all triangulations of M starting from any one. The case of interest to us here is $n = 3$, where M is a space-like hypersurface. The elementary move can be interpreted as gluing the 4-simplex σ to M, so that the new triangulation is that of the manifold which runs 'over the top of this', as in figure 2. The 4-simplex forms the space-time region which interpolates between the two hypersurfaces. By repeated application of such moves, one can build up an extensive region of space-time. (There are some conditions to check which are described below). In this way, finite element methods can be constructed which have the full flexibility to adjust the triangulation dynamically according to the needs of the data in each local neighbourhood. The idea is that the choice of elementary move which is to be performed should be made by examining the data. I make the rather bold claim that any numerical scheme which has such a dynamical grid refinement can be boiled down to essentially the same mathematics. This is because any other type of cell can be subdivided into simplexes without adding any extra vertices (see Rourke and Sanderson 1982), and any moves on cells can be achieved by the elementary moves on triangulations by the above theorem.

3 METRIC DATA

In a finite element method, one approximates the infinite dimensional space of continuous functions (or sections of the appropriate bundle) by a finite dimensional space of piecewise polynomial functions. The pieces on the simplexes of the triangulation are glued together in some appropriate way, to preserve some degree of continuity or differentiability. In relativity, the basic variable is the metric, a symmetric tensor field. The simplest choice for its discretisation is that of Regge calculus (Regge 1961). Here, the metric tensor is piecewise constant, with the condition that the induced metric tensor on the common face of any two simplexes should be common to the two. Under these circumstances, the simplex is by far the best choice of convex cell. This is because the metric tensor is uniquely determined by the squared length of each edge in the triangulation. In other words, simplexes are rigid. This relationship breaks down for other shapes of cell. Philip Tuckey (1989) investigated the use of prisms, and discovered that the relationship of the space of metrics to locally defined length measurements is complicated and can involve non-local constraints. These complications are to be avoided at all costs. However, if one is to use some other type of piecewise polynomial metric tensors, some other shapes of cells may well be appropriate. Alternatively, one might use simplexes but group them together and assign a metric polynomial to each cluster of simplexes. These possibilities have not been investigated to a great extent, and I shall stick to the piecewise-constant metric tensors of Regge calculus. The reader should bear in mind that this is due to the need for a simple and concrete example, and that many of my remarks will hold for a more general finite element method. I shall also confine myself to a discussion of the vacuum equations.

It is useful to spell out the precise relationship between the metric tensor and the edge lengths.

Proposition. The specification of the square of the length for each edge of a k-simplex uniquely determines a constant metric tensor.

Proof. Choose one vertex as the origin, so that the remaining vertices form a basis of the k-dimensional vector space in which the simplex is embedded. The square of each basis element is uniquely determined as the square of an edge length, and the inner product of two different basis vectors is determined from the three squared edge lengths of the triangle in which they lie by the formula $a.b = 1/2(a^2 + b^2 - (a - b)^2)$.

The Einstein-Hilbert action has a straightforward extension to the piecewise-constant metric tensors. The condition that this should be stationary with respect to variations of the metric determines the equations of motion for the metric. The exact form of the equations (see Regge 1961) is not important here. Variations of the metric are equivalent to variations of the edge lengths, and so there is one equation for each edge. This equation is a constraint on the geometry of a small neighbourhood of any point on the edge. Now suppose S is a simplicial spacetime manifold and M a simplicial space-like hypersurface. The equations which belong to the edges in M are constraints on the geometry immediately to the future and to the past of M. Thus in a Cauchy problem these are the evolution equations which allow one to constrain the geometry to the future of M, depending on a fixed value of the geometry of M and to the past of M. In favourable circumstances, one might be able to solve the equations uniquely, in which case one can speak of the evolution of the initial data. More realistically, there may be a class of solutions, or of solutions up to some predetermined tolerance of error.

Following the suggestion of Sorkin (1975), the initial data for the Cauchy problem consists of a solution to all of the equations which belong to the edges in S which are to the past of M. Let $P \subset S$ be the submanifold which consists of the complement of the future of M (the 'past'), so that $\partial P = M$. Now suppose that a larger set P', $P \subset P' \subset S$, is obtained from P by gluing a 4-simplex onto M, so that $M' = \partial P'$ is obtained from M by an elementary move. The equations for edges which are interior to P form the constraint equations for the initial data. The equations for edges which are interior to P' but not P are the evolution equations, and are also the extra constraint equations which need to be satisfied for the new initial data on (P', M'). These edges actually all lie on M, because only a single 4-simplex has been glued on.

The evolution equations are intended to constrain the geometry to the future of M, assuming fixed values for the geometry of P. This new geometry to the future is parameterised by the lengths of the edges which are in P' but not in P. One can count the number of new equations and the number of new variables.

order 1:	4 equations	no variables
order 2:	1 equation	no variables
order 3:	no equations	1 variable
order 4:	no equations	4 variables

This list is puzzling because all of the cases appear to be pathological in the sense that they either introduce new data without any constraint, or introduce new constraints on the already existing data. The regular case, where the number of new variables matches the number of new equations, is absent. However, this is not an indication that anything is wrong. To use the computer jargon, this is a feature rather than a bug.

In general, if one wishes to change the triangulation of space from M_1 to M_2 and M_2 contains more edges than M_1, then it will be necessary to create some of the data to match the more refined grid, by some process such as linear interpolation. Similarly, if M_2 contains less edges than M_1 some data must be lost in projecting onto a coarser grid. These are the processes which are occuring in the case of the elementary moves.

In fact, by examining the above list of elementary moves, it is clear that the change in the number of edges of M_1 as it evolves to M_2 through a number of elementary moves is equal to the total number of new variables minus the total number of new equations.

In order to obtain a regular evolution, where the number of new variables matches the number of new edges, several elementary moves of different types must be considered together. This means that $P' \supset P$ now differs by more than one 4-simplex, and that the equations for all of the edges interior to P' but not interior to P are considered together as one evolution step.

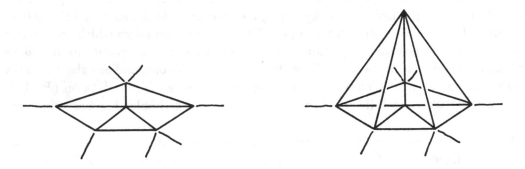

Figure 3.

An example of this is Sorkin evolution. Sorkin (1975) considered an example of the Cauchy evolution for a space containing just eight vertices. Phillip Tuckey realised that Sorkin's method generalises to any manifold, giving an evolution which does not change the triangulation of M. The extension P' is given as follows. Pick a vertex $v \in M$, and glue the cone over the star of v in M onto the star of v by its base (figure 3). (The star of v is the union of all simplexes containing v, while the cone over a space is a space of one higher dimension which contains one extra vertex, v'. The simplexes of the cone are of the form $(v', v, a, b, c, \ldots)$, where $(v, a, b, c, \ldots)$ is a simplex in M, and faces of these). One can show, for a 4-dimensional space-time, that P' is obtained by some sequence of elementary moves, and in fact this starts with one move of order 4, proceeds with an equal number of moves of orders 2 and 3, and ends with a move of order 1. The fact that there is a sequence of elementary moves is a little delicate, and would not always be true in a spacetime of 5 or more dimensions (see Goodrick (1968) for the complications).

4 CRITERIA FOR EVOLUTION

Sorkin evolution is an example of a cluster of elementary moves. A cluster means that the a sequence of elementary moves in the same neighbourhood are performed together. The big open question is what types of clusters are suitable for evolution. This will depend on the initial data, both the topology of the triangulation and the metric data. I shall outline some criteria which I think are important.

4.1 Topology

Let P be the past of an initial surface M, so that $M = \partial P$. However, let us not assume any more that the entire spacetime (previously called S) already exists, or is already determined. The problem is to glue a cluster onto P so that this defines an extension of the manifold which is actually a manifold. In general relativity, Cauchy development is always trivial topologically, and so the new manifold should have the same topology as P.

So let P' be obtained from P by gluing on a cluster of simplexes C, so that $P' = P \cup C$, $C \cap P = \partial C \cap M$. Then C should be topologically a 4-dimensional ball, and $\partial C \cap M$ a 3-dimensional ball. From these conditions, it follows that the evolution P' is a PL manifold, and is homeomorphic to P.

However it may not be true that $P' = P \cup C$ is a simplicial complex, if one just requires that the gluing is topological, as I specified above. For this to be true, the extra condition that has to be checked is that if σ_1 is a simplex of P, σ_2 a simplex of C, then $\sigma_1 \cap \sigma_2$ is a common face. This is a simple combinatorial condition which can be checked easily in practice, and I think that one would want the condition that all manifolds are simplicial complexes. For example, an elementary move of order p, $p \geq 1$, followed by its inverse, the move of order $4 - p$, does not satisfy the condition. In this example the cluster C is taken to be the single 4-simplex each time.

4.2 Signature

There is a large number of possibilities for the signatures of all the faces of a Lorentzian 4-simplex. However, all of the faces which are in one of the Cauchy hypersurfaces M should be spacelike. If the elementary moves are performed one at a time, then this means that all simplexes of dimension less than four are spacelike. This is the puff-pastry choice of signatures. If the equations for a cluster of elementary moves are solved together, the signatures of the faces interior to the cluster are not fixed, but will be constrained by other criteria. Kheyfets, LaFave and Miller (1990) used an evolution in which a large number of edges were fixed to be null. While this has certain advantages, it forced them away from simplexes. Also, they considered the time evolution of an entire Cauchy surface at one instant. If clusters are performed sequentially, then, as I have pointed out above, the faces on the boundary of each cluster have to be space-like, so these edges cannot be null. Thus, while maximally-null approaches are worth investigating, I shall not pursue them here.

4.3 Fatness

Let us, for the moment, consider Euclidean space with a positive definite signature metric. The fatness of a simplex is a parameter between 0 and 1 which measures the degree to which its shape departs from the equilateral simplex, with 1 being the equilateral case and 0 for any degenerate simplex, one such that the vertices lie in a hyperplane. There are several essentially equivalent ways one can define fatness. One can measure the volume v and the mesh η (the maximum edge length) and define fatness as

$$\Theta = \frac{v}{v(\eta)},$$

$v(\eta)$ being the volume of the equilateral simplex with edge length η. Alternatively, one can look at the eigenvalues of the operator which maps the equilateral simplex onto the given one, as in (Barrett and Parker 1992). Requiring the fatness of a simplex to be bounded below

$$\Theta \geq B > 0$$

is a regularity condition which is crucial if one examines the convergence of a sequence of approximations, as in (Barrett and Parker 1992). This condition also played a role in the approximation theory for the scalar curvature (Cheeger, Müller, Schrader, 1984) and in the linearised theory (Barrett 1988). The condition has also featured in many practical applications of the finite element method. In the first paper, Phil Parker and I examined examples of convergence without such a bound on the fatness. Various bizarre and counterintuitive things can happen, including the limit of the sequence of approximations being a metric with a different signature to the one being approximated. These examples indicate that the piecewise-constant metric tensors are bad approximations of smooth metrics unless the fatness is controlled. In a sense, allowing the fatness parameter to approach zero is akin to a coordinate singularity for a smooth metric. For this reason, I expect such a bound to be important in numerical relativity. Maintaining such a bound will influence both the use of gauge freedom (discussed below) and the choice of cluster to evolve.

The most fruitful definition of fatness for a Lorentzian signature simplex has to be determined. It is possible that it could rule out null faces, and null edges in particular, as too degenerate.

4.4 Stability
The numerical stability of the evolution is a key property. I expect that an analysis of the evolution equations will lead to restictions on the signature, topology and fatness bound of a cluster. As Mann (1989) points out, it is common in the application of finite element methods to hyperbolic systems that the weight functions have to be adjusted for a trade-off between accuracy and stability. Porter's (1987) implementation of a finite element code for a spherically-symmetric problem of relativity and matter using prisms appeared to be unstable. However, I should warn the reader that this remark is anecdotal, rather than based on analysis. Also, it applies to a scheme which uses prisms as the basic cells which I do not wish to consider.

4.5 Regularity of the Cauchy surface M
The surface $M = \partial P$ has an extrinsic curvature, whose definition is a simple extension of the usual smooth definition (Hartle and Sorkin 1981). Such a measure can be used to ensure that the boundary M does not become too irregular. For example, one could perform a local evolution at a vertex if the trace of the extrinsic curvature is negative there. This would provide a procedure analogous to the maximal slicing condition.

4.6 Magnitude of the curvature
A local measure of the amount of curvature per simplex can be obtained from the local holonomy elements in the piecewise-constant metric. In the positive signature case, this would reduce to the deficit angles of cones. This measure will be a guide to the accuracy of the approximation which is possible with this mesh size. Thus a large curvature indicates that the evolution should proceed by creating more vertices and edges in the Cauchy surface, and a small curvature indicates that it is safe to reduce the number of vertices and edges, increasing the mesh size. Since the evolution is local, both processes can occur in spatially separated regions at once. In fact, this idea should be sufficiently flexible to allow the finely triangulated portions of space to 'flow' with the data where the curvature is high, and where the high accuracy is required.

4.7 Use of the gauge freedom
The discrete equations of motion retain a gauge freedom. This means that the set of solutions to the equation |Einstein tensor| $< \epsilon$ forms a region in the space of metrics which is narrow in most directions, but very long in a few directions. The region

will be like a thickening of a part of a surface. If the space is flat, there is an exact continous family of symmetries.

Let x^μ be coordinates on the space of the new edge lengths to be determined in a cluster, and $I[x]$ the action. Denote the derivatives at x by $I_\nu[x]$, $I_{\mu\nu}[x]$, etc. In a Newtonian iteration for the stationary point of the action, one solves the linear equation

$$I_{\mu\nu}[x](x^\mu - x_0^\mu) = I_\nu[x]$$

for x_0 as a better approximation to a solution of the equation $I_\nu[x] = 0$ for x. As Hartle (1986) points out, some of the eigenvalues of $I_{\mu\nu}[x]$ are close to zero, depending on the accuracy ϵ being considered. Thus the solution for x_0 is approximately a linear space of some dimension rather than a point.

This indeterminacy has to be used carefully to satisfy the criteria outlined above. For example, it may be sufficient to pick an arbitrary choice of the lengths of the four edges introduced whenever a move of order 4 is made. In this case the choice could be made to make the new 4-simplex have a standard 'equilateral' Minkowskian geometry so that the fatness criterion is kept on track.

5 APPROXIMATION OF EXACT SOLUTIONS

Ultimately, one wants to be able to say that a computed solution is an approximation of an exact solution to the Einstein equations. This has not been carried out for any approach, but, due to the geometrical nature of the finite element approach, the prospects for results in this direction with this approach are good. A program for this is as follows:-

(1) The computed solution defines the metric tensor as a discontinous function, and the curvature as a tensor distribution.

(2) A smooth metric is obtained by convolution with a suitable kernel on $S \times S$.

(3) A calculation with generalised functions shows that the curvature of the smooth metric differs from the curvature of the computed solution by a small distribution. The estimate for this will depend on the mesh size, the convolution kernel and the magnitude of the computed curvature.

(4) The distributional Einstein tensor for the computed solution is small, as in (Barrett 1986, 1987, 1988) and so leads to the result that the Einstein tensor of the smooth metric is small, in a similar way to (Barrett 1988).

(5) A smooth metric with an Einstein tensor which is small in a suitable Sobolev norm, is close to a smooth metric which is vacuum. (I am indebted to James Vickers for discussing these points with me).

I should stress that this has not been carried through to the desired conclusion, but that various parts of this program have been investigated in isolation from each other. I look forward to progress in this direction.

REFERENCES

Alexander, J. W. (1930). *Ann. of Math.(2)*, **31**, 294–322.

Agistein, M. E. and Migdal A. A. (1991). *Mod. Phys. Lett. A.*, **6**, 1863–1884.

Barrett, J. W. (1986). *Class. Quant. Grav.*, **3**, 203–206.

Barrett, J. W. (1987). *Class. Quant. Grav.*, **4**, 1565–1576.

Barrett, J. W. (1988). *Class. Quant. Grav.*, **5**, 1187-1192.

Barrett, J. W. and Parker, P. E. (1992). *J. Approx. Theory*, to appear.

Barrett, J. W. and Westbury, B. W. (1992). "Invariants of piecewise-linear manifolds", preprint.

Cheeger, J., Müller, W. and Schrader, R. (1984). *Comm. Math. Phys.*, **92**, 405-454.

Goodrick, R. E. (1968). *Proc. Camb. Phil. Soc.*, **64**, 31–36.

Gross, M. and Varsted, S. (1991). "Elementary moves and ergodicity on D-dimensional simplicial quantum gravity", Niels Bohr Institute preprint, NBI-HE-91-33.

Hartle, J. B. (1986). *J. Math. Phys.*, **27**, 287–295.

Hartle, J. B. and Sorkin, R. (1981). *Gen. Rel. Grav.*, **13**, 541-549.

Kheyfets, A., LaFave, N. J. and Miller, W. A. (1990). *Phys. Rev. D*, **41**, 3628–3636, and *Phys. Rev. D.*, **41**, 3637–3651.

Mann, P. J. (1989). In *Frontiers in numerical relativity*, C.R. Evans, L.S. Finn and D.W.Hobill (eds.), Cambridge University Press, Cambridge, 230–238.

Pachner, U. (1991). *Europ. J. Combinatorics*, **12**, 129–145.

Porter, J. (1987). *Class. Quant. Grav.*, **4**, 391–410.

Regge, T. (1961). *Nuovo Cimento*, **19**, 558-571.

Rourke, C. P. and Sanderson, B. J. (1982). *Introduction to piecewise-linear topology*, Springer, Berlin.

Sorkin, R. (1975). *Phys. Rev. D*, **12**, 385–396.

Tuckey, P. A. (1989). *Class. Quant. Grav.*, **6**, 1–21.

MAKING SENSE OF THE EFFECTS OF ROTATION IN GENERAL RELATIVITY

J. C. Miller

Trieste Astronomical Observatory, Trieste, Italy

Abstract. It is argued that having a good conceptual understanding of relativistic effects is very important when undertaking large computations in numerical relativity. The radius of gyration (the square root of the ratio of the specific angular momentum to the angular velocity) is proposed as a useful quantity for the analysis of effects which are related to rotation.

1 INTRODUCTION

This paper is concerned with the effects of rotation in general relativity and parts of it draw heavily on joint work carried out together with M.A. Abramowicz of NORDITA and Z. Stuchlík of the Silesian University of Opava (see Abramowicz et al. 1992).

Among the areas of particular interest for current work in numerical relativity are the following:

   (i) collisions of neutron stars or black holes;
  (ii) realistic three-dimensional relativistic gravitational collapse;
 (iii) non-axisymmetric behaviour of compact objects;
  (iv) accretion onto compact objects;
   (v) processes in the early universe;
  (vi) behaviour of gravitational waves;
 (vii) formation of singularities.

In much of this, the calculations are intrinsically three-dimensional and rotation plays a crucial role. The question arises: how well do we *understand* the effects of rotation in general relativity? One could argue that this is unimportant; there is a system of equations to be solved for given initial conditions and boundary conditions and the process is a mechanical one which leads to predictions which could then be tested experimentally, at least in principle. On this view, conceptual understanding is not relevant and, indeed, attempting to achieve it may actually be counter-productive since our intuition is firmly based in Newtonian theory and may be misleading when applied to relativistic phenomena. Everything which one needs to know is there in the equations and so one should let them "do the work" without becoming involved with other unnecessary considerations. While this view has some merit, I do not agree with it for two reasons which are of rather different types. Firstly, I think that one of the main roles of science is precisely to give us this sort of conceptual understanding. Being able to predict the results of possible experiments is only part of the process. Secondly, and this is particularly relevant in the present context, conceptual understanding is directly useful and possibly even essential for people involved with numerical relativity. When working with large computer codes, it is very important to have some idea of what behaviour is to be expected as a guide to knowing what is real and what is not, and for being able to tell when some extraneous numerical effect is corrupting the solution. Error analyses and checks with test problems, while essential, are not sufficient on their own for knowing when results are reliable.

2 IMPORTANT DIFFERENCES BETWEEN GENERAL RELATIVITY AND NEWTONIAN THEORY WITH REGARD TO ROTATION

In strong-field general relativity, the effects of rotation differ considerably from those familiar in Newtonian theory. These differences are related to the following:

(i) dragging of inertial frames;
(ii) changes in centrifugal effects;
(iii) a change in the relation between angular momentum and angular velocity;
(iv) emission of gravitational radiation when the shape of a gravitating object changes with time.

An example which shows some of these differences in action is given by the case of the quasi-stationary sequence of slowly rotating, general relativistic Maclaurin spheroids as studied by Chandrasekhar and Miller (1974). This sequence is constructed in such a way that (J/M^2) is constant for all of the models (where J is the total angular momentum and M is the mass) but they have progressively decreasing values of (R/R_S) (where R is the radius and R_S is the Schwarzschild radius). (Throughout, we are using units for which $c = G = 1$ and we will adopt the space-like signature convention.) This sequence mimics the contraction of a Maclaurin spheroid, with mass and angular momentum being conserved. Some results of the calculations are shown in figures 1 and 2. Figure 1 shows the ellipticity ϵ, measured in units of $(J/M^2)^2$, plotted as a function of (R/R_S). In Newtonian theory, this curve would be a rectangular hyperbola; as contraction proceeds to progressively smaller values of (R/R_S), the ellipticity continually increases. In contrast, general relativity predicts that when neutron star dimensions are reached, the increase in ellipticity ceases and the object starts to become more spherical again. Centrifugal effects are being seriously weakened with respect to what would be expected from Newtonian theory for objects on the scale of neutron stars. Using Newtonian theory to calculate ϵ for $R/R_S \sim 2.5$ (a value typical for neutron stars) gives differences of order 100% with respect to

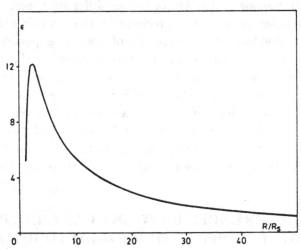

Figure 1. *Results for a sequence of slowly rotating Maclaurin spheroids calculated according to general relativity. Ellipticity ϵ [measured in units of $(J/M^2)^2$] is plotted as a function of (R/R_S).*

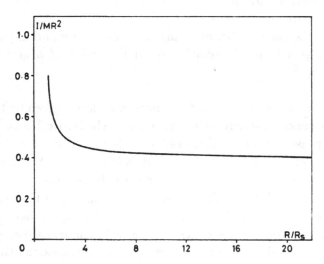

Figure 2. *Results for the same models as in figure 1. The ratio (I/MR^2) is plotted as a function of (R/R_S).*

results coming from general relativity. This is not a small difference and casts doubt on the use of weak-field approximation schemes for calculations involving *any* rotational properties of objects as compact as this. Figure 2 shows the ratio (I/MR^2) plotted as a function of (R/R_S) for the same sequence (I is the moment of inertia defined as (J/Ω) where Ω is the angular velocity as measured by an observer at infinity). For large values of (R/R_S), this ratio tends to its Newtonian limit of 0.4 but at smaller (R/R_S) it increases. The object is behaving as if its radius were larger than the actual value; its inertia is increased above the corresponding Newtonian value.

Our motivation here is to search for a conceptual framework within which behaviour such as that just described would fall naturally into place.

3 RADIUS OF GYRATION, CENTRIFUGAL FORCE AND THE PHOTON EFFECTIVE POTENTIAL

In Newtonian theory, the radius of gyration is a familiar concept in connection with the rotational properties of rigid bodies. It is defined as the radius of the circular path on which a point-like particle having the same mass and angular velocity as the rigid body, would also have the same angular momentum; in other words, it is the value of $\tilde{r}$ given by the equation

$$J = M\tilde{r}^2\Omega,\tag{1}$$

so that

$$\tilde{r} = \sqrt{\frac{J}{M\Omega}}.\tag{2}$$

For a point particle moving along a circle, the value of $\tilde{r}$ given by (2) is (trivially) just the ordinary radius of the circle.

In general relativity, things are more complicated because there is more than one satisfactory way of measuring the radius of a circle. Two standard measures are the *circumferential* radius (the proper circumference divided by 2π) and the *radial proper distance*. If a generalization of (2) is used as the definition of $\tilde{r}$ for a point particle moving along the circle, then this provides a third measure. We use the generalization

$$\tilde{r} = \sqrt{\frac{L}{E\Omega}},\tag{3}$$

where L is the angular momentum of the particle and E is its energy. (Note that the level surfaces of $\tilde{r}$ defined in this way are the *von Zeipel cylinders* which play an important role in discussions of rotating fluids.) Whereas, in Newtonian theory, these three ways of measuring radius give results which are identical, in general relativity they are distinct. We propose that $\tilde{r}$ is the right quantity to use for discussing rotational effects and that the direction of increase of $\tilde{r}$ defines the *local outward symmetric* with Killing vectors η^α (time-like) and ξ^α (space-like). Generalization to the case of a stationary space-time is straightforward, however, and has already been completed. Extension to completely general space-times is presently under investigation.

First, I will give invariant definitions of various quantities which will be used in the subsequent discussion. The proper circumferential radii of circles whose symmetry axis corresponds to that of the space-time, are given by

$$r = \sqrt{(\xi\xi)}, \tag{4}$$

and the "global" outward direction is given by the vector $\nabla_\alpha r$. The radius of gyration $\tilde{r}$ is given by

$$\tilde{r} = \sqrt{\frac{L}{E\Omega}} = \sqrt{-\frac{(\xi\xi)}{(\eta\eta)}}, \tag{5}$$

(it is defined to be strictly positive) and the associated "local" outward direction is that of the vector $\nabla_\alpha \tilde{r}$.

For circular particle motion with constant speed we have

$$v^\alpha = \frac{(\eta^\alpha + \Omega\xi^\alpha)}{\sqrt{-[(\eta\eta) + \Omega^2(\xi\xi)]}}, \tag{6}$$

and

$$a_\beta \equiv v^\alpha \nabla_\alpha v^\beta = \frac{1}{2}\frac{\nabla_\beta(\eta\eta) + \Omega^2\nabla_\beta(\xi\xi)}{(\eta\eta) + \Omega^2(\xi\xi)}, \tag{7}$$

where v^α and a_β are the particle four velocity and four acceleration respectively. A *static observer* with four velocity

$$u^\alpha = \frac{\eta^\alpha}{[-(\eta\eta)]^{1/2}}, \tag{8}$$

uniquely defines a global rest frame and a projected three-space. The orbital speed of the particle in this projected three-space is

$$\tilde{v} = \frac{\tilde{r}\Omega}{\sqrt{1 - \tilde{r}^2\Omega^2}}, \tag{9}$$

and the acceleration can be written as

$$a_\alpha = \frac{1}{2}\nabla_\alpha \ln[-(\eta\eta)] - \frac{\tilde{v}^2}{\tilde{r}}\nabla_\alpha \tilde{r}. \tag{10}$$

In order to keep the particle moving on its circular path, a force must in general be applied so as to produce the acceleration given by (10) except that in the special case of geodesic motion, this applied force goes to zero.

In Newtonian theory, where gravity appears as an applied force, we are used to thinking of the condition for motion on a free circular orbit as being given by the balance between gravity and the centrifugal force. Also, it is usual to describe the effect produced on passengers in a car which turns a sharp corner as being due to centrifugal force. There is no *necessity* for introducing the concept of centrifugal force in these cases (one could always talk in terms of the applied force producing an acceleration rather than of it acting against another force) but it has proved to be a very useful concept in analyzing physical situations. It is natural, therefore, to ask how an equivalent idea of centrifugal force could be introduced in general relativity. Here, there is the difference that gravity now appears as part of the acceleration, and so the question arises of how one should identify the "gravitational" and "centrifugal" parts of the acceleration vector. This question is controversial; in principle, there is not only one unique way to do it. Abramowicz (1990) argued that since the gravitational force is normally thought of as being independent of velocity, the natural splitting consists of identifying the first term on the right hand side of (10) as the "gravitational force per unit mass" and the second term as the "centrifugal force per unit mass". For a particle with rest mass m_0 the centrifugal force is then given by

$$C_\alpha = m_0 \frac{\tilde{v}^2}{\tilde{r}} \nabla_\alpha \tilde{r}. \tag{11}$$

A consequence of this definition is that the centrifugal force always acts in the local outward direction (as given by $\nabla_\alpha \tilde{r}$) even if this becomes misaligned with the global outward direction (as given by $\nabla_\alpha r$). In the vacuum Schwarzschild metric, the acceleration becomes independent of velocity on the circular photon orbit at $r = 3M$ ($\nabla_\alpha \tilde{r}$ is zero there) and interior to this the local and global outward directions are directly opposite for motion in the equatorial plane.

This approach has been criticized by de Felice (1991) on the grounds that (i) he doubts the value of introducing the idea of centrifugal force into general relativity at all; and (ii) he argues that if one *is* to introduce it then one should choose a definition which will keep it always pointing in a direction away from the axis of rotation. We take the view that the concept of centrifugal force *is* a valuable one to introduce into general relativity and that while it is certainly true that the definition is not unique, the simplicity and elegance of the formulae resulting from definition (11) and the formal unity to which it gives rise are powerful arguments in its favour. Also, the reversal of direction does not seem counter-intuitive when viewed in the light of the fact that C_α is always aligned with $\nabla_\alpha \tilde{r}$ which we have argued defines the local outward direction in the way relevant for consideration of rotational effects. Indeed, a definition of centrifugal force which did *not* have the property of always being aligned with $\nabla_\alpha \tilde{r}$ would appear to be counter-intuitive. In the remainder of this article, we will therefore use the term "centrifugal force" to mean the quantity defined by (11).

The connection between the behaviour of centrifugal force and the location of the circular photon orbit in the vacuum Schwarzschild geometry suggests that circular photon orbits might in general play an important role in connection with rotational effects. This provides motivation for investigating the behaviour of the *effective potential* for photon motion in general static space-times.

Using the condition $(vv) = 0$ for geodesic photon motion and defining V^α to be the component of the photon four velocity orthogonal to both η_α and ξ_α, one obtains

$$(VV) = -\frac{E^2}{(\eta\eta)} - \frac{L^2}{(\xi\xi)}, \tag{12}$$

which can be rearranged to give

$$-\frac{(\eta\eta)}{L^2}(VV) = \frac{E^2}{L^2} - V_{eff}, \tag{13}$$

where the left hand side is strictly non-negative and V_{eff} is the effective potential for photon motion, given by

$$V_{eff} = -\frac{(\eta\eta)}{(\xi\xi)}. \tag{14}$$

A remarkable consequence of (14) is that

$$V_{eff} = \frac{1}{\tilde{r}^2}, \tag{15}$$

giving

$$C_\alpha = -\frac{1}{2}m_0 \tilde{v}^2 \tilde{r}^2 \nabla_\alpha V_{eff}. \tag{16}$$

The equipotential surfaces for photon motion precisely coincide with the level surfaces of $\tilde{r}$ (the von Zeipel cylinders) and the direction of the centrifugal force C_α (which is the direction of *increase* of $\tilde{r}$) is the direction of *decrease* of V_{eff}. Circular photon orbits occur at local extrema of the effective potential, with maxima corresponding to unstable orbits and minima to stable ones. The centrifugal force is zero on circular photon orbits and, in the equatorial plane, always points *away* from unstable ones and *towards* stable ones. In the next section, we illustrate these features for a particular case.

4 EFFECTIVE POTENTIAL CURVES AND VON ZEIPEL CYLINDERS FOR THE INTERIOR AND EXTERIOR SCHWARZSCHILD METRICS

The equatorial effective potential curve for photon motion and the structure of the von Zeipel cylinders are already well-known for the case of Schwarzschild black holes. Here, we want to investigate the situation for objects which are not as extreme as black holes and which will provide a link between the familiar Newtonian conditions and the extreme relativistic ones. For this purpose, we have carried out a study of space-times generated by non-rotating spheres of matter with constant density and surrounded by vacuum (giving the linked Schwarzschild interior and exterior solutions). We have examined a sequence of models with progressively increasing degrees of compactness which can be seen as representing a quasi-stationary contraction. The models are very simple but provide a useful probe of the effects being considered and the results are exactly the lowest order ones for the sequence of slowly-rotating models studied by Chandrasekhar and Miller (1974).

Some results are shown in figures 3 – 6 which correspond to $R_*/2M = 2.0$, 1.5, 1.2 and 1.125 respectively (where R_* is the circumferential radius of the surface of the sphere). For each value of $R_*/2M$ we show both the photon effective potential curve for the equatorial plane and a cross-section through the von Zeipel cylinders in a plane containing the rotation axis (the drawings have been made using standard Schwarzschild coordinates). The dashed line in the von Zeipel diagrams marks the surface of the sphere; interior to this, there is matter with constant density; exterior to it, there is vacuum. Figure 7 shows equivalent plots for a Schwarzschild black hole with the dashed line now marking the position of the event horizon ($R_*/2M = 1.0$). For $R_*/2M = 2.0$ (figure 3), the effective potential curve differs only very slightly from the equivalent Newtonian one (a truncus) but a more obvious difference is seen for the von Zeipel cylinders. The ones with the largest radii are near to being straight cylinders but those with smaller radii are pulled significantly inwards in the vicinity of the equatorial plane. For $(R_*/2M) = 1.5$ (figure 4), the value at which the circular photon orbit of the vacuum Schwarzschild space-time is coincident with the surface of the sphere, the effective potential curve has developed a stationary point (an inflection) at $r = 3M$ and the centrifugal force has gone to zero there. The von Zeipel cylinder which touches the sphere at the equator has developed a cusp at which the normal vector has zero length.

For models which are more compact than this, a special role is played by the cylinder which intersects the equatorial plane at the unstable photon orbit ($r = 3M$; $\tilde{r} = \tilde{r}_c = 3\sqrt{3}M/2$) since this separates regions with different regimes of behaviour. Each of the von Zeipel diagrams shows cylinders with $\tilde{r}$ ranging from $\tilde{r}_c/3$ to $5\tilde{r}_c/3$

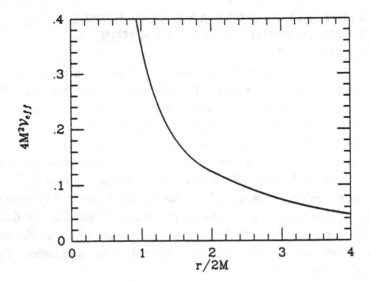

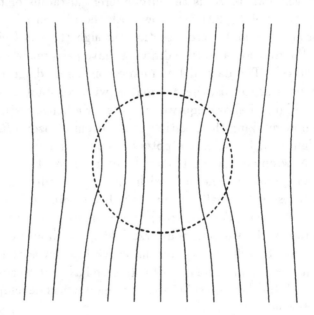

Figure 3. The equatorial photon effective potential curve for $R_*/2M = 2.0$ and a cross-section through the corresponding von Zeipel cylinders in a plane containing the rotation axis.

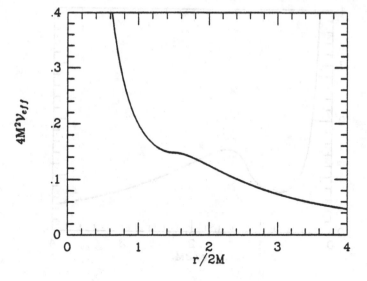

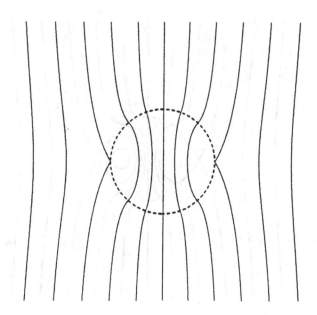

Figure 4. The equivalent of figure 3 for $R_\star/2M = 1.5$.

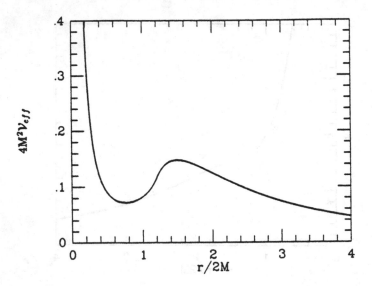

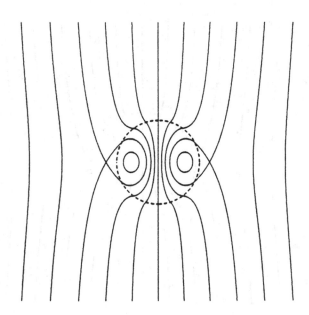

Figure 5. The equivalent of figure 3 for $R_*/2M = 1.2$.

in equal intervals of $\tilde{r}_c/3$. For $R_\star/2M = 1.2$ (figure 5), the effective potential curve has developed a clear minimum corresponding to a *stable* circular photon orbit internal to the matter. The existence of these stable orbits was first demonstrated and discussed by de Felice (1969). After their first appearance at the equator when $R_\star = 3M$, they move progressively inwards through the matter region (in the sense of decreasing $r/R_\star$) as $R_\star$ is further reduced. The *unstable* circular photon orbit in the vacuum region remains at $r = 3M$. Recalling that $\tilde{r}$ increases when V_{eff} decreases, one sees that, in the equatorial plane, the local and global outward directions become opposite in between the two circular photon orbits but coincide elsewhere. The centrifugal force, which is always aligned with the local outward direction becomes globally inward pointing in the equatorial plane between the two circular photon orbits. The central part of the corresponding von Zeipel diagram has become quite complicated by this stage with the cylinder for $4\tilde{r}_c/3$ having divided into two disconnected parts, one with cylindrical topology and the other with toroidal topology. The centrifugal force always points towards the the exterior of sections with cylindrical topology but towards the interior of toroidal sections. This is a general behaviour.

Figure 6 shows the situation for the most compact possible equilibrium model, which has $R_\star/2M = 9/8$. (The central pressure is infinite in this case.) Here, the minimum in the effective potential curve has moved to $r = 0$, *all* cylinders with $\tilde{r}_c \leq \tilde{r} < \infty$ have both outer and inner sections and the holes in the middle of the tori have closed up. (Note that only the cylinders with our five chosen values of $\tilde{r}$ have been illustrated here. The innermost torus shown is completely filled by a family of further tori which are the counterparts of the family of outer cylindrical sections stretching out to infinity.) Finally, in figure 7 we show the corresponding situation for a Schwarzschild black hole. All of the cylinders have been expelled from the interior of the event horizon and those with $\tilde{r} \geq \tilde{r}_c$ consist of an outer part with cylindrical topology and an inner part with spheroidal topology which is tangential to the event horizon at the poles. As $\tilde{r} \to \infty$, the inner spheroidal region becomes closer and closer to being exactly coincident with the event horizon. For $\tilde{r} < \tilde{r}_c$, the cylinders form two disconnected parts, one above the equatorial plane and one below it, with each part being closed off at $R = 2M$ where it becomes tangential to the event horizon at the poles in the same way as the spheroidal sections. Following the behaviour described earlier, the centrifugal force points towards the exterior of the topologically cylindrical sections (and also of the cylindrical sections which are broken in the middle for $\tilde{r} < \tilde{r}_c$) but towards the interior of the spheroidal sections.

A more detailed analysis of the behaviour described here is contained in the paper by Abramowicz et al (1992).

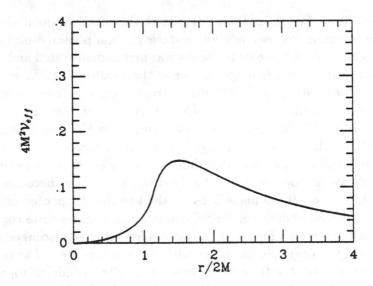

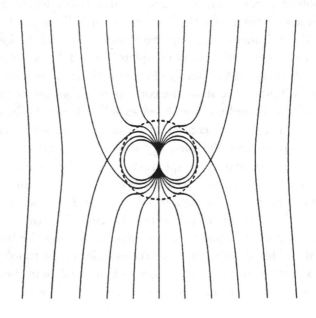

Figure 6. The equivalent of figure 3 for $R_*/2M = 9/8$.

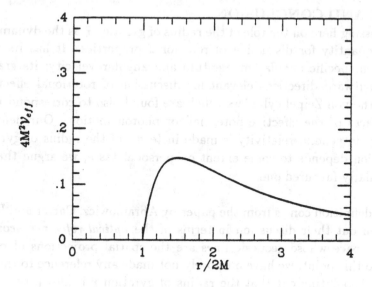

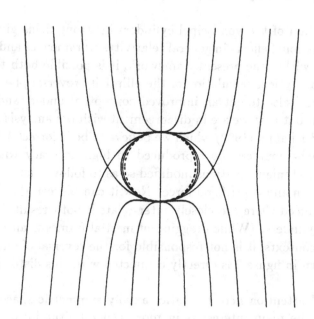

Figure 7. The equivalent of figure 3 for a Schwarzschild black hole.

5 DISCUSSION AND CONCLUSION

We have been focussing here on the role of the radius of gyration $\tilde{r}$ as the dynamically important radial quantity for discussion of rotational properties. It has its origin in the link between specific angular momentum and angular velocity; its gradient defines the local outward direction relevant for discussion of rotational effects; its level surfaces are the von Zeipel cylinders which are found also to correspond to the equipotential surfaces of the effective potential for photon motion. Our definition of centrifugal force in general relativity is made in terms of the radius of gyration. While this definition depends to some extent on personal taste, we argue that our choice is a particularly favoured one.

The origin of this definition comes from the paper by Abramowicz, Carter and Lasota (1988) who carried out their discussion in terms of the *optical reference geometry*, a projected three space whose geodesic lines are the spatial projections of photon trajectories. Up to this point, we have purposely not made any reference to this here but it is now worth pointing out that the radius of gyration $\tilde{r}$ is also precisely the proper circumferential radius in the optical reference geometry. Also, the fact that the von Zeipel cylinders and the equipotential surfaces for photon motion coincide, is an indication of the underlying interconnection which exists.

Study of the distortion of the von Zeipel cylinders in strong fields gives some valuable insight into rotational effects in general relativity. Abramowicz and Miller (1990) demonstrated how, within the present framework, it is possible both to make a surprisingly accurate analytical calculation of the ellipticity reversal phenomenon illustrated in figure 1 and also to get an improved conceptual understanding of why it occurred. (We point out that we are in disagreement with the analysis made recently by Chakrabarti and Khanna (1992) which we believe to be incorrect.) It was demonstrated that this behaviour can be reproduced to high accuracy with an analytic calculation using Newtonian equations modified so as to follow relativistic definitions of angular momentum and centrifugal force. Now it can be seen that the two types of correction introduced there are closely inter-related; both result from distortion of the von Zeipel cylinders. (While dragging of inertial frames is an important phenomenon in other contexts, it is not responsible for the reversal of ellipticity.) Also, the behaviour shown in figure 2 is directly connected with this distortion.

We have restricted attention here to static, axially symmetric space-times but, of course, eventually the main interest is in more general situations. As mentioned previously, the extension to stationary space-times has already been completed and extension to completely general space-times is currently under investigation.

ACKNOWLEDGEMENTS
This research is being carried out with financial support from the Italian Ministero dell'Università e della Ricerca Scientifica e Tecnologica.

REFERENCES
Abramowicz, M. A. 1990. *Mon. Not. R. Astr. Soc.*, **245**, 733.
Abramowicz, M. A., Carter, B. and Lasota, J. P. 1988. *Gen. Rel. Grav.*, **20**, 1173.
Abramowicz, M. A. and Miller, J. C. 1990. *Mon. Not. R. Astr. Soc.*, **245**, 729.
Abramowicz, M. A., Miller, J. C. and Stuchlík, Z. 1992. Submitted to *Phys. Rev. D.*
Chakrabarti, S. K. and Khanna, R. 1992. Submitted to *Mon. Not. R. Astr. Soc.*
Chandrasekhar, S. and Miller, J. C. 1974. *Mon. Not. R. Astr. Soc.*, **167**, 63.
de Felice, F. 1969. *Nuovo Cimento*, **63B**, 649.
de Felice, F. 1991. *Mon. Not. R. Astr. Soc.*, **252**, 197.

STABILITY OF CHARGED BOSON STARS AND CATASTROPHE THEORY

Franz E. Schunck

Institute for Theoretical Physics, University of Cologne, Cologne, Germany

Fjodor V. Kusmartsev

Department of Physics, University of Oulu, Linnanmaa, Finland

Eckehard W. Mielke

Faculty of Mathematics, University of Kiel, Kiel, Germany

Abstract. We investigate the stability of *charged* boson stars in the framework of general relativity. The constituents of these stars are scalar bosons which interact not only via their charge and mass but also via a short–range Higgs potential U. Our stability analysis is based on catastrophe theory which is capable of providing more information than perturbation theory. In fact, it predicts novel oscillation and collapse regimes for a certain range of the particle number.

1 INTRODUCTION

In the early universe, spin–zero particles, such as the scalar Higgs particles, may have played an important rôle [1]. At that early time it is conceivable that clouds of particles created stars which are kept together by their own gravitational field, the so–called *boson stars* [2]. These stars could make up a considerable fraction of the hypothetical dark matter.

The boson star consists of many particles and may have a very large mass comparable or larger than that of a neutron star. The latter depends upon the form of a self–interaction between the bosons [3]. Generally speaking, the boson star is in many ways analogous to the neutron star [4,5]. Both stars consist of one matter component. Recently, Higgs particles interacting with gauge field have been studied [6]. If we attribute charge to the bosons, they will interact also via electromagnetic forces. Because of the repulsive nature of this interaction there exists a critical total charge of these scalar particles beyond which the star becomes unstable [6].

Ultimately, these issues depend on the stability criteria for stars. Recently, we have studied the stability of boson and neutron stars [7] by transferring a method which was developed in the theory of solitons [8]. Therein, non–elementary catastrophe theory based on the classification of singularities of the smooth mappings is applied. In our case, the mapping is induced by the integrals of motion of the Einstein–Maxwell–Klein–Gordon equation, which are the mass and the charge of the star.

The charged boson star is stable in the region of small values of the central density. Instability will appear at some critical value, which corresponds to a maximum of the mass as a function of the central density. This result has also been obtained with the aid of perturbation theory [9]. In practice, the procedure of catastrophe theory reduces the problem to that of building up the *bifurcation diagram* which describes the functional dependence of the conserved quantities with respect to each other [7,8]. For the case of two conserved quantities M and N, as in the case of boson stars, the bifurcation diagram is represented by cuspoidal curves. Each of these cusps is associated with some surface, the so–called Whitney surface [10]. The critical points of this catastrophe manifold correspond to the bifurcation diagram. One of us [8] has shown that critical points of the minimum type correspond to the stable soliton and that the critical points of the maximum type correspond to the unstable soliton. This method has been applied to the case of boson stars [7] and was related to the stability analysis of neutron stars and white dwarfs [5].

2 EINSTEIN–MAXWELL SCALAR FIELD EQUATION

As a general–relativistic model of a charged boson star, we consider a self–interacting scalar field Φ coupled self–consistently to its own gravitational field and to $U(1)$–gauge fields having the Lagrangian

$$L = \frac{1}{2\kappa}\sqrt{|g|}\,R + \frac{1}{2}\sqrt{|g|}\,[g^{\mu\nu}(D_\mu\Phi)^*(D_\nu\Phi) - U(\Phi^*\Phi)] - \frac{1}{4}\sqrt{|g|}\,F_{\mu\nu}F^{\mu\nu}\,, \quad (2.1)$$

where $\kappa = 8\pi G$ is the gravitational constant in natural units, g the determinant of the metric $g_{\mu\nu}$, $\mu, \nu = (0, 1, 2, 3)$, and R the curvature scalar. The coupling between the scalar field and the Maxwell field $A = A_\mu dx^\mu$ ($U(1)$–valued 1–form), is introduced via the gauge and generally covariant derivative for the Higgs field

$$D_\mu\Phi = \partial_\mu\Phi + ieA_\mu\Phi\,, \quad (2.2)$$

where e is the $U(1)$ coupling constant. Furthermore, the electromagnetic field strength is given by

$$F_{\mu\nu} = \partial_\mu A_\nu - \partial_\nu A_\mu \quad (2.3)$$

and the polynomial self–interaction

$$U(\Phi^*\Phi) := m^2\Phi^*\Phi + \frac{1}{2}\alpha(\Phi^*\Phi)^2 \quad (2.4)$$

is written in the required local $U(1)$–symmetric form.

From the principle of minimal action, we obtain the *coupled* Einstein–Maxwell–Klein–Gordon equations:

$$R_{\mu\nu} - \frac{1}{2}g_{\mu\nu}R = -\kappa T_{\mu\nu}(\Phi, A) ,\tag{2.5}$$

$$\Box\Phi + \frac{\partial U}{\partial\Phi^*} = 0 ,\tag{2.6}$$

$$\partial_\nu\left(\sqrt{|g|}F^{\mu\nu}\right) = e\,j^\mu .\tag{2.7}$$

The energy–momentum tensor reads

$$T_{\mu\nu}(\Phi, A) = (D_\mu\Phi)^*(D_\nu\Phi) - F_\mu{}^\kappa F_{\nu\kappa} - \frac{g_{\mu\nu}}{\sqrt{|g|}}L(\Phi, A) ,\tag{2.8}$$

whereas

$$\Box\Phi = \frac{1}{\sqrt{|g|}}D_\mu\left(\sqrt{|g|}\,g^{\mu\nu}D_\nu\right)\Phi\tag{2.9}$$

is the generally covariant and $U(1)$–gauge–covariant d'Alembertian, and

$$\begin{aligned}j^\mu &= \frac{i}{2}\sqrt{|g|}\,g^{\mu\nu}[\Phi^*(D_\nu\Phi) - \Phi(D_\nu\Phi)^*]\\&= \frac{i}{2}\sqrt{|g|}\,g^{\mu\nu}[\Phi^*(\partial_\nu\Phi) - \Phi(\partial_\nu\Phi)^* + 2ieA_\nu\,|\,\Phi\,|^2]\end{aligned}\tag{2.10}$$

the Noether current density.

3 LOCALIZED SOLUTIONS VIA NUMERICAL INTEGRATION

In this paper, we restrict ourselves to a static, spherically symmetric metric

$$ds^2 = e^{\nu(r)}dt^2 - e^{\lambda(r)}dr^2 - r^2(d\theta^2 + \sin^2\theta d\phi^2),\tag{3.1}$$

in which the functions $\nu = \nu(r)$ and $\lambda = \lambda(r)$ depend on the Schwarzschild type radial coordinate r. For the boson field, we make the stationarity ansatz

$$\Phi(r,t) = P(r)e^{i\omega t} ,\tag{3.2}$$

which describes a spherically symmetric bound state with frequency ω. In order to have only electric charges for the scalar field, we choose $A_\mu(r,t) = (C(r), 0, 0, 0)$ for the gauge field. The resulting coupled system of ordinary differential equations reads

$$\nu' + \lambda' = \kappa(\rho + p_r)re^\lambda ,\tag{3.3}$$

$$\lambda' = \kappa\rho re^\lambda - \frac{1}{r}e^\lambda + \frac{1}{r} ,\tag{3.4}$$

$$P'' + \left(\frac{2}{r} + \frac{\nu' - \lambda'}{2}\right)P' = -e^{\lambda-\nu}(\omega + eC)^2\,P + e^\lambda\frac{dU(P^2)}{dP^2}P ,\tag{3.5}$$

$$C'' + \left(\frac{2}{r} - \frac{\nu' + \lambda'}{2}\right)C' = eP^2(\omega + eC)e^\lambda .\tag{3.6}$$

Because of the contracted Bianchi identity $\nabla^\mu \left(R_{\mu\nu} - \frac{1}{2} g_{\mu\nu} R \right) \equiv 0$, a further equation (involving $T_\theta{}^\theta = T_\phi{}^\phi = -p_\perp$) is identically satisfied. The energy–momentum tensor becomes diagonal, i.e. $T_\mu{}^\nu = diag\, (\rho, -p_r, -p_\perp, -p_\perp)$ with

$$\rho = \frac{1}{2} \left[(\omega + eC)^2 P^2 e^{-\nu} + P'^2 e^{-\lambda} + C'^2 e^{-\lambda-\nu} + U \right], \tag{3.7}$$

$$p_r = \rho - U - C'^2 e^{-\lambda-\nu}, \tag{3.8}$$

$$p_\perp = \rho - U - P'^2 e^{-\lambda} = p_r - P'^2 e^{-\lambda} + C'^2 e^{-\lambda-\nu}. \tag{3.9}$$

The form of $T_\mu{}^\nu$ is familiar from an ideal fluid, except that the radial and tangential pressure generated by the scalar field are generally different, i.e. $p_r \neq p_\perp$. This *fractional anisotropy* $a_f := (p_r - p_\perp)/p_r$ has already been noted by Ruffini and Bonazzola [2] for the uncharged boson star. For the charged boson star there also arises an anisotropy, provided that the radial distribution of the electric potential $A_0 = C$ does not satisfy $C' = \pm e^{\nu/2} P'$.

Equation (3.4) possesses a Schwarzschild–type solution

$$e^{-\lambda(r)} = 1 - \frac{\kappa M(r)}{4\pi r}, \qquad M(r) := 4\pi \int_0^r \rho x^2 dx, \tag{3.10}$$

where $M(r)$ is the mass function such that $M(\infty)$ yields the Newtonian mass at infinity.

The case of the uncharged boson star can be simply obtained by taking the limit $e \to 0$ of the coupling constant and requiring the electromagnetic field strength to vanish, i.e. $C' = 0$. For the polynomial self–interaction (2.4), these equations have then been solved numerically for non–singular, finite mass and zero–node solution [2, 3, 7]. Two and higher node solutions occurred already in Mielke and Scherzer in [2c]. For a massless scalar field with $U = 0$, an exact solution is known, cf. Ref. [12]. For a massless real scalar field, Christodoulou [13] could show that a spherically symmetric time–dependent field configuration must either disperse to infinity or, for non–vanishing Bondi mass, forms a black hole.

For a configuration with a localized charged scalar field, i.e. one for which $P(r) \simeq 0$ holds for $r \to \infty$, the metric (3.1) and the electric field strength will acquire asymptotically the Reissner–Nordstrøm form

$$e^{\nu(r)} = e^{-\lambda(r)} \simeq 1 - \frac{2GM}{r} + \frac{GQ^2}{4\pi r^2}, \qquad E(r) = C'(r) \simeq \frac{Q}{4\pi r}, \qquad \text{for } r \to \infty. \tag{3.11}$$

where M is the mass of the star, Q the charge, and G the gravitational constant. In this paper we numerically solve the full coupled set (3.3)–(3.6) for the charged boson star with the aid of the Runge–Kutta routine of the International Mathematical and Statistical Library (IMSL) packet and a self–developed shooting method [2g]; cf. Jetzer [9] for related results.

4 INTEGRALS OF MOTION

The concept of an energy–momentum 4–vector for a field configuration is a notoriously subtle one [11] in general relativity. In analogy to the case of the uncharged boson star, we can use the Tolman mass formula

$$M := \int (2T_0^0 - T_\mu^\mu)\sqrt{|g|}\, d^3x = 4\pi \int_0^\infty (\rho + p_r + 2p_\perp)\, e^{(\nu+\lambda)/2} r^2 dr$$

$$(4.1)$$

$$= 4\pi \int_0^\infty [2P^2(\omega + eC)^2 e^{-\nu} + C'^2 e^{-\lambda-\nu} - U] e^{(\nu+\lambda)/2} r^2\, dr\ .$$

The exponential decrease of the radial function $P(r) \simeq exp[-\sqrt{m^2 - \omega^2}\, r]$ for $|\omega| < m$ secures the applicability for such an *isolated*, static system. The formula can be derived from the *local* conservation law $\partial_\nu(T_\mu^{\ \nu} + \tau_\mu^{\ \nu}) = 0$, where $T_\mu^{\ \nu} = \sqrt{|g|}\, T_\mu^\nu$ and $\tau_\mu^{\ \nu}$ is the gravitational energy–momentum complex. For the charged boson star, the explicit expression (4.1) does involve derivatives, in contrast to the boson star [2g, 7].

A second "integral of motion" arises from the fact that the Lagrangian (2.1) is invariant under the *local* phase transformation $\Phi \to \Phi e^{-i\vartheta(x)}$ ($U(1)$–symmetry). Therefore the Noether current density j^μ, cf. (2.10), is *locally* conserved, i.e. $\partial_\mu j^\mu = 0$. The time–component j^0 integrated over space yields the *particle number* N or the *charge* $Q = eN$, respectively:

$$N = 4\pi \int_0^\infty P^2(\omega + eC)e^{(\lambda-\nu)/2} r^2\, dr\ .$$

$$(4.2)$$

Since the current density (2.10) is a "measure" for the radial distribution of charged "particles" in the charged boson star, its *effective radius* can be defined by [6,7]

$$R := \frac{1}{N} \int rj^\mu d\Sigma_\mu = \frac{4\pi}{N} \int_0^\infty P^2(\omega + eC)e^{(\lambda-\nu)/2} r^3\, dr\ .$$

$$(4.3)$$

5 STABILITY CRITERIA FROM CATASTROPHE THEORY

In order to investigate the stability of soliton–type solutions against radial perturbations, we consider the two–dimensional mapping

$$F\ :\ (k,\omega) \mapsto (M,N)\ ,$$

$$(5.1)$$

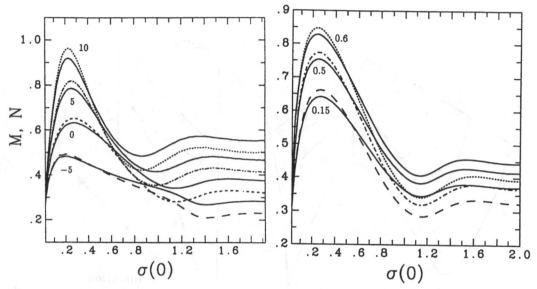

Figure 1. *The Tolman mass M in units of $(1/mG)$ and particle number N in dimensionless units of $(1/m^2G)$ as a function of the central density $\sigma(0) = \sqrt{\kappa/2} \mid \Phi(0) \mid$. Left: Uncharged boson star (e=0) for $\tilde{\alpha} := (2\alpha/\kappa m^2) = -5, 0, 5, 10$ in the potential U [2g, 7]. Right: Charged boson star for $\tilde{\alpha} = 0$ and $\tilde{e} := e/\sqrt{4\pi G}m = 0.15, 0.5, 0.6$.*

where k is a variational parameter which dilatates the effective radius R of the star and ω is the frequency. The parameter k induces a scaling of the metric, the frequency, and the scalar field in accordance with their normal physical dimensions $ds^2 \rightarrow k^2 ds^2$, $\omega \rightarrow \omega/k$, $P(r) \rightarrow kP(kr)$, such that N is kept fixed.

In order to classify the singularities of the mapping F, let us consider the Jacobi matrix

$$J = \begin{pmatrix} \partial M/\partial k & \partial M/\partial \omega \\ \partial N/\partial k & \partial N/\partial \omega \end{pmatrix} . \tag{5.2}$$

According to Whitney's theorem [8, 10], the singularities of the mapping F can be one of three types, depending on the rank $R_J = 2, 1$ and 0 of J, respectively. Since we require the soliton solution to be an extremal point of the Lagrange manifold, we have $\partial M/\partial k = \partial N/\partial k = 0$. For the *soliton* the rank of J is $R_J < 2$ and, consequently, the singularities of the mapping F may have either $R_J = 1$ or $R_J = 0$. In that case, our soliton solution corresponds to the extremal or critical points of the Whitney surfaces [8] which has a very definite form (see [10]). In our numerical examples (figure 1), the dependence $M = M(\omega)$, $N = N(\omega)$ is implicit via $\omega = \omega(\sigma(0))$. If the rank of J is zero, the critical points are degenerate. These are the extremal points of figure 1 which give rise to *cusps* in the diagram $M(N)$ presented in figure 2.

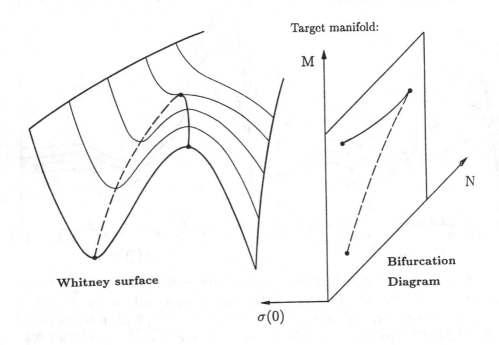

Figure 2. *This figure shows the projection of the Whitney surface (left) onto the bifurcation diagram (right). The drawn lines on the Whitney surfaces show that the manifold can be understood as the unfolding of a perturbed cubic function [10].*

The dependence $M = M(N)$ or, equivalently, the binding energy $E_B = M - mN$ represents the so–called *bifurcation diagram* (figures 2 and 3). Each cusp corresponds to some *Whitney surface*, which is a part of the total mass–energy surface called the *manifold of catastrophe* (figure 2). As shown in Ref. [8], the minimum of the Whitney surface corresponds to the stable soliton star, the maximum corresponds to the unstable soliton. At the cuspoidal point, the minimum coalesces with the maximum and the soliton loses its stability. Thus, the lower branch of the lowest cusp corresponds to the *absolutely stable soliton*. The upper branches of the first and second cusp correspond to unstable solitons. At the next cusp a new instability can appear or disappear, provided the next branch is higher or lower, respectively, than the previous one.

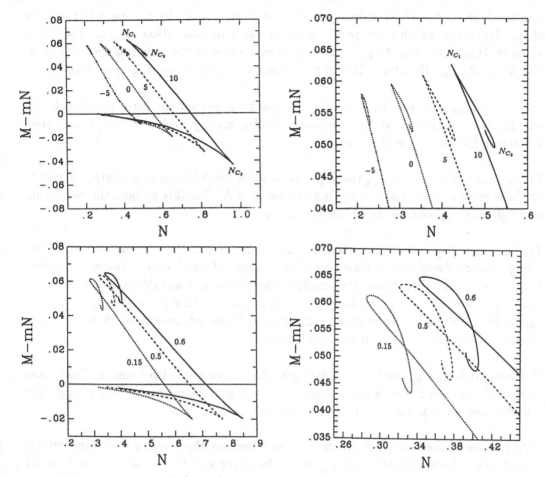

Figure 3. *Above left: Binding energy $M - mN$ as a function of N for different coupling constants in the self–interacting potential $\tilde{\alpha} = (2\alpha/\kappa m^2) = -5, 0, 5, 10$. Above right: Magnification of the second and third cusp regime [7]. Below: The bifurcation diagram for $\tilde{e} = 0.15, 0.5, 0.6$ and $\tilde{\alpha} = 0$. The boson star mass M and the particle number N are measured in units of $(1/mG)$ and of $(1/m^2 G)$, respectively.*

In our method, as in Arnold's classification of singularities of smooth mappings [10], the bifurcation diagram is a skeleton of the mass–energy surface. The extremal points of this surface correspond to the soliton solution. There exist also other types of solutions, with a different dynamical behaviour. Depending on the critical values $N_{C_1}, N_{C_2}, N_{C_3}$, of the particle number (figure 3), we obtain different cross–sections of the catastrophe manifold both for uncharged and charged boson stars (see Ref. [7] for details).

For fixed $N < N_{C_1}$ there exists only one minimum of $M(R)$ for the lower branch of the first cusp, which corresponds to the stable soliton solution $M_{soliton}$. For every value of $M_{soliton} < M < M_{collapse}$, there is an oscillation of the star near the minimum. For $M \geq M_{collapse}$ the star will stop oscillating and start collapsing to a black hole.

For $N > N_{C_3}$, the section has only the marginal extremum corresponding to an effectively free boson field with vanishing binding energy, and therefore, there exists a collapse for any value of the mass.

In the range $N_{C_2} < N < N_{C_3}$ the dependence of M on R has as many extremal points as there exist branches of cusps at a given value of N. For this section, the oscillating and collapse regimes described before coexist.

In the range $N_{C_1} < N < N_{C_2}$, the star's behaviour is also characterized by *pulsation*. The pulsation consists of at least two different types of oscillations. Numerical studies [14] of the evolution of boson star configuration show that a stable soliton (S–branch star), which is slightly perturbed, will oscillate with a fundamental frequency. The unstable soliton (U–branch star) will either collapse to form a black hole or will disperse. This coincides with our conclusion.

Subsequent to our paper [7a], oscillating soliton stars were also found by Seidel and Suen [15]. In their paper, a model with a real self–gravitating scalar field is suggested which, however, has no regular static solutions.

The picture of the star's behaviour, obtained here on the basis of (non–elementary) catastrophe theory [8, 10], has a general character and has an analogous form for neutron stars [4, 5], rotating stars [16], fermion Q–balls [17], dilaton stars [18], and gravitating magnetic monopoles [19]. In fact, for neutron stars the diagram $M(N)$, obtained from numerical integration of a certain equation of state, exhibit similar bifurcations as in the case of boson stars. For the first cusp, Harrison et al. 'saw' the Whitney surface already in 1965 (see figure 9 of Ref. [4]) and deduced the correct stability criteria from the analysis of the mass–energy surface $M(\rho_0, N)$. Following this idea of Harrison et al. (see p.60-66 in Ref. [4]), Thorne [20] has derived conditions under which a mass–radius–diagram for cold stars is valid. This represents another projection of the Whitney surface, so that Thorne obtains the same stability criteria as we do.

The earlier application of the elementary catastrophe theory to general relativity has been based on the modelling of stars and planets as solid or fluid continua very similar in form to those in engineering mechanics (Zeemann machine). This approach has been started in Thomson's group (see [20-21] and references therein). They studied the gravitational collapse, thermodynamical instabilities of hot stellar system, Lynden–Bell gravothermal catastrophe and instabilities of the rotating fluid mass together with its self–gravitation.

They demonstrated that the two–thirds–law cusp appears in the energy–mass–control projection. The other application, which we are not concerned with here, is the gravothermal catastrophe of hot stellar systems. Following Thompson and Hunt, Katz [22] has developed a conjugate theorem to find stability transitions at successions of folds along the equilibrium path. In this context, Sorkin [23] has developed further theorems for turning points which arise in such figures. Our method, which independently has been introduced by one of us in the theory of solitons [8], is just exactly the generalization of Thomson et al.'s approach to the system with infinite degrees of freedom, like classical field theory, where elementary catastrophe theory, used by Thomson et al., is not applicable.

ACKNOWLEDGMENTS

We are grateful to F. W. Hehl and R. Hecht for a useful discussion and J. A. Wheeler as well as J. Winicour for some important hints concerning the stability of stars and catastrophe theory. Furthermore, we thank Dieter Schunck for drawing Fig. 2. We acknowledge support form the Deutsche Forschungsgemeinschaft and the TFT Fellowship, Helsinki.

REFERENCES

[1] Thachev, I. I. (1991). *Phys. Lett. B*, **261**, 289.

[2] Kaup, D. J. (1968). *Phys. Rev.*, **172**, 1331; Ruffini, R. and Bonazzola, S. (1969). *Phys. Rev.*, **187**, 1767; Mielke, E. W. and Scherzer, R. (1981). *Phys. Rev. D*, **24**, 2111; Lee, T. D. (1987). *Comm. Nucl. Part. Phys.*, **17**, 225; Lee, T. D. (1987). *Phys. Rev. D*, **35**, 3637; Friedberg, R., Lee, T. D. and Pang, Y. (1987). *Phys. Rev. D*, **35**, 3640, 3658, 3678; Schunck, F. E. (1991). Diploma–Thesis, University of Cologne.

[3] Colpi, M., Shapiro, S. L. and Wasserman, I. (1986). *Phys. Rev. Lett.*, **57**, 2485.

[4] Zeldovich Ya. B. and Novikov, I. D. (1971). *Stars and Relativity*, Relativistic Astrophysics, Vol.1 University of Chicago Press, Chicago; Harrison, B. K., Thorne, K. S., Wakano, M. and Wheeler, J. A. (1965). *Gravitation Theory and Gravitational Collapse*, University of Chicago Press, Chicago; Shapiro, S. L. and Teukolsky, S. A. (1983). *Black Holes, White Dwarfs, and Neutron Stars. The Physics of Compact Objects*, New York, Wiley.

[5] Kusmartsev, F. V. and Schunck, F. E. (1992). *Physica B*, to appear.

[6] Jetzer, Ph. and van der Bij, J. J. (1989). *Phys. Lett. B*, **227**, 341.

[7] Kusmartsev, F. V., Mielke, E. W. and Schunck, F. E. (1991). *Phys. Rev. D*, **43**, 3895; (1991). *Phys. Lett. A*, **157**, 465.

[8] Kusmartsev, F. V. (1989). *Phys. Rep.*, **183**, 1.

[9] Jetzer, Ph. (1989). *Phys. Lett. B*, **231**, 433; University of Zürich preprint, ZU–TH 25/91 to be published in *Phys. Rep.*.

[10] Arnold, V. I., Gusein–Zade, S. M. and Varchenko, A. N. (1985). *Singularities of differentiable maps*, Birkhäuser, Boston; Poston T. and Stewart, I. (1978). *Catastrophe Theory and its Applications*, Pitman.

[11] Tolman, R. C. (1930). *Phys. Rev.*, **35**, 875 ; see also R. Penrose, (1986). In *Gravitational Collapse and Relativity*, H. Sato and T. Nakamura (eds.), World Scientific, Singapore, 43.

[12] Baekler, P., Mielke, E. W., Hecht, R. and Hehl, F. W. (1984). *Nucl. Phys.*, **288**, 800.

[13] Christodoulou, D. (1987). *Commun. Math. Phys.*, **109**, 613.

[14] Seidel, E. and Suen, W–M. (1990). *Phys. Rev. D*, **42**, 384.

[15] Seidel, E. and Suen, W–M. (1991). *Phys. Rev. Lett.*, **66**, 1659.

[16] Friedman, J. L., Ipser, J. R. and Sorkin, R. D. (1988). *Astr. J.*, **325**, 722.

[17] Bahcall, S., Lynn, B. W. and Selipsky, S. (1990). *Nucl. Phys. B*, **325**, 606 (1989); **331**, 67.

[18] Gradwohl B. and Kälbermann, G. (1989). *Nucl. Phys. B*, **324**, 215.

[19] Breitenlohner, P., Forgács, P. and Maison, D. (1991). *Gravitating monopole solutions*, MPI preprint, Munich.

[20] Thorne, K. S. (1966). In *High–Energy Astrophysics, Proceedings of Course 35 of the International Summer School of Physics "Enrico Fermi"*, Academic Press, New York.

[21] Saunders, P. T. (1980). *An Introduction to Catastrophe Theory*, Cambridge University Press, Cambridge.

[22] Thompson, J. M. T. (1979). *Phil. Trans. Roy. Soc. London*, **292**, 1.

[23] Katz, J. (1978). *Mon. Not. R. Astr. Soc.*, **183**, 765.

[24] Sorkin, R. (1982). *Astr. J.*, **249**, 254 (1981); **257**, 847.

PART B

PRACTICAL APPROACHES

NUMERICAL ASYMPTOTICS

R. Gómez and J. Winicour

Department of Physics and Astronomy, University of Pittsburgh, Pittsburgh, US

Abstract. We review the present status of the null cone approach to numerical evolution being developed by the Pittsburgh group. We describe the simplicity of the underlying algorithm as it applies to the global description of general relativistic spacetimes. We also demonstrate its effectiveness in revealing asymptotic physical properties of black hole formation in the gravitational collapse of a scalar field.

1 INTRODUCTION

We report here on a powerful new approach for relating gravitational radiation to its matter sources based upon the null cone initial value problem (NCIVP), which has been developed at the University of Pittsburgh . We are grateful to the many graduate students and colleagues who have made important contributions: Joel Welling (Pittsburgh Supercomputing Center), Richard Isaacson (National Science Foundation), Paul Reilly, William Fette (Pennsylvania State University at McKeesport) and Philipos Papadopoulous.

As will be detailed, the NCIVP has several major advantages for numerical implementation. (i) There are no constraint equations. This eliminates need for the time consuming iterative methods needed to solve the elliptic constraint equations of the canonical formalism. (ii) No second time derivatives appear so that the number of basic variables is half the number for the Cauchy problem. In fact, the evolution equations reduce to one complex equation for one complex variable. The remaining metric variables (2 real and 1 complex) are obtained by a simple radial integration along the characteristics. In null cone coordinates, Einstein's equations form a system of radial differential equations which can be integrated in hierarchical order for one variable at a time. We have been able to utilize this structure to construct a marching algorithm in which evolution to the next grid point is carried out with no extra computational baggage such as iterative procedures or inversion of matrices. (iii) The radiation zone can be idealized as a finite grid boundary using the Penrose compactification technique (Penrose 1963) for null infinity. No extraneous outgoing radiation conditions are required at null infinity, which both in theory and in practice acts as a perfectly absorbing boundary. This allows the rigorous description of radiation in terms of geometrical quantities such as the Bondi mass and news function (Bondi et. al. 1962), the angular momentum and supermomentum associated with the asymptotic symmetry group (Winicour 1989), and the Newman-Penrose conserved quantities (Newman & Penrose 1968). It supplies the waveform and the polarization incident on a distant antenna. Because of the singular time behavior of the compactified version of spatial infinity (Ashtekar & Hansen 1978), the analogous approach to the Cauchy problem is not practical numerically. Instead, the grid is terminated at some radius R, where an outgoing radiation condition must be imposed. There do not exist specific estimates of the effect of such boundary conditions on the interior physics, e.g. by reflection of waves off the boundary. Furthermore, combined with gauge ambiguities, use of a finite grid boundary complicates the extraction of the true wave profile seen by distant observers, who are essentially at null infinity. Although this has been effectively accomplished for the radiation from axially symmetric perturbations of relativistic stars (Abrahams & Evans 1990) it becomes much more problematical in the highly nonlinear and asymmetric case. (iv) The grid domain is exactly the region in which waves propagate, which is ideally efficient for radiation studies. Since each null cone extends from the source to null infinity, we see the radiation immediately with no need for numerical evolution to propagate it across the grid. Furthermore, in the case of black hole formation, the exterior region of spacetime which is of physical interest is itself bounded in the future by a null hypersurface which forms the horizon associated with the final black hole state. Here the use of null hypersurfaces again leads to an efficient choice of grid domain, although in highly asymmetric systems, such as two coalescing black holes, the caustic structure of the horizon is expected to be much more complicated than that of a null cone.

There are also disadvantages of the NCIVP. (i) One is the issue of caustics. A null cone is a special type of null hypersurface in which the caustics consist of a single spacetime point. In gravitational systems with large asymmetry, the focusing effect can lead to more complicated caustic structure and preclude the existence of null cones. Although there are methods to include arbitrary caustics in the characteristic initial value problem (Friedrich & Stewart 1983), at the present developmental stage we prefer to avoid this issue. In analogy with geometric optics, it is not the lensing effect of strong curvature by itself which leads to caustics but also the location of the lens with respect to the vertex. In a spacetime with negligible curvature containing, say, two peanuts, no global null cones exist if the peanuts are sufficiently far apart (approximately 10^{10} light years). On the other hand, in spacetimes with strong curvature, global null cones exist (or end on true physical singularities) in the case of near spherical symmetry and even for a binary neutron star system with orbital separation of less than 5 neutron star radii. It is known from the geometric optics approximation that gravitational lensing by an intermediary object between the source and observer can enhance the detectability of gravitational radiation in the same manner as for electromagnetic radiation, although this is likely to be too fortuitous to be of practical value. However, there is also another qualitatively different form of lensing which occurs when the "focal length" is matched to the size of the dynamical radius of the source. This occurs in the case of binary neutron stars at the minimal orbital radius that admits global null cones and it is generic of binary black holes. It is not known to what extent this dynamical lensing might enhance the detectability of gravitational radiation from these binary systems. This is an effect for which standard perturbation theory based upon harmonic light cones cannot be trusted and which has not yet been investigated numerically. (ii) The Courant stability condition requires that the physical domain of dependence be smaller than the domain of dependence determined by the numerical algorithm. For an explicit finite difference algorithm, this places a stronger restriction on the size of the time step near the vertex of the null cone than occurs for the Cauchy problem (Gómez, Isaacson & Winicour 1992). This can be circumvented by using either an implicit algorithm or a variable grid but again, at the present developmental stage, we prefer to keep the algorithm as simple as possible in finite difference form for purposes of calibration. (iii) Although the technique of shooting along characteristics is common in computational mathematics, there is very little history for the numerical implementation of the characteristic initial value problem. Outside of work done in general relativity, the literature only treats systems with one essential spatial dimension. This makes it prohibitive to try to develop an astrophysically realistic hydrodynamic code along with the gravitational code. (iv) There is a paucity of exact solutions that can be expressed in null coordinates for use in code calibration and debugging. One is the Oppenheimer-Snyder solution. In collaboration with W. Fette, we have developed a spherically symmetric code for Einstein's equations coupled to dust which tracks Oppenheimer-Snyder collapse, to second order accuracy in grid size, through the horizon up to the formation of the singularity. However, until recently, there were no nonspherically symmetric metrics known in null cone coordinates, other than the Minkowski metric, that were sufficiently global to serve as a test bed. This requirement is intrinsically more difficult than in the Cauchy problem, where the domain of dependence of a small portion of the initial spacelike hypersurface is nonempty, so that evolution can be tested locally by avoiding the singular regions of the exact solution. In the case of a null cone, the domain of dependence is empty for any portion not containing the vertex. Following a suggestion of J. Bicak, we developed a null cone formulation of the boost-rotation symmetric spacetimes (Bicak, Reilly & Winicour 1988). In further work with Reilly, we have used this formalism to find the only known nonflat vacuum spacetime that can be analytically expressed in null coordinates with a nonsingular vertex. This provides an important test bed for code development.

2 THE NULL CONE FORMALISM

Figure 1 illustrates how null cone coordinates $x^\mu = (u, r, x^A)$ are uniquely determined (up to a trivial angular coordinate freedom) by a point O and a timelike vector T^μ in its tangent space. These determine a timelike geodesic which serves as the origin worldline for the vertices of a family of null cones. Let the coordinate u be the proper time along this geodesic, with $u = constant$ on the outgoing null cones. Let x^A ($A = 2, 3$) be coordinates for the outgoing null rays, consistent with parallel propagation along the origin worldline. Let r be a surface-area distance on the null cones. Then, in the corresponding Bondi coordinate system, the line element takes the form

$$ds^2 = H_\mu dx^\mu du - r^2 h_{AB} dx^A dx^B, \tag{1}$$

where, for numerical purposes we choose $x^A = [-cos(\theta), \phi]$ in terms of the usual polar coordinates. Then $det(h_{AB}) = 1$. Also, the numerical grid is based upon the compactified radial coordinate $x = r/(1 + r)$, so that points at future null infinity J^+ are included in the grid at $x = 1$. The coordinate conditions at the origin imply that the metric reduces to a Minkowski (null polar) form along the central worldline. However, the resulting metric does not take an asymptotic Minkowski form at J^+.

Initial null data for the gravitational field consists of the 2-metric h_{AB}. Because of the unimodular condition, this entails two degrees of freedom describing the conformal 2-geometry of the surfaces of constant r on the initial null cone. It is convenient to introduce a complex polarization dyad

$$h_{AB} = 2m_{(A}\bar{m}_{B)}. \tag{2}$$

Then these dynamical degrees of freedom can be efficiently described in terms of the single complex function $\zeta = m_3/m_2$. There is a one-to-one correspondence between choices of ζ and symmetric unimodular matrices h_{AB}. There are no constraints on this data except that it be consistent with smoothness at the origin and asymptotic flatness.

The vacuum field equations consist of four hypersurface equations for H_μ and one complex evolution equation for ζ. They take the symbolic form

$$H_1 = \int_0^r dr J_1[\zeta] \tag{3}$$

$$H_A = \int_0^r dr J_A[\zeta, H_1] \tag{4}$$

$$H_0 = \int_0^r dr J_0[\zeta, H_1, H_A] \tag{5}$$

$$\partial \zeta / \partial u = \int_0^r dr J[\zeta, H_\mu]. \tag{6}$$

Here the the J-operators consist of explicit operations intrinsic to the null cone. These equations form a hierachy which leads from the initial data ζ to its time derivative by means of a series of radial integrations.

Our first attempt at a numerical code for axisymmetric vacuum space-times based upon the null cone algorithm led to unexpected difficulties. The numerical grid included null infinity as a compactified boundary and yielded the first successful numerical calculations of the Bondi mass and news function for gravitational waves (Isaacson, Welling & Winicour 1983). However, near the vertex of the null cone, instabilities arose which destroyed the accuracy of the code over long time scales. We felt that this problem was too complicated to analyze in the context of general relativity, considering that the numerical analysis of the characteristic initial value problem had not yet been carried out even for simplest linear axisymmetric systems.

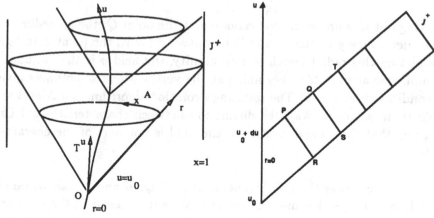

Fig. 1 Null cone coordinates.   Fig. 2 Scheme for the marching algorithm.

3 THE FLAT SPACE WAVE EQUATION

This warranted an investigation of the basic mathematical properties of the numerical evolution of the flat space scalar wave equation using a null cone initial value formulation (Gómez, Isaacson & Winicour 1992). Consider the scalar wave equation

$$\Box \Phi = NL + S, \tag{7}$$

where the terms on the right hand side represent a nonlinear potential, such as a Φ^4

potential, and an external source. This can be reexpressed in the form

$$\Box^{(2)}g = -\frac{L^2 g}{r^2} + r(NL + S), \tag{8}$$

where $g = r\Phi$ and L^2 is the angular momentum operator. Integration over a null parallelogram in the (u, r)-plane as depicted in figure 2, leads to the integral equation

$$g_Q = g_P + g_S - g_R + \frac{1}{2} \int_A du\, dr [-\frac{L^2 g}{r^2} + r(NL + S)]. \tag{9}$$

This identity gives rise to an explicit marching algorithm for evolution. Let the null parallelogram span null cones at adjacent grid values u_0 and $u_0 + \Delta u$, as shown in figure 2, for some θ and ϕ. Imagine for now that the points P, Q, R and S lie on the grid, so that $x_Q - x_P = x_S - x_R = \Delta x$. If g has been determined on the entire u_0 cone and on the $u_0 + \Delta u$ cone radially outward from the origin to the point P, then (9) determines g at the next radial grid point Q in terms of an integral over A. The integrand can be approximated to second order, i.e. to $O(\Delta x \Delta u)$, by evaluating it at the center of A. To the same accuracy, the value of g at the center equals its average between the points P and S, at which g has already been determined.

After carrying out this procedure to evaluate g at the point Q, the procedure can be repeated to determine g at the next radially outward point, the point T in figure 2. After completing this radial march to null infinity, the field g is then evaluated on the next null cone at $u_0 + 2\Delta u$, beginning at the vertex where smoothness gives the start up condition that $g = 0$. The resulting evolution algorithm is a 2-level scheme which reflects, in a natural way, the distinction between characteristic and Cauchy evolution, i.e. that the time derivative of the field is not part of the characteristic initial data.

In practice, the corners of the null parallelogram, P, Q, R and S, cannot be chosen to lie exactly on the grid because the velocity of light in terms of the compactified coordinate x is not constant even in flat space. As a consequence, the field g at these points is approximated to second order accuracy by linear interpolation between grid points. However, cancellations arise between these four interpolations so that (9) is satisfied to fourth order accuracy. The net result is that the numerical version of (9) steps g radially outward one cell with an error of fourth order in grid size. Second order global accuracy is indeed confirmed by convergence tests of the code.

For sufficiently large r, we found from an analysis of domains of dependence that the Courant limit on the step size is the same as for a standard Cauchy evolution in spherical coordinates,

$$\Delta u < 2\Delta r \tag{10}$$

and

$$\Delta u < r\Delta\theta. \tag{11}$$

However, near the origin, this analysis gives a much stricter limit

$$\Delta u < Kr(\Delta\theta)^2, \tag{12}$$

where $K \approx 1$. These stability limits were confirmed by numerical experimentation.

Operating within this Courant limit, the algorithm has been implemented as a stable, calibrated, globally second order accurate evolution code on a compactified grid (Gómez, Isaacson & Winicour 1992). Numerical evolution accurately satisfies the mass-energy flux conservation law. Furthermore, null infinity behaves as a perfectly absorbing boundary so that no radiation is reflected back into the system. This algorithm offers a powerful new approach to generic wave type systems. By constructing an exact nonspherical solution for a Φ^4 potential, we were able to calibrate the algorithm in the nonlinear case. It tracked the solution with the predicted second order accuracy right up to the formation of physical singularities. By other choices of potential, we were able to study approximate axisymmetric versions of solitary wave phenomena. The basic algorithm is applicable to any of the hyperbolic systems occurring in physics.

4 SELF-GRAVITATING SCALAR WAVES.

We subsequently extended this algorithm to self-gravitating, spherically symmetric, zero-rest-mass scalar waves, as described by the Einstein-Klein-Gordon equation (Gómez & Winicour 1992a).

$$G_{\mu\nu} = 8\pi[\nabla_\mu\Phi\nabla_\nu\Phi - \frac{1}{2}g_{\mu\nu}\nabla_\alpha\Phi\nabla^\alpha\Phi]. \tag{13}$$

In null cone coordinates, the line element takes the Bondi form

$$ds^2 = e^{2\beta}du(\frac{V}{r}du + 2dr) - r^2(d\theta^2 + sin^2\theta d\phi^2). \tag{14}$$

Let $H(u) = \beta(u,\infty)$. Then Bondi time $\tilde{u}$, measured by inertial observers at null infinity, is related to central time by

$$\frac{d\tilde{u}}{du} = e^{2H}. \tag{15}$$

Horizon formation occurs at a finite central time $u = u_H$ but at an infinite Bondi time $\tilde{u}_H \approx \infty$.

In the case of spherical symmetry, $g = r\Phi$ obeys a two dimensional wave equation intrinsic to the (u,r) plane. In two dimensions, the geometry is conformally flat, the wave operator has conformal weight -2 and the surface area element has conformal weight +2, so that the surface integral of $\Box^{(2)}g$ over a null parallelogram gives exactly the flat space result. This allows use of the same basic evolution algorithm already described. The only new feature is that the radial integration of the hypersurface equations must be worked into the algorithm to determine $\beta(u,r)$ and $V(u,r)$.

4.1 Asymptotic Properties

Christodoulou (Christodoulou 1986a, 1986b, 1987a, 1987b) has made a penetrating analysis of the existence and uniqueness of solutions describing gravitational collapse of a scalar field, in the spherically symmetric scalar case, and has established a rigorous version of a no-hair theorem. He proved that weak initial data evolves to Minkowski space asymptotically in time but that sufficiently strong data forms a horizon, with nonzero final Bondi mass M_H. The geometry is asymptotically Schwarzschild in the approach to I^+ (future timelike infinity) outside the sphere $r = 2M_H$. Figure 3 depicts the spacetime of such a field beginning at initial retarded time u_0 and forming a horizon at u_H. The situation differs from Oppenheimer-Snyder collapse in that the backscatter of radiation causes the $r = 2M_H$ curve to intersect the horizon only in the asymptotic limit at I^+. In that respect, it is more akin to the spacetime of a dust distribution whose interior collapses but whose exterior escapes to infinity, as depicted in figure 4.

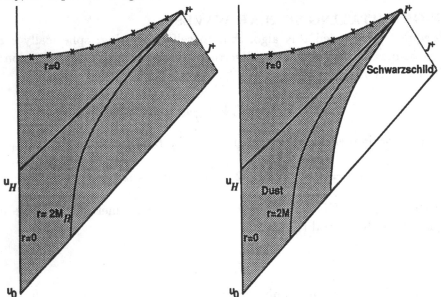

Fig. 3 Gravitational collapse of a scalar field.

Fig. 4 Collapsing-expanding dust.

When a horizon forms, Christodoulou's no hair theorem states that the geometry has the asymptotic step function behavior

$$e^{2(\beta-H)} \rightarrow \begin{cases} 0 & \text{for} \quad r < 2M_H \\ 1 & \text{for} \quad r > 2M_H \end{cases} \tag{16}$$

in the limit $u \rightarrow u_H$. From the hypersurface equation

$$\beta_{,r} = 2\pi r (\Phi_{,r})^2 , \tag{17}$$

it follows that $\Phi \rightarrow 0$ as $u \rightarrow u_H$ for $r > 2M_H$. The compactified grid allows the accurate tracking of the scalar radiation field right up to time infinity I^+. In the curved space case, the Courant condition requires that the ratio of the time step size to radial step size be limited by $(V/r)\Delta u \leq 2\Delta r$, where $\Delta r = \Delta[x/(1-x)]$. The strongest restriction arises at J^+, just before the formation of a horizon. In this limit, $V/r \rightarrow \infty$ so that the conformal singularity at I^+ freezes the numerical evolution. The code becomes unreliable when the red shift between central time and Bondi time is of the order of 10^9. In order to evolve across the horizon, exterior radial points must be dropped from the domain of the grid. Figure 5 illustrates the formation of a step function just before horizon formation during a typical numerical evolution, in confirmation of Christodoulou's theorem.

One static solution for a spherically symmetric, self gravitating zero rest mass scalar field is $\Phi = constant$, which is pure gauge and not by itself physically interesting. Another is the analog of the static solution $\Phi = 1/r$ in a Minkowski background (Janis, Newman & Winicour 1968). Pasting these together gives rise to initial data whose evolution is not static because of the jump discontinuity in the curvature at their interface. This gives rise to a shock front along a radially incoming characteristic. Results of a numerical evolution of $g(x, u)$ are shown in figure 6 for the case of initial amplitude large enough to form a horizon.

To the past of the shock front, Φ remains constant. Although $g(x, u) = r\Phi$ has a curved profile in this region, the evolution is manifestly static. The numerical code clearly handles the inward propagation of the shock front without difficulty. Outside the shock front, backscattering distorts the initial profile. This distinctly illustrates the breakdown of Huyghen's principle due to curvature. The numerical code handles the propagation of the shock front without any substantial difficulty. It introduces some slight high frequency numerical noise just outside the shock front, but too small to be perceptible in the figure.

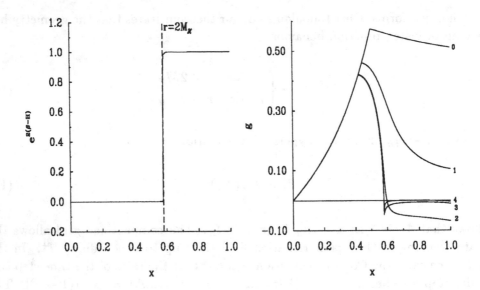

Fig. 5 Formation of a step function.   Fig. 6 Evolution of static-static data.

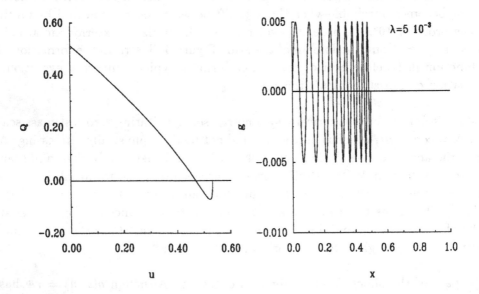

Fig. 7 Decay of the monopole mopment.   Fig. 8 The initial data for a wave train.

The early stage of the evolution appears innocuous but then the outer region of the profile flattens. This marks the beginning of the process of "shedding hair" as a black hole starts to develop. As this process intensifies in the late stage, a cusp begins to form near $r = 2M_H$, which corresponds to the conformal singularity developing at I^+. Note that the slope of g at J^+ is conserved during the evolution. It is an example of a Newman-Penrose conserved quantity. Our results demonstrate how the Newman-Penrose constant is indeed conserved while remaining consistent with the no hair scenario.

The time dependence of the monopole moment $Q(u) = g(u, \infty)$ is graphed in figure 7. After a long dive to a negative value, it bobs up abruptly to zero just before the horizon forms. The numerical evolution is stopped just short of the horizon due to the inability of the grid to resolve the cusp. The redshift factor at this time is $d\tilde{u}/du \approx 10^7$. Careful inspection of the numerical results shows that its final decay has time dependence $\sqrt{u_H - u}$. From recent discussions with P. Rabier and W. Rheinboldt, we have learned that this corresponds to a canonical singularity in the general theory of Differential Algebraic Equations (DAE's) (Rabier 1989; Rheinboldt 1984), in which a quantity has a well defined limit but its time derivative does not. We find that central time and Bondi time are related asymptotically by

$$\frac{u_H - u}{4 M_H} \sim e^{-\tilde{u}/4 M_H}, \tag{18}$$

analogous to the relation between Kruskal and Schwarzschild times. This implies that the monopole moment decays exponentially in Bondi time, in contrast to the power law predicted by perturbation theory in an Oppenheimer-Snyder background (Price 1972). This is surprising since in this final phase the metric is very close to a Schwarzschild metric in the region exterior to $r = 2M_H$.

Another illustrative example is provided by the strong amplitude initial data given in figure 8. Figure 9 graphs g just before horizon formation in a strong amplitude case; and figure 10 graphs what would be the linear evolution of g at a suitable time for comparison with the strong amplitude graph. The qualitative difference between these graphs highlights the nonlinear effects of self gravitation. The most striking feature of the strong amplitude case is that the horizon forms quite insensitively to the detailed structure of the field in the inner region $r < 2M_H$. The chief difference, between the strong and weak case, in the evolution of g inside this inner region arises from the way in which the outgoing wave from the origin interferes with the incoming signal. In the weak case this interference lowers the entire profile in figure 10 by a constant determined by the amplitude of the outgoing wave leaving the origin at that time. In the strong case, backscattering couples the the incoming and outgoing waves. The linear slope modulating the wave profile in figure 9 is a prime illustration of backscattering depleting the outgoing wave.

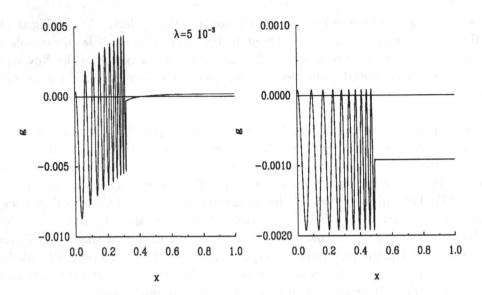

Fig. 9 Strong amplitude evolution.   Fig. 10 Weak amplitude evolution.

4.2 Scaling Limits

The Bondi mass of this Einstein-Klein-Gordon system has some remarkable asymptotic behavior with respect to the one parameter family of data obtained from the amplitude rescaling $\Phi(u_0, r; \lambda) = \lambda\Phi(u_0, r)$, which preserves asymptotic flatness. By an application of the method of Laplace, we have derived the following asymptotic formula for large λ which holds when the monopole moment is nonvanishing,

$$M(u_0; \lambda) \sim \frac{\pi}{\sqrt{2}}|Q(u_0; \lambda)|. \tag{19}$$

In this regime, the mass is essentially the magnitude of the monopole moment and scales linearly with λ! Compare these results to the small amplitude regime, in which M depends quadratically on λ. In the strong field case, a redshift type effect weakens the dependence of M on the the inner region of the matter distribution and the dominant contribution comes from the far field monopole moment. The details of the transition from the low amplitude to high amplitude asymptotic regimes can be obtained from numerical calculations of the mass using our code. The result for a typical choice of data is illustrated in figure 11. The transition from quadratic to linear λ dependence occurs around the critical value λ_c, at which the evolution bifurcates between forming or not forming a horizon.

For data of compact support within a radius R, the method of Laplace leads to the high amplitude asymptotic dependence

$$M(u_0; \lambda) \sim R/2. \tag{20}$$

In this case, redshifting completely saturates the λ dependence and the null data approaches that for a horizon at $R = 2M$. This is illustrated in figure 12. The study of a specific analytic model (Winicour 1990) with compact null data also shows that the news function is completely redshifted away, $N(\lambda) \to 0$ as $\lambda \to \infty$ holding R constant. Amplitude scaling, does not commute with evolution, $\Phi(u_1, r; \lambda) \neq \lambda \Phi(u_1, r)$, except in the small λ linearized limit. This lies behind the somewhat magical way in which high amplitude noncompact data, with large M and Q, rapidly sheds its monopole moment, along with its exterior field, without significant mass loss as a black hole forms.

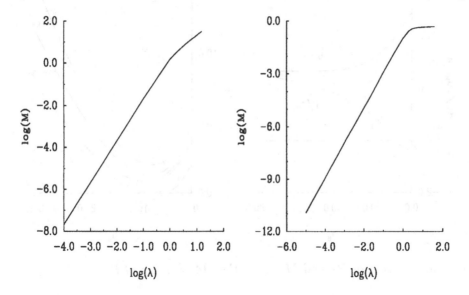

Fig. 11. Scaling of the Bondi mass
for nonvanishing monopole moment.

Fig. 12. Scaling of the Bondi mass
for compact data.

Another global quantity with interesting properties is

$$P = 4\pi \int_0^\infty r(g_{,r})^2 dr. \tag{21}$$

$P^{1/2}$, like the square root of the Bondi mass, provides a norm on the vector space of asymptotically flat initial data. P scales quadratically with respect to λ under amplitude rescaling. It is implicit from the results of Christodoulou that its derivative with respect to Bondi time can be expressed in the form

$$P_{,\bar{u}} = M - \lambda \partial_\lambda M. \tag{22}$$

In the linear regime, the Bondi mass scales quadratically so that $\lambda \partial_\lambda M \sim 2M$ and (22) reduces to $P_{,\tilde{u}} \sim -M$. Thus P must monotonically decrease in the weak field regime. This explains how flat spacetime arises as a basin of attraction for the weak field case. Since during the formation of a horizon the field develops compact support within a radius $2M_H$, one would expect that $\partial_\lambda M \to 0$ and therefore that

$$P_{,\tilde{u}} \sim M_H, \tag{23}$$

These late time asymptotics can also be confirmed by numerical evolution, as illustrated in figure 13.

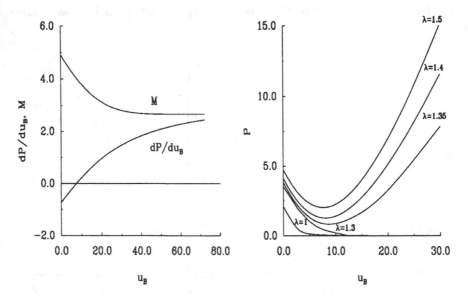

Fig. 13. Comparison of $P_{,\tilde{u}}$ and M.   Fig. 14. Scaling of P.

Figure 14 plots $P(\tilde{u})$ for several representative values of λ. Many qualitative features of these graphs can be explained in terms of asymptotic results previously discussed. For the two graphs with $\lambda < \lambda_c$, P decays to zero as the system decays to flat space, in keeping with the role of P as a norm. For the three graphs with $\lambda > \lambda_c$, P exhibits interesting asymptotic behavior. First, at the initial time $\tilde{u} = 0$, these three graphs of P all have approximately the same slope. This is a manifestation of the high amplitude scaling properties which can be established for $dP/d\tilde{u}$ (Gómez & Winicour 1992a). The numerical results show that the high amplitude limit remains a good approximation to 10% accuracy even for values $\lambda \approx \lambda_c$. At large Bondi time, these three graphs again tend toward straight lines exhibiting the asymptotic relation $dP/d\tilde{u} \to M_H$. From the asymptotic slope of these graphs, it is evident that systems which start out at higher amplitude develop horizons with greater mass.

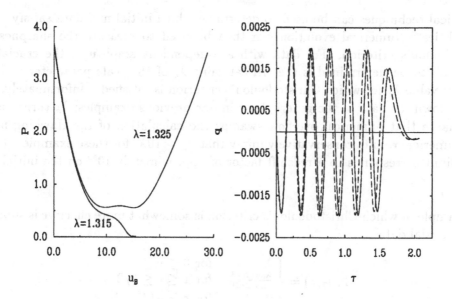

Fig. 15. Behavior of P near critical data. Fig. 16. Comparison of profiles at $R = 25M$ (dashed line) and J^+ (solid line) for compact initial data.

Figure 15 plots $P(\tilde{u})$ for values of λ just above and just below the critical value of the amplitude. The bumpy features in the graph may be a rough version of the intricate structure that appears in the neighborhood of the critical value, which has been described by Choptuik (Choptuik 1989). The graph also clearly shows the value of P as a diagnostic tool for determining the fate of the system at large times.

4.3 Collapse Criteria

Christodoulou (*to appear*) has established a condition on the initial data for the scalar field which is sufficient to guarantee collapse to a black hole. It is based upon pairs of spheres along the initial null cone with radii (r_1, r_2) (measured by surface area) and upon the Hawking masses (m_1, m_2) of these spheres. For a given pair, let

$$\delta := \frac{r_2 - r_1}{r_1} \tag{24}$$

and

$$\eta := \frac{2(m_2 - m_1)}{r_2}. \tag{25}$$

Then a horizon forms if there is a pair which satisfies the inequalities

$$0 < \delta < \frac{1}{2} \quad \text{and} \quad \eta \geq \frac{\delta}{(1+\delta)^2}[5 - \delta - log(2\delta)]. \tag{26}$$

Numerical techniques can be used to determine which initial null data satisfy these inequalities. Numerical evolution can then be used to measure the sharpness of Chritodoulou's criterion. For data with a λ-dependent amplitude, the crucial information to be determined is the critical value λ_c of the scale parameter and the smallest value λ_x for which Christodoulou's criterion is satisfied. Unfortunately, this breaks down for data of the generic type in our previous examples; λ_x turns out to be so large that numerical overflow swamps the calculation of the Hawking mass. The numerical results conservatively imply that $\lambda_x > 10\lambda_c$ for these examples. (This lower limit corresponds to a redshift factor of approximately 10^{30} on the initial null cone.)

An example in which Christodoulou's criterion is somewhat more effective is supplied by the initial data

$$\Phi(u_0, r) = \begin{cases} \lambda & \text{for } 0 \leq r \leq 1 \\ \frac{\lambda \log(2r/3)}{\log(2/3)} & \text{for } 1 \leq r \leq 3/2 \\ 0 & \text{for } r \geq 3/2 \,. \end{cases} \tag{27}$$

This is a special case of data for which the hypersurface equations can be integrated analytically (Winicour 1990). The Hawking mass is

$$m = \begin{cases} 0 & \text{for } 0 \leq r \leq 1 \\ \frac{r(z-1)(1-r^{-z})}{2z} & \text{for } 1 \leq r \leq 3/2 \\ 0 & \text{for } r \geq 3/2 \,. \end{cases} \tag{28}$$

Here $z = 1 + 4\pi\lambda^2[log(3/2)]^{-2}$. Analysis of the inequality (26) then leads to the value $\lambda_x \approx .33$ (corresponding to $r_1 = 1$ and $r_2 \approx 1.055$), compared with the value $\lambda_c \approx .097$ obtained from numerical evolution of this data.

Are there other more physically intuitive collapse conditions? The numerical results strongly support the hypothesis that once the redshift attains a sufficiently high value the consequences are irreversible and the system must form a horizon. Can then a collapse criterion be formulated in terms of an inequality $H > H_c$, where H_c is independent of the choice of data?

4.4 Waveforms at Finite Distances
We have also used the spherically symmetric, massless Einstein-Klein-Gordon system to investigate the discrepancies that arise if waveforms are observed at a grid boundary with finite radius R, as opposed to null infinity (Gómez & Winicour 1992b). For a system of mass $M \ll R$, it has tacitly been assumed that the waveform at the grid boundary approximates the waveform at infinity, after compensating for the $1/r$ falloff, with error of order of magnitude M/R. We have performed some numerical

experiments which show that this is not true for radiation consisting of a long wave-train. Discrepancies close to 100% can arise at large observation distances $R >> M$ for sufficiently periodic systems. They are most pronounced for radiation losses between one quarter and one half of the initial mass. This falls within the expected regime of the spiral infall of a relativistic binary system. The predominant contribution to this discrepancy stems from a time dependent redshift arising from radiative mass loss.

For gravitational waves, there are severe complications which can potentially affect the accuracy of waveforms based upon a finite worldtube. Gauge ambiguities make it unclear which components of the metric or curvature tensor to use. Some method of selecting the components transverse to the propagation direction is necessary but there is no unique means of defining this propagation direction locally. The choice of time coordinate can introduce further gauge effects. There are additional physical complications. Time dependent versions of redshifting occur. The nonlinear gravitational self-source is noncompact and introduces backscattering which blurs the distinction between incoming and outgoing fields. A finite grid boundary also leads to some backscatter. Furthermore, in a case such as a binary black hole system, there is no practical scheme for eliminating incoming waves from the initial data. In spite of these complications, techniques used by numerical relativists with spacelike codes have led to consistent and sensible results in the range of problems where they have been applied (neutron star oscillations, supernovae).

The spherically symmetric scalar model has no ambiguity in the local radial direction and no other gauge ambiguities. Furthermore, initial incoming waves are eliminated in the test region by choosing initial data with compact support. In the nonspherical case, this remedy is not possible for the initial data describing gravitational degrees of freedom because of the constraint equations. Elimination of these effects allows isolation of the nonlinear effects of backscattering and time dependent redshifting. Furthermore, spherical symmetry allows the luxury of a very fine grid so that these effects can be studied without contamination by numerical noise.

The radiation amplitude at J^+ is $Q(\tilde{u}) = g(\tilde{u}, \infty)$, where the Bondi time $\tilde{u}$ plays the role of the proper time used by an observer at infinity. As the counterpart of $Q(\tilde{u})$ based upon the worldtube, we choose $q(\tau; R) = g(\tilde{u}(\tau, R), R)$, where $\tau = \tau(\tilde{u}, R)$ is the proper time on the worldtube. There is a relative redshift between observers on the worldtube and at J^+. In order to synchronize the two time coordinates we set $\tau_0 = \tilde{u}_0 = 0$ on the initial null cone. We have tested how accurately $q(\tau; R)$ serves as a substitute for $Q(\tilde{u})$. By construction, the test is automatically satisfied in the linear weak field case throughout any region where the wave is purely outgoing. The waveform at a finite radius can be very misleading when incoming waves are present.

In strongly nonlinear fields, there is no clean way to decompose the wave into incoming and outgoing parts. But to make the test clear cut, we will only consider initial data with support $r < 1$ and test radii $R \geq 1$ at which $q(\tau; R)$ measures only the outgoing radiation and the backscattering. In order to quantify the discrepancy between the waveform q and the radiative waveform Q we use the l_2 norm and the figure of merit

$$E = \frac{||q - Q||}{||Q||}. \tag{29}$$

Since the l_2 norm is independent of basis, the same error applies to the fourier transform of the signal.

The strategy here is to choose initial data characterized by two parameters, representing amplitude and wavelength, and to investigate the resulting waveforms over a comprehensive range. E is small for small amplitudes, in accord with the weak field limit. It is also small at high amplitudes for which a black hole forms very rapidly. This stems from a rigorous version of the no-hair theorem (Christodoulou 1987b) which establishes that the scalar field must vanish in the limit $\tau \to \infty$, $r > 2M_H$. As a result, in the region of observational interest, the field is zero initially by construction and, in the high amplitude regime, it never builds up any appreciable amplitude before the interior region collapses to a black hole. Systems in this regime would not be readily detectable because of their extreme redshift. We focus our attention on the intermediate amplitude region of greatest physical relevance.

As an example of our test, consider the initial data

$$g(u_0, r) = \begin{cases} \Lambda \sin(2\pi N(r - 1)) & \text{for } r \leq 1 \\ 0 & \text{for } r \geq 1, \end{cases} \tag{30}$$

where Λ controls the amplitude and N controls the wavelength. The results are graphed in figure 16 for the choices $N = 20$ and $\Lambda = 9.5 \times 10^{-4}$, for which the initial mass is $M_0 = 0.041$ and the system forms a black hole with $M_H = 0.029$. The discrepancy in the waveform at $R = 25M_0$ is $E = 0.91$. The amplitudes shown in the figure are in close agreement, which indicates very little backscatter in the intervening region. However, there is considerable phase shifting, which results from a time dependent redshift effect, and this is the prime source of the discrepancy between the waveforms.

In the same manner, we also find a large discrepancy, $E = 0.80$, at $R = 100M_0$ by increasing N to 100. High accuracy computer simulations for $N >> 100$ would require too much computer time to be practical. However the trend we have already seen for large N can be understood and extrapolated in terms of a rough analytic model which leads to the approximation $E \approx 2\pi n M/\sqrt{3}R$, where n is the number of wavetrains radiated before horizon formation. For the previous case with $N = 20$ (for which $n = 8$), this gives better than 10% agreement with the value of E from the numerical evolution. For $N = 100$ (for which $n = 24$), there is 2.5% agreement between the formula and the numerical value obtained for E. This provides strong evidence that this formula gives a reliable estimate of E in the large N regime. Thus large waveform discrepancies, $E \approx 1$, can arise at any radius for a system of sufficiently high frequency, i.e. $n \approx R/M$.

5 GRAVITATIONAL WAVES

Our aim is to be able to study asymptotic properties of gravitationally radiating spacetimes with the high level of accuracy achieved for scalar radiation. We are currently incorporating the lessons learned from the scalar algorithm into the evolution code for axisymmetric spacetimes. For axisymmetry, the conformal geometry of the spheres of constant u and r is

$$h_{AB}dx^A dx^B = e^{2\gamma}d\theta^2 + sin^2\theta e^{-2\gamma}d\phi^2, \tag{31}$$

so that γ represents the entire null datum for the gravitational field. The evolution equation (6) can be reexpressed in the form

$$\Box^{(2)}\gamma = \hat{J}[\gamma, H_\mu], \tag{32}$$

where $\Box^{(2)}$ is the D'Alembertian with respect to the induced geometry of the (u, r) submanifold and $\hat{J}$ again consists only of operations intrinsic to the null cone $u = const$.

This has the identical structure as the scalar wave equation (8), thereby leading to the analogue of the null parallelogram identity (9) and to an explicit marching algorithm for the evolution of γ. The same cancellations in the numerical error from the four sides of the null parallelogram that arise in the scalar case can also be arranged here so that the algorithm should in principle yield a globally second order accurate solution. We expect, subject to the Courant limit (12), that it will be free of instabilities at the origin. It is reassuring to know that progress has also been made in this direction at Southampton (Bishop, Clarke & d'Inverno 1990) and at Cambridge (Stewart, *this volume*). Although the analysis leading to (12) was based on a 2-level evolution algorithm, the results of the Southampton group seem to imply that (12) also applies to 3-level Adams-Bashworth and Predictor-Corrector algorithms.

ACKNOWLEDGEMENTS
This work was supported by NSF Grant PHY-8803073. Computer time was provided by the Pittsburgh Supercomputing Center under Grant PHY860023P.

REFERENCES

Abrahams, A. M. and Evans, C. R. (1990). *Phys. Rev D*, **42**, 2585.

Ashtekar, A. and Hansen, R. O. (1978). *J. Math. Phys.*, **19**, 1542.

Bicak, J., Reilly, P. and Winicour, J. (1988). *Gen. Rel. and Grav.*, **20**, 171.

Bishop, N. T., Clarke, C. J. S. and d'Inverno, R. A. (1990). *Class. Quantum Grav.*, **7**, L23. See also the articles by Bishop and d'Inverno in this volume.

Bondi, H., van der Burgh, M. G. J. and Metzner, A. W. K. (1962). *Proc. R. Soc. A*, **270**, 103.

Choptuik, M. W. (1989). In *Frontiers in Numerical Relativity*, C. R. Evans, L. S. Finn and D. W. Hobill (eds.), Cambridge University Press, Cambridge. Also, see article in this volume.

Christodoulou, D. (1986a). *Commun. Math. Phys.*, **105**, 337.

Christodoulou, D. (1986b). *Commun. Math. Phys.*, **106**, 587.

Christodoulou, D. (1987a). *Commun. Math. Phys.*, **109**, 591.

Christodoulou, D. (1987b). *Commun. Math. Phys.*, **109**, 613.

Christodoulou, D. *Commun. Math. Phys.*, to appear.

Friedrich, H. and Stewart, J. M. (1983). *Proc. R. Soc. A*, **385**, 345.

Gómez, R., Isaacson, R. A. and Winicour, J. (1992). *J. Comp. Phys.*, **98**, 11.

Gómez, R. and Winicour, J. (1992a). *J. Math. Phys.*, **33**, 1445.

Gómez, R. and Winicour, J. (1992b). "Gravitational Waveforms at Finite Distances and at Null Infinity", *Phys. Rev. D*, in press.

Isaacson, R. A., Welling, J. S. and Winicour, J. (1983). *J. Math. Phys.*, **24**, 1824.

Janis, A., Newman, E. T. and Winicour, J. (1968). *Phys. Rev. Lett.*, **20**, 878.

Newman, E. T. and Penrose, R. (1968). *Proc. R. Soc. A*, **305**, 175.

Penrose, R. (1963). *Phys. Rev. Lett.*, **10**, 66.

Price, R. H. (1972). *Phys. Rev. D*, **5**, 2419

Rabier, P. J. (1989). *J. Math. Anal. and Appl.*, **144**, 425.

Rheinboldt, W. C. (1984). *Math. of Comp.*, **43**, 473.

Stewart, J. M. This volume.

Winicour, J. (1989). In *Highlights in Gravitation and Cosmology*, A. Khembavi, J. Narlikar, C. Vishveshwara and B. Iyer (eds.), Cambridge University Press, Cambridge.

Winicour, J. (1990). In *Proc. of the 3rd Canadian Conf. on Gen. Rel. and Rel. Astrophysics*, A. Coley, F. Cooperstock B. and Tupper (eds.), World Scientific, Singapore, 94.

INSTABILITIES IN RAPIDLY ROTATING POLYTROPES

Scott C. Smith and Joan M. Centrella

Department of Physics and Atmospheric Science, Drexel University, Philadelphia, US

Abstract. We review the classical and modern work on the stability of rotating fluid configurations with particular interest in astrophysical scenarios likely to produce gravitational radiation. We describe a hybrid method for numerically generating axisymmetric equilibrium models in rapid differential rotation based on the self consistent field approach. We include a description of the 3-D hydrodynamics code that we have developed to model the production of gravitational radiation, and present the results of a 3-D test case simulating the growth of the dynamical bar mode instability in a rapidly rotating polytrope.

1 INTRODUCTION

1.1 Overview

The study of the effects of rotation on equilibrium fluid bodies was begun by Newton in Book III of the *Principia* where he investigated the consequences of rotation on the figure of the earth, and concluded that the result would be a flattening at the poles to give the earth a slightly oblate shape. Much of the classical work accomplished since that time has been concerned with equilibrium configurations for fluids with uniform density and/or rigid rotation. Recent advances in computing technology, however, have allowed more detailed investigations involving differential rotation, various equations of state, and dynamical evolution of self gravitating systems. This work has fostered a variety of astrophysical applications, notably in the study of the formation of single and binary stars from collapsing gas clouds, and of the structure of compact objects, such as white dwarfs and neutron stars.

Rapidly rotating axisymmetric systems exhibit both secular and dynamical instabilities to the growth of nonaxisymmetric modes (Tassoul 1978), and the growth of such structure in compact objects or during stellar core collapse could lead to the release of significant amounts of gravitational radiation (Thorne 1987). We have developed a 3-D hydrodynamics code with Newtonian gravity to model these events. Gravitational radiation is calculated by the quadrupole formalism, while gravitational radiation losses and the secular instability will be handled by adding gravitational radiation reaction terms. The code is currently being run on a variety of test-bed calculations to verify its accuracy. Once this testing is complete, we will apply it to the calculation of the gravitational wave signature from nonaxisymmetric instabilities. Such calculations are becoming increasingly important due to progress towards the goal of detecting gravitational waves from astrophysical sources, with detectors of sufficient sensitivity expected to be operational by the end of the decade (Vogt 1991).

We begin with an historical overview of the classical work on incompressible fluids, followed by a summary of more recent work involving polytropes and neutron stars. We then present a method for generating axisymmetric equilibrium configurations for differentially rotating fluids. These models serve as the starting point for investigating the secular instability induced by dissipative forces and the dynamical instability which occurs spontaneously under small perturbations away from axisymmetry. Finally, after a brief description of our code, we present some results of a 3-D simulation of the growth of a dynamical bar instability that we are using to verify the performance of our code.

1.2 Classical results - incompressible fluids

The history of the study of rotating fluids is well reviewed in the texts by Chandrasekhar (1969) and Tassoul (1978), and also in the review article by Durisen and Tohline (1985). The bulk of this section is taken from these sources.

The problem of calculating the equilibrium configuration of rotating, self- gravitating systems was first solved in the case of an incompressible fluid undergoing rigid rotation by Maclaurin in the early 1700's. Maclaurin assumed axisymmetry, and the resultant family of uniform density oblate spheroids are known as the Maclaurin spheroids. For a given fluid density, these form a single parameter family, which is frequently parametrized by the eccentricity of the ellipsoid or the total angular momentum. We shall find it useful to parametrize them by the ratio

$$\beta \equiv T/|W| \tag{1}$$

where T is the total kinetic energy and W is the gravitational potential energy. In terms of this parameter, Maclaurin spheroids exist for $0 \leq \beta \leq 0.5$, the full range allowed by the virial theorem.

The first nonaxisymmetric solutions to this problem were found nearly a century later by Jacobi, who demonstrated the existence of a family of triaxial ellipsoids that satisfied the equilibrium conditions for uniform density fluids. Unlike the Maclaurin spheroids, there is a lower bound on β for the Jacobi ellipsoids at $\beta = 0.1375$, at which point the sequence of Jacobi ellipsoids bifurcates from the sequence of Maclaurin spheroids. These ellipsoids rotate rigidly around the short axis.

The first work that relaxed the restriction of rigid rotation was accomplished by Dirichlet, and published posthumously by Dedekind. Dirichlet addressed the problem of the conditions under which an equilibrium configuration would at all times present the figure of an ellipsoid. The general equations were worked out by Dirichlet, who only solved them in detail for the axisymmetric case. Dedekind, however, worked out a case that was related to the Jacobi ellipsoids by a simple transformation. These Dedekind ellipsoids represent a family of triaxial ellipsoids for which the surface figure remains fixed in an inertial reference frame, but within which there is a uniform, non-zero vorticity ζ. The Dedekind ellipsoids bifurcate from the sequence of Maclaurin spheroids at the same point as the Jacobi ellipsoids.

The general solution to Dirichlet's problem was carried out by Riemann, who showed that a general uniform density equilibrium ellipsoid is formed by superposing a uniform rotation with internal motions having uniform vorticity. These configurations are known collectively as the Riemann ellipsoids. Those configurations for which the rotation and vorticity are both directed along the short axis of the figure are known as the Riemann S-type ellipsoids. The Jacobi and Dedekind ellipsoids represent the special cases of uniform rotation of the surface with no internal motions and pure internal motion with no overall rotation of the envelope, respectively. In general, a Riemann ellipsoid presents the surface figure of a triaxial ellipsoid in uniform rotation, although the internal motions may be more complicated.

There are also equilibrium configurations with geometries other than ellipsoidal. These show more complicated symmetries corresponding to higher order harmonics and have not been as thoroughly studied.

The co-existence of different equilibria having the same energy leads naturally to the question of stability. The stability of the uniform density equilibrium configurations has been well studied. The Maclaurin spheroids are stable as long as $\beta < 0.1375$, the same value at which the Jacobi and Dedekind sequences bifurcate. At this point, the Maclaurin spheroids become subject to a secular instability in the presence of dissipative mechanisms that allows the growth of nonaxisymmetric structure. This is a bar mode instability, for which growth occurs in nonaxisymmetric Kelvin modes of the form $e^{\pm im\phi}$ with m=2. At somewhat higher energies, $\beta \geq 0.2738$, a true dynamical bar mode instability sets in.

The secular instability occurs on the timescale of the dissipative mechanism that triggers it, and terminates as either a Jacobi or Dedekind ellipsoid. In the presence of viscosity, a Maclaurin spheroid will evolve into the rigidly rotating Jacobi ellipsoid having the same angular momentum. Under the action of gravitational radiation reaction, the end result will be the stationary Dedekind ellipsoid having the same circulation. The latter case is commonly referred to as the Chandrasekhar-Friedman-Schutz (CFS) instability, after the researchers who initially studied it (Chandrasekhar 1970; Friedman and Schutz 1978). When both viscosity and gravitational radiation reaction are operating, the two mechanisms can compete with each other, leading to increased stability (Lindblom and Detweiler 1977). In the presence of dissipative mechanisms the general Riemann S-type ellipsoids are no longer true equilibria, and will evolve onto the Jacobi, Dedekind, or Maclaurin sequences.

Evolution away from the Maclaurin sequence under the secular instability occurs though a series of these Riemann S-type quasi-equilibrium configurations.   For $\beta \geq 0.2738$, however, no such equilibria exist. At this point the Maclaurin spheroids become subject to a dynamical bar mode instability that grows on a timescale comparable to a rotation period. The Riemann S-type ellipsoids are themselves subject to a dynamical m=3 "pear" mode instability. This occurs at the same point at which a sequence of pear shaped figures bifurcates from the S-type ellipsoids. For the specific case of the Jacobi ellipsoids, this bifurcation point occurs at $\beta = 0.1628$.

1.3 Rotating polytropes and compact objects

The uniform density configurations have been well studied classically. More recently, work has been focused on more centrally condensed models such as polytropes, which can provide more realistic approximations to actual stars and compact objects. Work in this area has been greatly facilitated by the advances in computer technology over the last few decades, although some results were derived classically much earlier. The equation of state (EOS) for a polytrope is given by

$$P = k\rho^{(1+\frac{1}{n})}, \tag{2}$$

where k and n are constants, and n is commonly termed the polytropic index. Note that smaller values of n produce "stiffer" equations of state, with $n = 0$ corresponding to the case of an incompressible fluid.

Early work has shown that rigidly rotating polytropes with soft equations of state are unable to achieve high rotation rates. Jeans (1919) showed that, for polytropes with polytropic index $n \geq 0.8$, the uniformly rotating sequence terminates due to mass shedding at the equator before any nonaxisymmetric structures bifurcate. James (1964) later refined this result to $n \geq 0.808$. Thus, these high index polytropes remain stably axisymmetric under uniform rotation. More recently, three groups (Vandervoort and Welty 1981; Ipser and Managan 1981; Hachisu and Eriguchi 1982, 1983) have constructed sequences of low index polytropes under rigid rotation. These have been found to bifurcate into Jacobi-like triaxial ellipsoids for $\beta > 0.137(\pm 0.003)$

for all values of n. Only for extremely low polytropic indices ($n \leq 0.1$), however, do the sequences extend significantly beyond the bifurcation point before they terminate due to equatorial mass loss.

If the constraint of uniform rotation is relaxed, polytropes can be constructed with much higher values of β. Differential rotation laws for which the angular velocity is reduced far from the rotation axis can allow values of β as high as the maximum permitted by the virial theorem. Several groups (Lucy 1977; Gingold and Monaghan 1978, 1979; Durisen and Tohline 1980; Tohline et al. 1985; Durisen et al. 1986; Williams and Tohline 1987,1988; Hachisu et al. 1987) have investigated differentially rotating polytropes using 3-D numerical simulations. Their work, although still limited in the number of initial models treated, seems to indicate that the onset of the dynamical bar mode instability for initially spheroidal configurations occurs at close to the Maclaurin value of $\beta = 0.27$, regardless of the form of the rotation law used or the polytropic index.

Rotating neutron stars, *e.g.* pulsars, can emit interesting amounts of gravitational radiation due to nonaxisymmetric instabilities. In general, neutron stars do not rotate fast enough for the dynamical bar mode instability to operate, although the secular instability can be important. Under the CFS instability, the higher order modes can be excited at lower values of β, but the timescales for the growth of these modes are longer than for modes with smaller m. The CFS instability is believed to limit the rotation rate of pulsars since the emission of gravitational waves carries away angular momentum. The stiffness of the neutron star EOS further limits the rotation rate that can be achieved, and hence which modes will operate under the CFS instability (Friedman, Ipser, and Parker 1986). In addition, the CFS instability can be damped by neutron star viscosity (Ipser and Lindblom 1991). Overall, it is believed that the modes with $m = 3, 4$, and 5 are the most important for neutron stars (Weber, Glendenning, and Weigel 1991; Friedman, Ipser, and Parker 1986), whereas the $m = 2$ bar mode may operate in pulsars composed of strange quark matter (Colpi and Miller 1991). In addition, the secular instability may play an important role during stellar core collapse and neutron star coalescence if a rapidly rotating object forms. Wagoner (1984) has estimated the gravitational radiation from this instability for a neutron star accreting matter from a disk. To date, however, the CFS instability and the resulting gravitational wave emission have not been modelled using fully dynamical numerical simulations with realistic neutron star models. We hope to accomplish this in the future using our 3-D code.

2 GENERATION OF AXISYMMETRIC EQUILIBRIA

2.1 Background

Several approaches to the generation of rotating Newtonian equilibrium configurations have been proposed (see Hachisu 1986a for a review). The earliest was that of James (1964), who utilized an expansion of the density and gravitational potential in terms of Legendre polynomials and integrated the solution directly. This method, however, could not achieve highly flattened configurations and was not useful with polytropes having $n > 3$. More recently, a powerful method was proposed by Eriguchi and Müller (1985) that utilizes an integral representation to solve Poisson's equation and directly obtains the density distribution using Newton-Raphson iteration. However, this method requires a massive investment in computer time and memory, and thus has practical limitations.

An approach that has proven to be versatile while remaining computationally efficient is the self-consistent field (SCF) method first introduced by Ostriker and Mark (1968) and later refined by Hachisu (1986a). This method has been used to study rapidly rotating polytropes (Mark 1968; Bodenheimer and Ostriker 1970,1973; Ostriker and Bodenheimer 1973) and white dwarfs (Ostriker and Bodenheimer 1968; Ostriker and Tassoul 1969), as well as to investigate the fragmentation of gas clouds, rings and tori (Hachisu et al. 1987, 1988; Tohline and Hachisu 1990). Hachisu (1986b) has extended the method to produce non-axisymmetric structures, although limited to the case of uniform rotation. We use a variation of Hachisu's method in generating our initial models.

2.2 Self-consistent field approach

The SCF method is based on an integral formulation of the equations of hydrodynamic equilibrium. This simplifies the problem somewhat because it automatically incorporates the boundary conditions into the integral equations. Integrated quantities also tend to be less sensitive to discontinuities and grid-based effects than differential quantities, so the method can be expected to be more stable than methods relying on a differential formulation. The technique begins by choosing an initial guess density distribution, calculating gravitational and rotational potential fields for that distribution, using those fields to obtain an improved density distribution, and iterating until convergence is achieved. In this section we will examine the SCF method, specialized to the case of rotating polytropes, as implemented by Ostriker and Mark (1968, hereafter O&M) and by Hachisu (1986a; subsequent references to Hachisu throughout this section refer to this paper unless otherwise specified).

Although the approaches of O&M and Hachisu differ in some details, they share many features common to the generic SCF method. Both are based on the same basic set of equations, and rely on iterative relations between the potential fields and the density distribution. In addition, this is a numerical technique performed on a grid, and both authors have chosen to utilize spherical coordinates in defining their grids.

Consider an isolated, axisymmetric fluid rotating in hydrodynamic equilibrium about the z axis in cylindrical coordinates (ϖ, φ, z). Euler's equation gives (Tassoul 1978)

$$\frac{1}{\rho}\frac{\partial P}{\partial \varpi} = -\frac{\partial \Phi}{\partial \varpi} + \Omega^2 \varpi \tag{3}$$

and

$$\frac{1}{\rho}\frac{\partial P}{\partial z} = -\frac{\partial \Phi}{\partial z}, \tag{4}$$

where the angular velocity $\Omega = \Omega(\varpi, z)$ and Φ is the gravitational potential, given by

$$\Phi = -G \int \frac{\rho(\vec{r'})}{|\vec{r} - \vec{r'}|} d^3 r'. \tag{5}$$

By integrating (3) with respect to ϖ and (4) with respect to z, and adding the two we obtain the integral form of the equilibrium condition,

$$\int \frac{dP}{\rho} + \Phi - \int \Omega^2 \varpi d\varpi = C, \tag{6}$$

where C is a constant of integration. The enthalpy is defined by

$$H = \int \frac{dP}{\rho}. \tag{7}$$

Note that this can be integrated directly for any specified barotropic EOS, $P = P(\rho)$ (cf. (2)).

Next, we require that the angular velocity be constant on cylindrical surfaces, $\Omega = \Omega(\varpi)$. This allows us to consider the centrifugal acceleration term in (6) to be derived from a rotational potential, given by

$$\phi_{rot} = -\int \Omega^2(\varpi)\varpi d\varpi = -\int \frac{j^2(\varpi)}{\varpi^3} d\varpi, \tag{8}$$

where $j(\varpi)$ is the specific angular momentum. Equation (6) then becomes

$$H + \Phi + \phi_{rot} = C. \tag{9}$$

If we further assume that the EOS is barotropic, then Lichtenstein's theorem shows that the star must have an equatorial plane of symmetry (Tassoul 1978). If we consider specifically a polytrope of index n, then the enthalpy is given by the simple *algebraic* equation

$$H = (1+n)P/\rho, \tag{10}$$

and the surface boundary condition that $P = 0$ becomes $H = 0$. With the addition of an equation defining the rotation law, $\Omega = \Omega(\varpi)$, this completes the basic set of equations needed to carry out the SCF procedure for a rotating polytrope.

The SCF prescriptions of both Hachisu and O&M require the choice of a rotation law (discussed in more detail below) and an initial guess density distribution. Equations (8) and (5) are then used to determine the two potentials ϕ_{rot} and Φ. These are substituted into equation (9), which is solved to give the enthalpy at all points on the numerical grid. Equation (10) is then inverted to find the corresponding density distribution.

Although the actual methods employed by O&M and Hachisu for solving (5) differ, this should not effect the solutions themselves. The differences in their approach to (8), however, result in significant differences in the applicability of their algorithms to different situations, and thus should be considered here. O&M specify their rotation laws explicitly as functions of the mass interior to the cylinder of radius ϖ. Thus they write

$$j = j(m) = j(m(\varpi)), \tag{11}$$

where $m(\varpi)$ specifies the mass interior to the cylinder of radius ϖ on which the angular momentum is defined. This implies that the explicit relation between Ω and ϖ varies according to the mass distribution, so ϕ_{rot} must be recalculated on each iteration. The rotational potential (8) is integrated by quadrature using an expansion approximation to the mass density. The mixed nature of this integral, performed on a spherical grid but involving a quantity that is specified naturally in cylindrical coordinates, may be in part responsible for the fact that O&M's method is unable to produce models with high values of β. Hachisu eliminates this difficulty by further restricting himself to rotation laws that can be integrated analytically to yield a solution of the form

$$\phi_{rot} = h_0^2 \psi(\varpi), \tag{12}$$

where h_0^2 is a constant. This limits him to the three cases that he terms rigid, v-constant, and j-constant rotation, which are summarized in table 1.

	Rigid	v-constant	j-constant
Angular velocity, Ω	Ω_0	$v_0/\sqrt{\varpi^2 + d^2}$	$j_0/(\varpi^2 + d^2)$
h_0^2	Ω_0^2	v_0^2	j_0^2
$\psi(\varpi)$	$-\varpi^2/2$	$-\frac{1}{2}ln(\varpi^2 + d^2)$	$\frac{1}{2}(\varpi^2 + d^2)^{-1}$

Table 1. *Hachisu's rotation laws. d is a free parameter, and as $d \to 0$ the v-constant and j-constant rotation laws correspond to constant linear velocity in the angular direction and constant specific angular momentum, respectively.*

The major difference between the approaches of O&M and Hachisu lies in their choice of fixed quantities. We have already pointed out that to implement the SCF technique one must choose an EOS and a rotation law to complete the set of equations to be solved. In order to obtain a particular solution, one must then choose two parameters that will determine the final configuration. O&M make the seemingly natural choice of total mass and total angular momentum. In practice, however, this requires them to continually rescale their grid, and Hachisu claims that this choice is responsible for the difficulty they encounter in achieving convergence for high values of β. In addition, in cases where two equilibruium solutions coexist, this method provides no means of distinguishing which solution will be achieved. Hachisu instead chooses to fix the maximum density and the axis ratio of the final configuration. The axis ratio is determined by choosing two points on the surface of the configuration, one on the equator and one on the pole. Hachisu's method can also produce toroidal mass distributions by choosing the second surface point on the inner surface of the toroid. Since there are values of mass and angular momentum for which both flattened and ring-like solutions exist, this provides a means for distinguishing between the two, and Hachisu is able to converge to either solution.

2.3 Our SCF approach

We have seen that the O&M SCF formulation is fundamentally limited in its ability to distinguish between flattened and ring-like configurations with the same overall parameters. It is further practically limited by its difficulty in achieving convergence for rapidly rotating ($\beta > 0.25$) configurations (Bodenheimer and Ostriker 1973; Clement 1974). These limitations are avoided in Hachisu's formulation, but only at the expense of restricting the choice of rotation laws. This turns out to be a severe restriction if one is interested in studying the dynamical instability, because for these rotation laws only ring-like solutions exist for high β. We have thus adopted a variation of Hachisu's formulation that allows us to return to an arbitrary rotation law that can be written in the O&M form of (11). Our algorithm is illustrated schematically by the flowchart in figure 1, and is described in detail below.

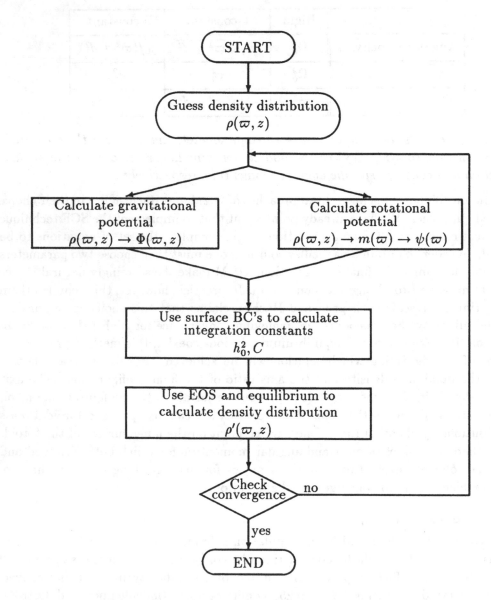

Figure 1. Flowchart for modified Hachisu SCF method of obtaining axisymmetric equilibria. The rotation law, EOS, maximum density, and axis ratio are fixed at the start of the calculation.

We formulate our equations on a cylindrical, rather than a spherical, coordinate grid. This allows us to numerically integrate (8) directly. We can cast (8) in the form of (12) by making the identifications

$$h_0^2 = J^2/M^2 \tag{13}$$

$$\psi(\varpi) = \int_0^\varpi \frac{h^2(m(\varpi'))}{\varpi'^3} d\varpi' \tag{14}$$

where J and M are the total mass and angular momentum, respectively, and $h = \frac{M}{J} j$ is the dimensionless form of the specific angular momentum. If we are using a general rotation law of the form (11), we must re-calculate ψ on each iteration as the mass distribution changes. For Hachisu's rotation laws in table 1, ψ does not vary and need only be calculated once. J and M are, of course, unknown at the start of the procedure, but the constant h_0^2 is determined during the solution process by applying the boundary condition, $H = 0$, at the two chosen surface points A and B. From (9) and (12) this gives

$$h_0^2 = \frac{\Phi(B) - \Phi(A)}{\psi(A) - \psi(B)}. \tag{15}$$

The vanishing enthalpy at one of the surface points can then be used to detemine the integration constant C from

$$C = \Phi(A) + h_0^2 \psi(A). \tag{16}$$

We then substitute these constants into (12) and (9) to find the enthalpy at all points on the grid, which in turn yields the density distribution through (10). With this density we then recalculate the potentials and iterate the procedure, as shown schematically in figure 1. We use Hachisu's criteria to determine convergence.

By formulating Hachisu's method on a cylindrical grid we can achieve the robustness of his formulation for producing high β configurations, while still maintaining the flexibility in choosing rotation laws afforded by O&M's method. We also obtain the numerical advantage of being able to choose the limits of our grid in the z direction closer to the surface of the configuration, thus reducing calculations for empty grid zones. We follow Hachisu in using a Legendre polynomial expansion to solve Poisson's equation for the gravitational potential, which requires additional memory and computer time on the cylindrical grid, but this has not yet presented any serious limitation. In addition, since our hydrodynamics code also utilizes a cylindrical grid, we gain the practical advantage of being able to transfer the solutions directly into our hydro code without having to perform any interpolation from spherical to cylindrical geometry.

3 NUMERICAL SIMULATION OF THE DYNAMICAL INSTABILITY IN POLYTROPES

3.1 Previous work

A great deal of effort has been devoted to modelling the dynamical instability in rapidly rotating polytropes, primarily motivated by an interest in the fission hypothesis for the formation of binary star systems; for a review see Durisen and Tohline (1985). Although we believe the secular instability to be more relevant for the problems we are interested in, the dynamical instability provides an excellent 3-D code test. We therefore present a brief summary of some of the relevant work before proceeding to show a sample of our test results.

Williams and Tohline (1987) investigated the evolution of polytropes of varying indices $n = 0.8, 1.0, 1.3, 1.5,$ and 1.8, with a fixed value of $\beta = 0.310$. In a related paper, Tohline, Durisen and McCollough (1985, hereafter TDM) examined the evolution of polytropes of fixed index $n = 1.5$, with different values of $\beta = 0.28, 0.30, 0.33,$ and 0.35. Both of these studies used a 3-D Eulerian hydrodynamics code and began with equilibrium models of polytropes with differential rotation. They found that an open, two armed spiral pattern evolved for all models with $\beta \geq 0.3$, with more open and barlike spirals developing for lower polytropic indices. Comparison of the simulations with the predictions of a tensor virial equation analysis showed good agreement with the expected growth of the bar mode instability, as well as higher order modes.

Durisen et al. (1986) published a comparison of the results produced by three different codes for an $n = 1.5$ polytrope with $\beta = 0.33$ and 0.38. Two of the codes were Eulerian, finite difference codes, and the third was a smoothed particle hydrodynamics (SPH) code. All three used the same initial models and produced similar results. Qualitatively, all three codes showed growth of the bar mode instability leading to a trailing spiral structure. For both values of β, this eventually leads to mass being ejected through the spiral arms, leaving behind a remnant with much lower angular momentum. Quantitative comparison of the growth of the structure and the amount of mass ejected was reasonable within the limits of the simulations. The central remnant for $\beta = 0.33$ was a stable Dedekind-like bar, while that for $\beta = 0.38$ exhibited more complicated symmetry. The agreement of three programs using two very different numerical approaches provides a useful benchmark for future work.

3.2 Hydrodynamics code

We are currently studying the dynamical instability in rotating polytropes as a means of testing our 3-D Newtonian hydrodynamics code. The code was initially developed by Clancy (1989) for the study of stellar core collapse. We have completed the debugging and testing of the code and improved the handling of rotating configurations. In addition, we have added the calculation of gravitational radiation, and hope to soon include the effects of the gravitational radiation reaction. A full description of the original code and the tests performed on it can be found in Clancy (1989), and a more thorough discussion of our modifications and code tests, including a more detailed analysis of the dynamical bar instability, can be found in Smith, Centrella and Clancy (1992).

Our code is a fully 3-D implementation of the hydrodynamical equations with Newtonian gravity on a cylindrical coordinate grid (ϖ, φ, z) that allows nonuniform zoning in the ϖ and z directions. We utilize an Eulerian formulation, allowing the fluid to flow freely through the grid, but allow the grid to move in the ϖ and z directions to permit better resolution of evolving configurations. The plane $z = 0$ is assumed to be a plane of symmetry, so only the upper half of the distribution, $z \geq 0$, is modelled.

The Eulerian hydrodynamic equations are implemented using an explicit finite difference method similar to that described by Bowers and Wilson (1991). Advection is accomplished using the monotonic scheme of LeBlanc (Clancy 1989), with the consistent advection algorithm of Norman and Winkler (1986) being used to insure local angular momentum conservation. Poisson's equation for the gravitational potential is solved by the preconditioned conjugate gradient method using diagonal scaling (Meijerink and Van Der Vordst 1981). To date, this code has been tested on the Riemann shock tube and Noh shock problems (Clancy 1989); as well as on the stability of rotating and non-rotating polytropes, the homologous collapse of polytropes, and the collapse of dust ellipsoids (Smith, Centrella and Clancy 1992).

The calculation of gravitational radiation has been implemented using the quadrupole approximation. In addition to a straight forward implementation of the quadrupole formula, we have implemented the momentum divergence (QF2) and first moment of momentum (QF1) formulations introduced by Finn and Evans (1990). These are expected to eliminate much of the noise associated with repeated numerical time derivatives. Gravitational radiation reaction is not currently treated, but will be added shortly following Blanchet, Damour, and Schäfer (1990).

3.3 Results
As a stringent test of our hydrodynamics code in three dimensions, we are repeating some of the calculations of TDM. This section will present an example of such a calculation. The specific case shown here corresponds to the case of a polytrope of index $n = 1.5$ with $\beta = 0.327$.

The model is actually specified by the axis ratio $R_{eq}/R_p = 5.96$, where R_{eq} is the equatorial radius and R_p is the polar radius. Following TDM, the rotation law is chosen to give a specific angular momentum distribution identical to that for a Maclaurin spheroid, given by

$$h(m) = \frac{5}{2}[1 - (1 - m)^{2/3}].$$ (17)

Our SCF code then produces an equilibrium model with $\beta = 0.3256$, which differs only slightly from the value quoted by TDM. The discrepancy is presumably due to the difference in the grids used. The initial model is scale free, but our code requires us to specify the maximum density ρ_{max} and constant k in the polytropic EOS. We choose $\rho_{max} = 8.44$ g/cm^3 and $k = 2.48 \times 10^{-14}$, which are identical to the values for a spherical $n = 1.5$ polytrope of mass $M_\odot$ and radius $R_\odot$. At this rotation rate, this yields a highly flattened configuration with total mass $5.83 M_\odot$ and equatorial radius $3.38 R_\odot$. This provides conversion factors for the time units used by TDM of 1 polytropic unit (p.u.) = 75.5 seconds and 1 central initial rotation period (cirp) = 7918.5 seconds. The density contours in the equatorial plane for the initial model are illustrated in figure 2.

To break the axisymmetry, we introduced a random density perturbation with an amplitude of 10^{-3} and followed the evolution of the system for more than 9 cirps (over 950 p.u. or 7×10^4 sec.). Although we have not yet performed a quantitative comparison with the results of TDM, we see qualitative agreement. We observed the growth of the bar mode and the development of trailing spiral arms on the same timescale as TDM. Some representative plots are shown in figures 3–6. These plots show density contours in the equatorial plane with the velocity field superimposed on them. Rotation is in the counterclockwise sense. In figure 3, at about 755 p.u., slight deviations from axisymmetry are noticeable. Figure 4, at 818 p.u. shows a growing bar mode distortion; and the figures 5 and 6, at 871 p.u. and 891 p.u., respectively, clearly exhibit spiral structure. Note that the plots in figures 5 and 6 are on an expanded grid, and that velocities have been suppressed in low density (empty) grid zones for these figures.

These preliminary results give us confidence that our code is behaving correctly in this 3-D situation. We are now making a quantitative comparison with the TDM and tensor virial equation results. Once this testing is successfully completed, we will begin using our code to study the generation of gravitational radiation in 3-D astrophysical scenarios.

CYCLE=   0 DENSITY AT TIME=0.00E+00

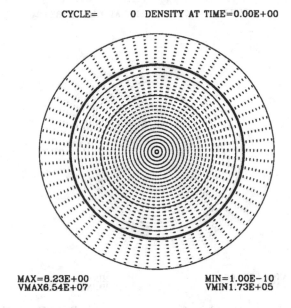

MAX=8.23E+00
VMAX6.54E+07

MIN=1.00E−10
VMIN1.73E+05

Figure 2. Initial model: density contours are shown for an equatorial section, with fluid velocities superimposed.

CYCLE=   11000 DENSITY AT TIME=5.70E+04

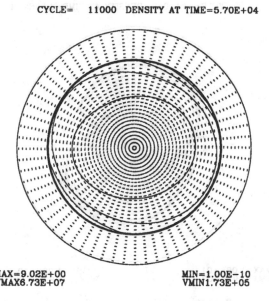

MAX=9.02E+00
VMAX6.73E+07

MIN=1.00E−10
VMIN1.73E+05

Figure 3. Density and velocity plots for an equatorial slice. Evolution time, in TDM units, is 755 p.u. (7.2 cirps). Rotation is counterclockwise.

CYCLE=   12000 DENSITY AT TIME=6.18E+04

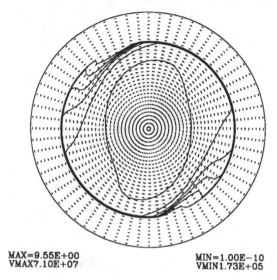

MAX=9.55E+00   MIN=1.00E−10
VMAX7.10E+07   VMIN1.73E+05

Figure 4. Density and velocity plots for an equatorial slice. Evolution time, in TDM units, is 818 p.u. (7.8 cirps). Rotation is counterclockwise.

CYCLE=   13050 DENSITY AT TIME=6.58E+04

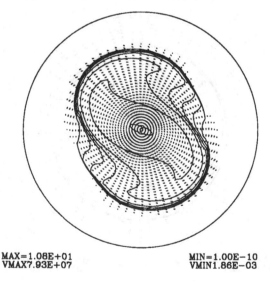

MAX=1.08E+01   MIN=1.00E−10
VMAX7.93E+07   VMIN1.86E−03

Figure 5   Same as figure 3 at time 871 p.u. (8.3 cirps). (Note that the grid has been expanded in this plot as compared with that figure, and that velocity has been suppressed in low density zones.)

CYCLE=   13550   DENSITY AT TIME=6.73E+04

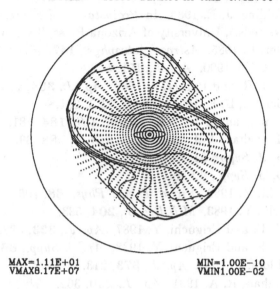

MAX=1.11E+01   MIN=1.00E−10
VMAX8.17E+07   VMIN1.00E−02

Figure 6. Same as figure 5 at time 891 p.u. (8.5 cirps).

ACKNOWLEDGEMENTS

We are pleased to acknowledge stimulating discussions with John Miller, Steve McMillan, Silvano Bonazzola, and Jean-Alain Marck. This work was supported by NSF grant PHY90-12383. The numerical simulations were perfomed on a Cray-2 computer at the National Center for Supercomputing Applications.

REFERENCES

Blanchet, L., Damour, T. and Schäfer, G. 1990. *M.N.R.A.S.*, **242**, 289.
Bodenheimer, P. and Ostriker, J. P. 1970. *Ap. J.*, **161**, 1101.
Bodenheimer, P. and Ostriker, J. P. 1973. *Ap. J.*, **180**, 159.
Bowers, R. L. and Wilson, J. R. 1991. *Numerical Modelling in Applied Physics and Astrophysics*, Jones and Bartlett, Boston.
Chandrasekhar, S. 1969. *Ellipsoidal Figures of Equilibrium*, Yale University Press, New Haven.
Chandrasekhar, S. 1970. *Phys. Rev. Lett.*, **24**, 611.
Clancy, S. P. 1989. Ph.D. thesis, University of Texas, Austin.
Clement, M. J. 1974. *Ap. J.*, **194**, 709.
Colpi, M. and Miller, J. C. 1992. *Ap. J.*, in press.
Durisen, R. H., Gingold, R. A., Tohline, J. E. and Boss, A. P. 1986. *Ap. J.*, **305**, 281.

Durisen, R. H. and Tohline, J. E. 1980. *Space. Sci. Rev.*, **27**, 267.

Durisen, R. H. and Tohline, J. E. 1985. In *Protostars and Planets, II*, D. C. Black and M. Matthews (eds.), University of Arizona Press, Tucson.

Eriguchi, Y. and Müller, E. 1985. *Astron. Astrophys.*, **147**, 161.

Finn, L. S. and Evans, C. R. 1990. *Ap. J.*, **351**, 588.

Friedman, J. L., Ipser, J. R. and Parker, L. 1986. *Ap. J.*, **304**, 115.

Friedman, J. L. and Schutz, B. F. 1978. *Ap. J.*, **222**, 281.

Gingold, R. A. and Monaghan, J. J. 1978. *M.N.R.A.S.*, **184**, 481.

Gingold, R. A. and Monaghan, J. J. 1979. *M.N.R.A.S.*, **188**, 39.

Hachisu, I. 1986a. *Ap. J. Supp.*, **61**, 479.

Hachisu, I. 1986b. *Ap. J. Supp.*, **62**, 461.

Hachisu, I. and Eriguchi, Y. 1982. *Progr. Theor. Phys.*, **68**, 206.

Hachisu, I. and Eriguchi, Y. 1983. *M.N.R.A.S.*, **204**, 583.

Hachisu, I., Tohline, J. E. and Eriguchi, Y. 1987. *Ap. J.*, **323**, 592.

Hachisu, I., Tohline, J. E. and Eriguchi, Y. 1988. *Ap. J. Supp.*, **66**, 315.

Ipser, J. R. and Lindblom, L. 1991. *Ap. J.*, **373**, 213.

Ipser, J. R. and Managhan, R. A. 1981. *Ap. J.*, **250**, 362.

James, R. A. 1964. *Ap. J.*, **140**, 552.

Jeans, J. H. 1919. *Problems of Cosmogony and Stellar Dynamics*, Cambridge University Press, Cambridge.

Lindblom, L. and Detweiler, S. L. 1977. *Ap. J.*, **211**, 565.

Lucy, L. 1977, *A.J.*, **82**, 1013.

Mark, J. W-K. 1968. *Ap. J.*, **154**, 627.

Meijerink, J. A. and Van Der Vordst, H. A. 1981. *J. Comp. Phys.*, **44**, 271.

Norman, M. L. and Winkler, K-H. A. 1986. In *Astrophysical Radiation Hydrodynamics*, K-H. A. Winkler and M. L. Norman (eds.), Reidel, Dordrecht.

Ostriker, J. P. and Bodenheimer, P. 1968. *Ap. J.*, **151**, 1089.

Ostriker, J. P. and Bodenheimer, P. 1973. *Ap. J.*, **180**, 171.

Ostriker, J. P. and Mark, J. W-K. 1968. *Ap. J.*, **151**, 1075.

Ostriker, J. P. and Tassoul, J. L. 1969. *Ap. J.*, **155**, 987.

Smith, S. C, Centrella, J. M. and Clancy, S. P. 1992. In preparation.

Tassoul, J. L. 1978. *Theory of Rotating Stars*, Princeton University Press, Princeton.

Tassoul, J. L. and Ostriker, J. P. 1969. *Ap. J.*, **154**, 613.

Thorne, K. 1987. In *300 Years of Gravitation*, S. W. Hawking and W. Israel (eds.), Cambridge University Press, Cambridge.

Tohline, J. E., Durisen, R. H. and McCollough, M. 1985. *Ap. J.*, **298**, 220.

Tohline, J. E. and Hachisu, I. 1990. *Ap. J.*, **361**, 394.

Vandervoort, P. O. and Welty, D. E. 1981. *Ap. J.*, **248**, 504.
Vogt, R. E. 1991. LIGO Project Report 91-7, to be published in *Proceedings of the Sixth Marcel Grossmann Meeting on General Relativity, Kyoto.*
Wagoner, R. V. 1984. *Ap. J.*, **278**, 345.
Weber, F., Glendenning, N. K. and Wiegel, M. K. 1991. *Ap. J.*, **373**, 579.
Williams, H. A. and Tohline, J. E. 1987. *Ap. J.*, **315**, 594.
Williams, H. A. and Tohline, J. E. 1988. *Ap. J.*, **334**, 449.

GRAVITATIONAL RADIATION FROM COALESCING BINARY NEUTRON STARS

Ken-Ichi Oohara

National Laboratory for High Energy Physics, Oho, Japan

Takashi Nakamura

Yukawa Institute for Theoretical Physics, Kyoto University, Kyoto, Japan

Abstract. We present three-dimensional Newtonian and post-Newtonian codes, including the gravitational radiation damping effect, using a finite difference method. We follow the emission of gravitational radiation using the quadrupole approximation. Using these codes we calculate the coalescence of a neutron star binary. For Newtonian calculations the initial configuration is given as a hydrostatic equilibrium model of a close neutron-star binary. Calculations were performed for neutron stars of different masses as well as of the same masses. In order to evaluate general relativistic effects, we compare the results of the calculation of the coalescence of a binary comprising two spherical neutron stars using the post-Newtonian code with results using the Newtonian code.

1 INTRODUCTION

The most promising sources for laser-interferometric gravitational-wave detectors are catastrophic events such as the gravitational collapse of a star or the coalescence of a black-hole or neutron-star binary. We need to know the characteristics of the waves for design of detectors. It requires general relativistic calculations of stellar collapse and binary coalescence. In the last decade, 2 dimensional (2D) calculations were successfully performed for a head-on collision of two black holes (Smarr 1979) and axisymmetric collapse of a rotating star (Stark and Piran 1986). They found that the efficiency of gravitational wave emission (the ratio of the energy emitted in gravitational radiation to the total rest mass) is less than 0.1%. Nakamura, Oohara and Kojima (1987), on the other hand, pointed out that the efficiency may be much greater in non-axisymmetric black-hole collision. They estimated the efficiency of the gravitational wave emission from a particle falling into a black hole using a perturbation technique and found that the efficiency for axially symmetric system is much less than that for 3D system because of phase cancellation effects. Extrapolation of their results to black-hole collision leads to an efficiency of as much as 9%, though it depends on the orbital angular momentum and the spin angular momentum of the black hole. (See also J.A. Marck in this volume.) These results encouraged us to consider evolutions of more realistic non-axisymmetric events and gravitational radiation from them. Thus we have performed 3D calculations of the coalescence of a neutron-star binary using a Newtonian or post-Newtonian (PN) hydrodynamics code (Oohara and Nakamura 1989, 1990, Nakamura and Oohara 1989a, 1991). Gravitational waves are calculated using the quadrupole approximation and the gravitational radiation damping effects are included there, while they come from 2.5 PN terms.

In the following we report results of numerical simulations of coalescing binary neutron stars. In section 2 we describe basic equations and our numerical method. In section 3 initial data sets and results of Newtonian calculation are shown. Results of post-Newtonian calculation is described in section 4.

2 BASIC EQUATIONS AND NUMERICAL METHODS

2.1 (1+2.5) Post-Newtonian Hydrodynamics Equations
The equations of motion including general relativistic effects up to order $(v/c)^2$ are called the first post-Newtonian (1PN) approximation. However, gravitational damping effects start at order $(v/c)^5$, second-and-a-half post-Newtonian (2.5PN) terms, and the formulation for adding the effects to the equation of motion depends on the gauge condition. For example, the radiation reaction is represented by adding to the standard Newtonian potential a "radiation-reaction" potential proportional to the fifth time derivatives of the quadrupole moments (Misner, Thorne and Wheeler 1973). In view of numerical calculation, however, it is hard to calculate the fifth time derivatives with satisfactory accuracy. Recently Blanchet, Damour and Schäfer (1990) have presented (1+2.5)PN hydrodynamics equations including the radiation damping effects, where we need up to the third time derivatives of the quadrupole moments. Following them, we use the evolution equations given by

$$\partial_t \, \rho + \partial_j \left(\rho v^j \right) = 0, \tag{1}$$

$$\partial_t \left(\rho w_i \right) + \partial_j \left(\rho w_i v^j \right) = F_i^{\text{press}} + F_i^{\text{1PN}} + F_i^{\text{reac}}, \tag{2}$$

$$\partial_t \left(\rho \varepsilon \right) + \partial_j \left(\rho \varepsilon v^j \right) = -p \, \partial_j v^j, \tag{3}$$

where ρ, ε and p are the coordinate rest-mass density, the internal energy density and the pressure, respectively. In order to express a hard equation of state for a neutron star, we use a polytropic equation of motion

$$p = (\gamma - 1)\rho \tag{4}$$

with $\gamma = 2$. The quantity v^i denotes the 3-velocity and w_i is the "momentum per unit rest-mass." In terms of w_i, the 3-velocity v^i is given by

$$v^i = \left(1 - \frac{\beta}{c^2}\right) w_i + \frac{1}{c^2} A_i + \frac{4G}{5c^5} w_j Q_{ij}^{[3]}. \tag{5}$$

The forces are given by

$$F_i^{\text{press}} = -\partial_i \left(1 + \frac{\alpha}{c^2}\right) p, \tag{6a}$$

$$F_i^{\text{1PN}} = -\rho \left[\left(1 + \frac{\delta}{c^2}\right) \partial_i \psi + \frac{1}{c^2} \partial_i \psi_2 + \frac{1}{c^2} w_j \partial_i A_j\right], \tag{6b}$$

$$F_i^{\text{reac}} = -\frac{1}{c^5} \rho \, \partial_i \psi_r. \tag{6b}$$

Here ψ is the Newtonian potential, α, β, δ, ψ_2 and A_i are 1PN quantities and ψ_r is the radiation-reaction potential (a 2.5PN quantity)

$$\alpha = (3\gamma - 2)\psi - \frac{1}{2}\gamma w^2, \tag{7a}$$

$$\beta = \frac{1}{2}w^2 + \gamma\varepsilon - 3\psi, \tag{7b}$$

$$\delta = \frac{3}{2}w^2 + (3\gamma - 2)\varepsilon + \psi, \tag{7c}$$

$$A_i = U_i - \frac{1}{2}x^i \psi_t, \tag{7d}$$

$$\psi_r = \frac{2}{5}G \left[R - Q_{ij}^{[3]} x^i \partial_j \psi\right], \tag{7e}$$

where $w^2 = \delta^{ij} w_i w_j$ and ψ_t is the time derivative of ψ with the addition of corrections of order $O(c^{-2})$. For ψ, ψ_t, ψ_2, R and U_i, we should solve seven Poisson equations at each time step

$$\Delta\psi = 4\pi G\rho, \tag{8}$$
$$\Delta\psi_t = -4\pi G\partial_j (\rho w_j), \tag{9}$$
$$\Delta\psi_2 = 4\pi G\rho\delta, \tag{10}$$
$$\Delta R = 4\pi G Q_{ij}^{[3]} x^i \partial_j \rho, \tag{11}$$
$$\Delta U_i = -4\pi G \left(4\rho w_i + \frac{1}{2} x^i \partial_j (\rho w_j)\right). \tag{12}$$

The quantity $Q_{ij}^{[3]}$ is given by

$$Q_{ij}^{[3]} = \text{STF} \left\{2 \int \left[p\partial_i w_j - 2\rho w^i \partial_j \psi + x^i \partial_j \psi \partial_k(\rho w^k) - \rho x^i \partial_j \psi_t\right] dV\right\}, \tag{13}$$

which differs from the third time derivative of the reduced quadrupole moment of the system by corrections of order $O(c^{-2})$ (Finn 1989). The notation STF means the operator taking the symmetric, trace-free part of any two-index object A^{ij}

$$\text{STF}\{A^{ij}\} = \frac{1}{2}A^{ij} + \frac{1}{2}A^{ji} - \frac{1}{3}\delta^{ij} A^{kk}. \tag{14}$$

The energy flux of the gravitational waves are given by

$$\frac{dE}{dt} = -\frac{G}{5c^5} Q_{ij}^{[3]} \frac{d}{dt} I_{ij},$$

(15)

where

$$I_{ij} = \text{STF} \left\{ 2 \int \rho \left[w_i w_j - x^i \partial_j \psi \right] dV \right\},$$

(16)

which is the second time derivative of the reduced quadrupole moment with the addition of corrections of order $O(c^{-2})$. The standard quadrupole formula gives the amplitude h_{ij}^{TT} of the gravitational wave in the transverse-traceless gauge (Misner, Thorne and Wheeler 1973)

$$h_{ij}^{TT} = \frac{2}{r} \left(P_{im} P_{jn} - \frac{1}{2} P_{ij} P_{mn} \right) \frac{d^2 Q_{mn}}{dt^2},$$

(17)

where Q_{mn} is the reduced mass quadrupole moment

$$Q_{mn} = \text{STF} \left\{ \int \rho x^m x^n dV \right\}$$

(18)

and $P_{ij} = \delta_{ij} - n_i n_j$ is the projection operator onto the plane transverse to the outgoing wave direction $n_i = x^i / r$. For consistency with post-Newtonian accuracy, we must take into account the relativistic corrections up to the relative order $(v/c)^2$. It requires the time derivatives of mass octupole, mass 2^4-pole, current quadrupole and current octupole (Blanchet, Damour and Schäfer 1990). However we neglect these corrections since the gravitational wave amplitude is evaluated with quantities at each time step and small truncation errors above 'Newtonian' accuracy do not accumulate. With this accuracy, we can use I_{ij} defined by (16) in place of $\ddot{Q}_{ij}$. From (17), two polarizations are given by

$$h_+ = \frac{1}{r} \left(I_{\hat{\theta}\hat{\theta}} - I_{\hat{\phi}\hat{\phi}} \right)$$

(19a)

$$h_\times = \frac{2}{r} I_{\hat{\theta}\hat{\phi}}$$

(19b)

where I_{ij} is the quantity in the orthonormal basis

$$I_{\hat{\theta}\hat{\theta}} = \left(I_{xx} \cos^2 \phi + I_{yy} \sin^2 \phi + 2 I_{xy} \sin \phi \cos \phi \right) \cos^2 \theta$$
$$+ I_{zz} \sin^2 \theta - 2 \left(I_{xz} \cos \phi + I_{yz} \sin \phi \right) \sin \theta \cos \theta$$

(20a)

$$I_{\hat{\phi}\hat{\phi}} = I_{xx} \sin^2 \phi + I_{yy} \cos^2 \phi - 2 I_{xy} \sin \phi \cos \phi$$

(20b)

$$I_{\hat{\theta}\hat{\phi}} = (I_{yy} - I_{xx}) \cos \theta \sin \phi \cos \phi + I_{xy} \cos \theta (\cos^2 \phi - \sin^2 \phi)$$
$$+ I_{xz} \sin \theta \sin \phi - I_{yz} \sin \theta \cos \phi.$$

(20c)

2.2 Numerical Methods

The equations are discretized by a finite difference method (FDM) with a uniform Cartesian grid. The evolution equations are integrated using van Leer's scheme (van Leer 1977) with second-order accuracy in space. In order to make the scheme stable, the monotonicity condition, the so-called TVD (Total Variation Diminishing) limiter (Harten 1983), is imposed on this scheme (Oohara and Nakamura 1989). In order to achieve second-order accuracy in time we adopt a two-step procedure. To describe our code, we write (1)–(3) as

$$\partial_t u + \partial_j(uv^j) = F, \tag{21}$$

where u is a 5-dimensional vector defined by

$$u = (\rho, \rho w_i, \rho \varepsilon) \tag{22}$$

and elements of the 5-vector F are given by the right hand sides of (1)–(3). We calculate u^{n+1} at $t^{n+1} = t^n + \Delta t^n$ from u^n at t^n as

$$u^{n+\frac{1}{2}} = u^n + Q(u^n, \Delta t^n/2) \tag{23a}$$
$$u^{n+1} = u^n + Q(u^{n+\frac{1}{2}}, \Delta t^n), \tag{23b}$$

where $Q(u, \Delta t)$ is given at each grid point by

$$Q_{i,j,k}(u, \Delta t) = \left(\Delta u_{i-\frac{1}{2},j,k} - \Delta u_{i+\frac{1}{2},j,k}\right) + \left(\Delta u_{i,j-\frac{1}{2},k} - \Delta u_{i,j+\frac{1}{2},k}\right)$$
$$+ \left(\Delta u_{i,j,k-\frac{1}{2}} - \Delta u_{i,j,k+\frac{1}{2}}\right) + F_{i,j,k}\Delta t. \tag{24}$$

The quantity $\Delta u_{i-\frac{1}{2},j,k}(\Delta t)$ is

$$\Delta u_{i-\frac{1}{2},j,k}(u, \Delta t) =$$

$$\begin{cases} \left[u_{i-1,j,k} + \nabla u_{i-1,j,k}\left(\Delta x - v^x_{i-\frac{1}{2},j,k}\right)\dfrac{v^x_{i-\frac{1}{2},j,k}\Delta t}{\Delta x}\right] & \text{if } v^x_{i-\frac{1}{2},j,k} > 0, \\[4mm] \left[u_{i,j,k} + \nabla u_{i,j,k}\left(\Delta x - v^x_{i-\frac{1}{2},j,k}\right)\dfrac{v^x_{i-\frac{1}{2},j,k}\Delta t}{\Delta x}\right] & \text{otherwise;} \end{cases} \tag{25}$$

$\nabla u_{i,j,k}$ is the limited slope

$$\nabla u_{i,j,k} = \begin{cases} 0 & \text{if } \partial_x u_{i-\frac{1}{2},j,k}\partial_x u_{i+\frac{1}{2},j,k} \leq 0 \\[3mm] \text{sign} \times \min\left(|\partial_x u_{i-\frac{1}{2},j,k}|, \dfrac{1}{2}|\partial_x u_{i,j,k}|, |\partial_x u_{i+\frac{1}{2},j,k}|\right) & \text{otherwise,} \end{cases} \tag{26}$$

where sign $= 1$ if $\partial_x u_{i-\frac{1}{2},j,k} > 0$ and $= -1$ otherwise. The velocity and the gradient of u on each cell face are

$$v^x_{i-\frac{1}{2},j,k} = \frac{\rho_{i-1,j,k} v^x_{i-1,j,k} + \rho_{i,j,k} v^x_{i,j,k}}{\rho_{i-1,j,k} + \rho_{i,j,k}} \tag{27a}$$

$$\partial_x u_{i-\frac{1}{2},j,k} = \frac{u_{i,j,k} - u_{i-1,j,k}}{\Delta x}, \tag{27b}$$

and the gradient of u at each grid point is given by

$$\partial_x u_{i,j,k} = \frac{u_{i+1,j,k} - u_{i-1,j,k}}{2\Delta x}. \tag{28}$$

The quantities $\Delta u_{i,j-\frac{1}{2},k}$ and $\Delta u_{i,j,k-\frac{1}{2}}$ are given similarly.

In order to express shock waves, we use the tensor artificial viscosity given by

$$p_{ij} = \begin{cases} \rho \ell^2 (\partial_k v^k) \text{STF}\{2\partial_i v^j\} & \text{if } \partial_k v^k < 0, \\ 0 & \text{otherwise,} \end{cases} \tag{29}$$

where ℓ is an appropriate number with the unit of length. The gas pressure p is replaced by $P_{ij} = p\delta_{ij} + p_{ij}$. Poisson equations are solved using the MICCG (Modified Incomplete Cholesky-decomposition and Conjugate Gradient) method (Meijerink and van der Vorst 1977, Gustafsson 1978, Ushiro 1984, van der Vorst 1989, Murata, Oguni and Karaki 1985), which is fully vectorized using the hyperplane method proposed by Ushiro (Murata, Oguni and Karaki 1985).

We have performed various tests for this code (Oohara and Nakamura 1989, Nakamura and Oohara 1989b), which include: (1) free transportation of a dust cube of a homogeneous or Gaussian density distribution, (2) 1D Riemann shock tube, (3) point explosion in the air, (4) local conservation of specific angular momentum for an axially symmetric collapse and (5) collapse of a homogeneous dust ellipsoid. Our code passed these tests with sufficiently good accuracy.

3 NEWTONIAN CALCULATION

First we performed an extensive series of numerical calculations using Newtonian hydrodynamics including the radiation damping effects, where terms of order $O(c^{-2})$ in (5) and (6) are neglected while terms of order $O(c^{-5})$ are kept to include the radiation damping effects. In this case we have only to solve three Poisson equations, namely (8), (9) and (11).

3.1 Initial Data

We consider two kinds of initial data: (1) two spherical polytropes and (2) a binary system in rotational equilibrium. In both cases, two neutron stars just touch each other and are rigidly rotating around the center of mass with angular velocity Ω.

In order to obtain hydrostatic equilibrium models of close neutron star binaries, we set $\partial_t = 0$, $F_i^{\text{reac}} = 0$ and $v^i = w_i$ in the Newtonian version of (1)–(3). Here the gravitational radiation damping effects are neglected since a binary system evolves quasi-stationary up to the onset of coalescence. Assuming the pressure p is given by $p = K\rho^\gamma$ and the two stars are in synchronized circular orbit around the center of mass $(x_0, y_0, 0)$, that is, the velocity is given by $v^i = (-(y - y_0)\Omega, (x - x_0)\Omega, 0)$, we have the equation that equilibrium models should satisfy

$$\nabla \left[\psi + H - \frac{1}{2} \{(x - x_0)^2 + (y - y_0)^2\} \Omega^2 \right] = 0, \tag{30}$$

where H is the enthalpy,

$$H = \frac{\gamma}{\gamma - 1} K\rho^{\gamma-1} . \tag{31}$$

A solution of (30) for each star is given by

$$\psi + H - \frac{1}{2} \{(x - x_0)^2 + (y - y_0)^2\} \Omega^2 = C_i \quad \text{(for } i = 1, 2) \tag{32}$$

where C_1 and C_2 are different constants in general. Since the position of the center of mass can be set freely, we set $y_0 = 0$. Instead of setting x_0, however, we fix the position x_1^s and x_2^s where the surface of each star intersects the x-axis between two stars: $\rho(x_1^s, 0, 0) = \rho(x_2^s, 0, 0)$. In addition, we fix the center of each star x_1^c and x_2^c and the density there: $\rho(x_1^c, 0, 0) = \rho_1^c$ and $\rho(x_2^c, 0, 0) = \rho_2^c$. A self-consistent solution is determined by an iterative method. First, setting constants γ and K of the equation of state, we make an initial guess of the density distribution, usually as two spherical polytropes. We then repeat the following steps until convergence.

(1) The potential ψ is calculated as the solution of (8).
(2) C_1, C_2, x_0 and Ω are determined from

$$\psi_1^c + H_1^c - \frac{1}{2} (x_1^c - x_0)^2 \Omega^2 = C_1, \tag{33a}$$

$$\psi_1^s \qquad - \frac{1}{2} (x_1^s - x_0)^2 \Omega^2 = C_1, \tag{33b}$$

$$\psi_2^c + H_2^c - \frac{1}{2} (x_2^c - x_0)^2 \Omega^2 = C_2, \tag{33c}$$

$$\psi_2^s \qquad - \frac{1}{2} (x_2^s - x_0)^2 \Omega^2 = C_2; \tag{33d}$$

or explicitly

$$x_0 = \frac{A_2\{(x_1^c)^2 - (x_1^s)^2\} - A_1\{(x_2^c)^2 - x(\substack{s\\2})^2\}}{2\{A_2(x_1^c - x_1^s) - A_1(x_2^c - x_2^s)\}} , \tag{34a}$$

$$\Omega^2 = \frac{A_1}{(x_1^c - x_1^s)(x_1^c + x_1^s - 2x_0)} , \tag{34b}$$

$$C_i = \psi_i^s - \frac{1}{2} (x_i^s - x_0)^2 \Omega^2, \tag{34c}$$

where $\psi_a \equiv \psi(x_a, 0, 0)$, $H_a \equiv H(x_a, 0, 0)$ and $A_i \equiv \psi_i^c - \psi_i^i + H_i^c$.

(3) Using these values, we determine a new density distribution from

$$H = \frac{\gamma}{\gamma - 1} K\rho^{\gamma-1} = C_i - \psi + \frac{1}{2}\{(x - x_0)^2 + y^2\}\Omega^2. \tag{35}$$

Here we use the constant C_1 for $x > x_0$ and C_2 for $x < x_0$.

To be more realistic, we consider the infalling velocity due to the gravitational radiation. Assuming the two stars to be point masses with separation ℓ in a circular orbit, ℓ decreases at a rate

$$\partial_t \ell = -\frac{64 m_1 m_2 (m_1 + m_2)}{5\ell^3}. \tag{36}$$

Thus we add the infalling velocity given by this equation to initial stars.

Parameters of initial data with which we started the numerical simulations are shown in table 1. For all of them, we use $\gamma = 2$ to express the hard equation of state of neutron-star matter. As for the coefficient K, we use the value for a spherical star of radius r_0:

$$K = \frac{2r_0^2 G}{\pi}. \tag{37}$$

3.2 Numerical Results

We take typically a $141 \times 141 \times 131$ grid. We assume reflection symmetry with respect to the $z = 0$ plane and consider the region of $z \geq 0$ only, since the system becomes rather flat and therefore a finer grid is needed in the z-direction. Calculations were performed on a supercomputer HITAC S820/80 at the National Laboratory for High Energy Physics (KEK). Each calculation requires 480 Mbytes of the main memory and 450 Mbytes of the ES (the ES, Extended Storage, of S820 is a device equivalent to the SSD of Cray Y-MP). A typical CPU time required is 100–200 hours for 50,000–90,000 time steps, which corresponds to an event of 4–8 milliseconds. Details of the numerical results with initial data sets in table 1 are given in Oohara and Nakamura 1989, 1990 and Nakamura and Oohara 1989a, 1991.

Here we show the results for equation (8) and TD1 in table 1. Figure 1 shows the evolution of the density and the velocity on the x-y plane for equation (8), where the mass and the radius of each star are $1.49 M_\odot$ and 8km, the separation between stars ℓ is 16km, the angular velocity Ω is $6.7 \times 10^3 \text{sec}^{-1}$ and the total angular momentum J_t is $5.7 G M_\odot^2/c$ initially. In each figure only the central part is represented, while the computational grid covers $[-31\text{km}, 31\text{km}]$ in the x and y directions and $[0\text{km}, 21\text{km}]$ in the z direction. Figure 2 shows the emitted energy L and the central density ρ_c as functions of time. There are three important time scales at the initial stage of

	initial data			radiation
Name	S/E	mass	infall vel.	damping
BNx	S	$1.40 M_\odot + 1.40 M_\odot$	NO	NO
BK1	S	$1.40 M_\odot + 1.40 M_\odot$	NO	YES
EQ0	E	$1.49 M_\odot + 1.49 M_\odot$	NO	YES
EQ8	E	$1.49 M_\odot + 1.49 M_\odot$	YES	YES
SN1	E	$0.85 M_\odot + 0.85 M_\odot$	YES	YES
TD1	E	$1.70 M_\odot + 1.28 M_\odot$	YES	YES
TD2	E	$1.83 M_\odot + 0.97 M_\odot$	YES	YES

Table 1. Numerical simulations we performed. The symbol S and E in the second column means initial data including two spherical stars and a binary system in rotational equilibrium, respectively.

coalescence; the rotational period $P = 2\pi/\Omega$, the coalescence time due to the initial infall velocity $\tau_{col} = \ell/\partial_t \ell$ and the time scale for 10% loss of the total angular momentum $\eta_{loss} = 0.1 J_t \Omega/L$. The gravity becomes 120% above the centrifugal force when 10% of the angular momentum is lost and consequently the initial configuration will be changed considerably. For equation (8), η_{loss}=0.5msec is not so different from τ_{col}=0.7msec, while P=0.9msec; therefore coalescence is made to begin before one rotational period, not only by the radiation reaction but also by the initial infalling velocity (figures 1a–d). Subsequently the central part begins to expand and the two stars appear again (figure 1e), since the infall velocity causes a rapid coalescence before the angular momentum is sufficiently lost and then the centrifugal force increased above the gravity. Consequently a coalescence-expansion oscillation arises. The oscillation lasts several cycles, and afterwards a small quasi-steady binary is formed in the central region (figures 1e–g). The small binary coalesces very slowly in consequence of the angular momentum loss and finally it becomes a ring, as shown in figure 1h. Finally the ring evolves to a disk (figure 1i); the system is almost axisymmetric and the luminosity of the gravitational radiation is very low, 2 orders of magnitude smaller than the peak luminosity (figure 2). The system is almost in equilibrium at this time but an oscillation in the central density remains as shown in figure 2; the oscillation can be seen on the video movie, too. This is a radial oscillation in the deep central region excited by the rapid coalescence and the

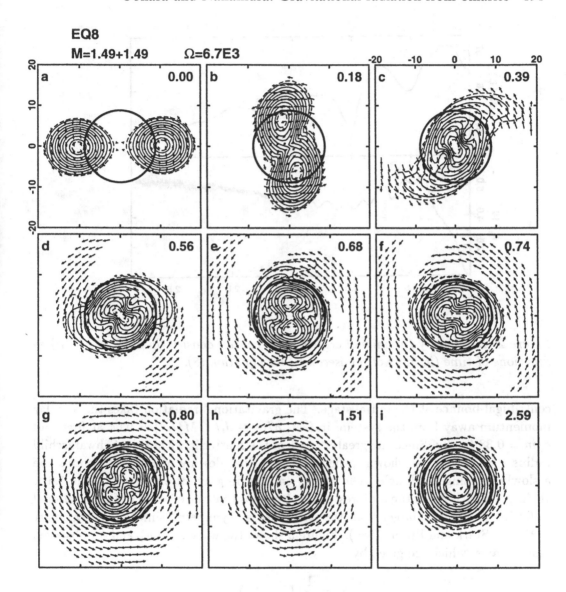

Figure 1. Density and velocity on the x-y plane for equation (8). Time is given in units of milliseconds. Solid lines are drawn by step of a tenth of the maximum density and the inner and the outer dashed lines indicate 19/20 and 1/20 of the maximum density. Arrows indicate the velocity vectors of the matter. A thick line shows a circle of radius $2GM_t/c^2$, to show the size of a spherical black hole for comparison.

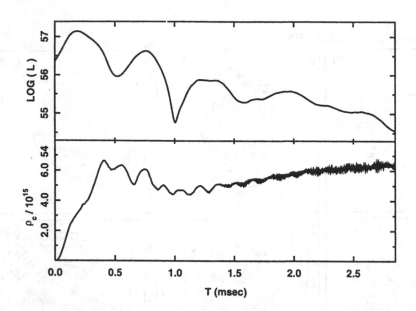

Figure 2. Luminosity (in units of erg/sec) and central density (in units of g/cm³) as functions of time (in units of milliseconds) for equation (8).

centrifugal bounce at the initial stage. The gravitational radiation takes the angular momentum away from the system; in fact, $a/m \equiv J_t/(GM^2/c^3) = 0.64$ at first and $a/m = 0.38$ finally. Since the greatest part of the matter is within the Schwarzschild radius $r = GM_t/c^2$ as shown in figure 1i, the final destiny of the system must be a slowly rotating black hole. The energy emitted in gravitational radiation is 1.6×10^{53}erg, which is 3% of the total rest mass. It means that the efficiency of gravitational radiation is 30 times larger than the results for axisymmetric simulations by Smarr (1979) or Stark and Piran (1986). Figure 3 shows the wave form h_+ and $h_\times$ observed along z-axis, which are given by

$$h_+ = \frac{1}{r}\left(I_{xx} - I_{yy}\right), \tag{38a}$$

$$h_\times = \frac{2}{r}I_{xy}, \tag{38b}$$

where r is the distance between the event and the observer. If the event occurs at 10Mpc from the earth, the maximum amplitude of h is 4.7×10^{-21}.

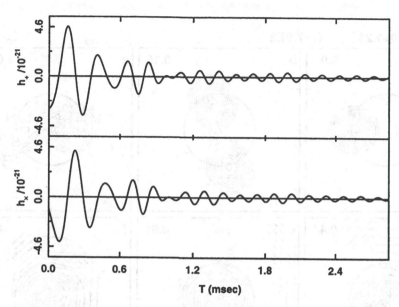

Figure 3. Wave forms of h_+ and $h_\times$ observed on the z-axis at 10Mpc for equation (8).

Figure 4 is the same as figure 1 for TD1, where the masses of stars are $1.70 M_\odot$ and $1.28 M_\odot$, the radii are 7km and 8km, $\Omega = 7.9 \times 10^3 \text{sec}^{-1}$ and $J_t = 5.1 GM_\odot^2/c$.

In this case, τloss=0.3msec is greater than τcol=1.2msec and P=0.9msec; therefore tidal disruption of the lower mass star proceed very rapidly and then it falls into the other one (figures 4a–e). In contrast to the evolution of (8), two stars do not appear after coalescence. The system is getting almost axially symmetric soon losing the angular momentum due to the gravitational radiation. However a ring-like profile appears before the coalescence is completed (figure 4g). Figure 5 shows a rapid decrease in the luminosity. The energy emitted in gravitational radiation is 1.6×10^{53}erg and which is 2.9% of the total rest mass.

The gravitational radiation in coalescence of neutron stars is summarized in table 2. In view of the proposed sensitivity of a large laser-interferometric gravitational wave detector, coalescence of binary neutron stars is likely to be so strong a source that we may observe events up to 50Mpc from the earth. The Fourier analysis of the waves shown in figure 3 shows a strong peak at f=7kHz. However if the red-shift effect, which is not included here, is considered, the frequency of the emitted waves may be a few kHz or lower.

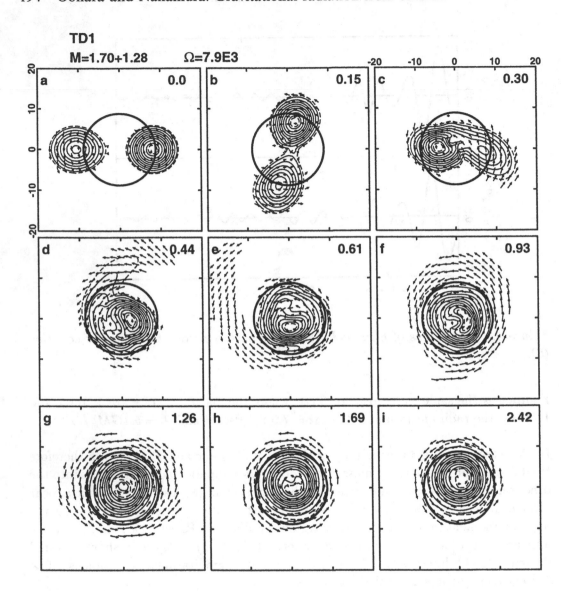

Figure 4. Density and velocity on the x-y plane for TD1. Notations are the same as figure 1.

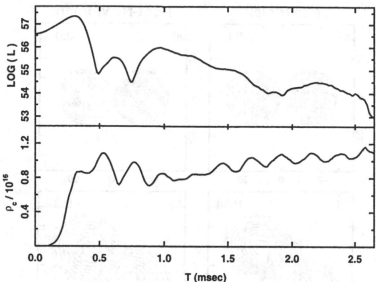

Figure 5. *Luminosity (in unit of erg/sec) and central density (in unit of g/cm³) as functions of time (in unit of millisecond) for TD1.*

Name	energy (erg)	efficiency	maximum amplitude at 10Mpc
EQ8	1.6×10^{53}	3.0%	4.7×10^{-21}
SN1	2.3×10^{52}	0.8%	1.8×10^{-21}
TD1	1.6×10^{53}	2.9%	4.4×10^{-21}
TD2	1.1×10^{53}	2.3%	3.4×10^{-21}

Table 2. *Gravitational radiation in coalescence of neutron stars.*

4 POST-NEWTONIAN CALCULATION

Now we show the results of a (1+2.5) post-Newtonian calculation we performed in order to study general relativistic effects. We performed also a Newtonian calculation with the same initial data for comparison. The initial data includes two spherical neutron stars, namely two relativistic polytropes with $\gamma = 2$ touching each other on the surface. The mass and the radius of each star is $0.62 M_\odot$ and 15km, respectively. Two stars are rotating around the center of mass with angular velocity $\Omega = 2.0 \times 10^3 \mathrm{sec}^{-1}$. The total angular momentum of the system J_t is $1.6 GM_\odot^2/c$.

Figure 6 shows the evolution of the density and the velocity on the x-y plane and figure 7 shows the emitted energy L and the central density ρ_c as functions of time. The result of post-Newtonian(PN) calculation is shown in the right of figure 6 and

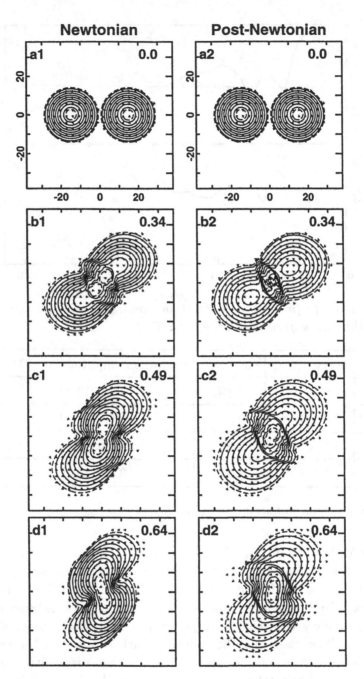

Figure 6. Density and velocity on the x-y plane. The left and right figures are for Newtonian (N) and post-Newtonian (PN) calculation, respectively. Notations are the same as figure 1.

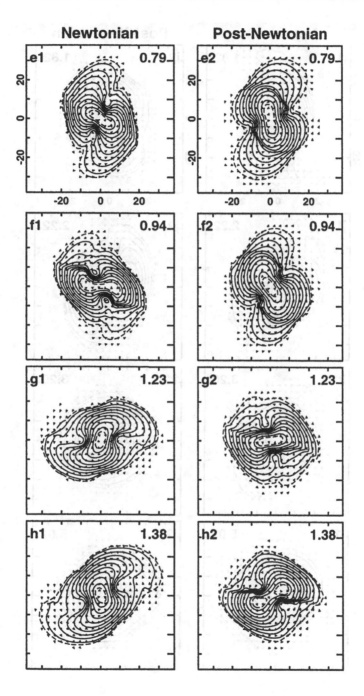

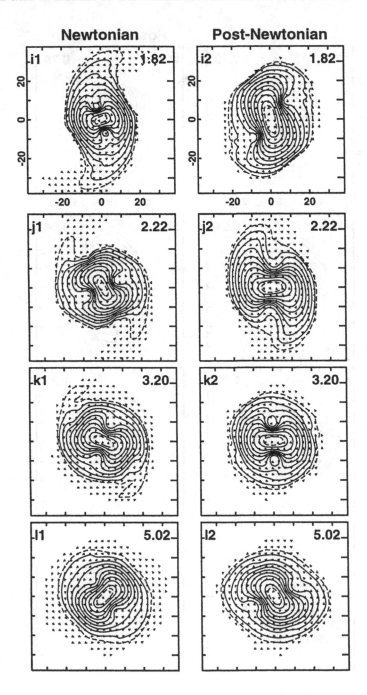

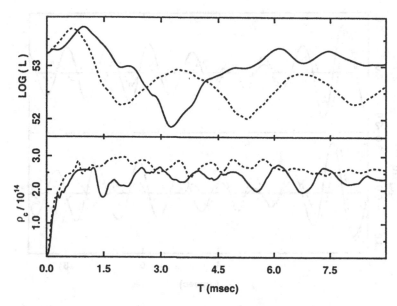

Figure 7. Luminosity (in units of erg/sec) and central density (in units of g/cm³) as functions of time (in units of milliseconds). The solid and dashed lines are for PN and N, respectively.

as solid lines in figure 7 while the result of Newtonian(N) calculation is in the left and as dashed lines. In PN, coalescence begins more rapidly since the general relativity makes an effective gravity force increase. Then a strong shock appears and it makes the matter expand again. Consequently a radial oscillation with large amplitude is excited. The coalescence-expansion oscillation lasts longer in PN case than in N case. In result the total energy emitted in the gravitational radiation increases in comparison with N case; 3.4×10^{50}erg in PN and 2.2×10^{50}erg in N up to 8.6msec. Figure 8 shows the wave form observed along z-axis, where solid lines are for PN and dashed lines are for N. Although h is damped faster in N, the maximum amplitudes are almost the same. Appearance of the strong shock in PN affects little the wave form.

The difference between PN and N is caused principally by the strong shock appearing in the initial stage. Thus if initial data includes two stars in rotational equilibrium and coalescence proceeds more slowly, the difference between PN and N may be more slight. However we don't consider the red-shift effect and excitation of the quasi-normal modes of the resultant black hole or neutron star, which may be essential in the gravitational radiation (Stark and Piran 1986, Nakamura, Oohara and Kojima

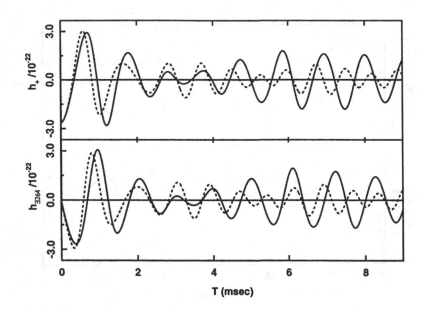

Figure 8. Wave forms of h_+ and $h_\times$ observed on the z-axis at 10Mpc. The solid and dashed lines are for PN and N, respectively.

1987). To complete the program of studying a final destiny of a coalescing binary system and computing emission of gravitational waves in such a event, we should solve a full set of Einstein equations. Therefore it is urgent to construct a fully general relativistic 3D numerical code for the evolution of matter and gravitational field.

This work was in part supported by Grant-in-Aid for Scientific Research (A) of the Ministry of Education, Science and Culture (02452051) and Grant-in-Aid for Scientific Research on Priority Areas of the Ministry of Education, Science and Culture (03250104).

REFERENCES

Blanchet, L., Damour, T. and Schäfer, G. (1990). *Mon. Not. R. Astron. Soc.*, **241**, 289.

Finn, L.S. (1989). *Frontiers of Numerical Relativity*, C. R. Evans, L. S. Finn and D. W. Hobill (eds.), Cambridge University Press, Cambridge, 126.

Gustafsson, I. (1978). *BIT*, **18**, 142.

Harten, A. (1983). *J. Comput. Phys.*, **49**, 357.

Meijerink, J. A. and van der Vorst, H. A. (1977). *Math. Comp.*, **31**, 148.

Misner, C. W., Thorne, K. S. and Wheeler, J. A. (1973). *Gravitation*, Freeman, San Francisco, 993.

Murata, K., Oguni, T. and Karaki, Y. (1985). *Supercomputer*, Maruzen, Tokyo, (in Japanese).

Nakamura, T. and Oohara, K. (1989a). *Prog. Theor. Phys.*, **82**, 1066.

Nakamura, T. and Oohara, K. (1989b). *Frontiers of Numerical Relativity*, C. R. Evans, L. S. Finn and D. W. Hobill (eds.), Cambridge University Press, Cambridge, 254.

Nakamura, T. and Oohara, K. (1991). *Prog. Theor. Phys.*, **86**, 73.

Nakamura, T., Oohara, K. and Kojima, Y. (1987). *Prog. Theor. Phys. Suppl.*, No. 90.

Oohara, K. and Nakamura, T. (1989). *Prog. Theor. Phys.*, **82**, 535.

Oohara, K. and Nakamura, T. (1990). *Prog. Theor. Phys.*, **83**, 906.

Smarr, L. (1979). *Sources of Gravitational Waves*, L. Smarr (ed.), Cambridge University Press, Cambridge, 245.

Stark, R.F. and Piran, T. (1986). *Proceedings of the Fourth Marcel Grossmann Meeting on General Relativity*, R. Ruffini (ed.), Elsevier, Amsterdam, 327.

Ushiro, Y. (1984). *Kokyuroku of RIMS*, **514**, 110, (in Japanese).

van Leer, B. (1977). *J. Comput. Phys.*, **23**, 276.

van der Vorst, H.A. (1989). *Comput. Phys. Comm.*, **53**, 223.

"CRITICAL" BEHAVIOUR IN MASSLESS SCALAR FIELD COLLAPSE

M. W. Choptuik

Center for Relativity, University of Texas at Austin, Austin, US

Abstract. The results of a detailed numerical investigation of the strong-field, dynamical behaviour of a collapsing massless scalar field coupled to the gravitational field in spherical symmetry are summarized. A variety of non-linear phenomena suggestive of a type of universality in the model have been discovered using a finite-difference approach combined with an adaptive mesh algorithm based on work by Berger & Oliger. A derivation of the equations of motion for the system is sketched, the adaptive algorithm is described, and representative examples of the strong-field behaviour are displayed.

1 INTRODUCTION

The problem of the collapse of a massless scalar field coupled to the Einstein gravitational field in spherical symmetry has been studied in considerable detail, both analytically (Christodoulou 1986a, 1986b, 1987a, 1987b), and through numerical work (Choptuik 1986, 1989, 1991, Goldwirth & Piran 1987, Goldwirth et al 1989, Gómez & Winicour 1989, 1992, Gómez et al 1992). In many ways, the system provides an ideal model for addressing a variety of basic issues in numerical relativity. The scalar field provides the model with a radiative degree of freedom which is necessarily absent from any "dynamics" of the Einstein (or Maxwell) field in spherical symmetry. At the same time, by suitable choice of initial data, the self-gravitation of the scalar field can be made arbitrarily strong, so that processes such as curvature scattering of radiation and black-hole formation can be studied. Operationally, the restriction to spherical symmetry makes the actual tasks of setting up the equations of motion for the model, and then solving the resulting set of partial differential equations (PDE's) numerically, *much* easier than for general calculations in numerical relativity. In particular, with currently available computing resources, and using appropriate algorithms, it is possible to construct genuine "computational laboratories" for this problem wherein one can *interactively* generate individual numerical solutions — a process which greatly aids in the exploration of the phenomenology of the model's solution space.

The main purpose of this paper is to summarize the results from such an exploration. I have focussed my study on the strong-field dynamics of collapsing shells of scalar radiation, and in particular, have examined the behaviour of one-parameter families of initial data where variation of the parameter, p, results in an sequence of geometries which are flat at one extreme and contain black holes at the other. For each such family, there is a "critical value", p^* at which black hole formation "turns on", and a detailed study of the parameter space regime, $p \approx p^*$, for various families has revealed a variety of interesting phenomena suggestive of a form of universality in the strong-field regime. These phenomena were first discovered for a *minimally* coupled scalar field, and in an attempt to assess the extent to which the results are dependent on the form of the coupling, I have recently extended the work to the case of *non-minimal* coupling. As the non-minimally coupled model is relatively unstudied in numerical relativity, I review the derivation of the equations of motion (in a particular coordinate system) in the next section. For this, I assume that the reader is familiar with the 3+1 approach to general relativity as described, for example, in Chapter 21 of Misner, Thorne & Wheeler (1973) — my notation, units, and nomenclature follows that source closely. Section 3 deals with a description of the adaptive mesh-refinement algorithm which, following Berger & Oliger (1984), I have implemented to solve the model. A key characteristic of the strong-field dynamics is the potential for development of structure on arbitrarily small spatio-temporal scales. The adaptive algorithm was absolutely crucial in revealing this behaviour so it is worthwhile understanding the basic principles of its operation. The results themselves are summarized in Section 4 which is followed by a few concluding remarks.

2 DERIVATION OF THE EQUATIONS OF MOTION

Adopting geometric units ($G = c = 1$), the (total) Lagrange function for a massless, non-self-interacting scalar field, ϕ, which is non-minimally coupled to the gravitational field is

$$L^{\text{Total}} \equiv L^{\text{Einstein}} + 16\pi L^{\phi} \sim R - 16\pi \left(\frac{1}{2}\phi^{;\mu}\phi_{;\mu} + \frac{1}{2}\xi R\phi^2 \right). \tag{1}$$

where ξ is the usual (non-minimal) coupling constant. It is convenient to introduce additional constants η_1, η_2 and η_{12} given by

$$\eta_1 = 1 - 6\xi \qquad \eta_2 = 8\pi\xi \qquad \eta_{12} = 8\pi\xi\,(1 - 6\xi) = \eta_1\eta_2.$$

The stress-energy tensor for the scalar field is

$$T^{\phi}{}_{\mu\nu} = T^0{}_{\mu\nu} + 2\xi \left(g_{\mu\nu}\Sigma - \phi_{;\mu}\phi_{;\nu} + g_{\mu\nu}\phi\Box\phi - \phi\phi_{;\mu\nu} \right) + \xi G_{\mu\nu}\phi^2, \tag{2}$$

where

$$\Sigma \equiv \phi_{;\alpha}\phi^{;\alpha}, \tag{3}$$

$G_{\mu\nu}$ is the Einstein tensor, and

$$T^0_{\ \mu\nu} \equiv \phi_{;\mu}\phi_{;\nu} - \frac{1}{2}g_{\mu\nu}\Sigma, \tag{4}$$

is the stress tensor for a *minimally* coupled scalar field. Observe that in contrast to the minimally coupled case, the more general stress tensor (2) involves *second* derivatives of ϕ, and in particular, second *temporal* derivatives. As I will discuss further at the end of this section, this complicates the process of putting the equations into "true" 3+1 form.

Given that ϕ is the only matter field on the spacetimes under consideration, Einstein's equations are

$$G_{\mu\nu} = 8\pi T^{\phi}_{\ \mu\nu}, \tag{5}$$

which, using (2) can be written as

$$T^{\phi}_{\ \mu\nu} = \left(1 - \eta_2\phi^2\right)^{-1}\left(T^0_{\ \mu\nu} + 2\xi\left(g_{\mu\nu}\Sigma - \phi_{;\mu}\phi_{;\nu} + g_{\mu\nu}\phi\Box\phi - \phi\phi_{;\mu\nu}\right)\right). \tag{6}$$

The equation of motion for ϕ is just

$$\phi_{;\mu}^{\ ;\mu} \equiv \Box\phi = \xi R\phi = -\xi G\phi = -\eta_2 T^{\phi}\phi, \tag{7}$$

where G and T^{ϕ} are the traces of the Einstein and stress-energy tensors, respectively, and the contracted form of (5) has been used to obtain the final relation. Now, from (2), the trace of the stress-tensor is

$$T^{\phi} = -\eta_1\Sigma + 6\xi\phi\Box\phi + \xi G\phi^2. \tag{8}$$

Using (7) to eliminate $\Box\phi$, and then applying the contracted Einstein equation, $R = -G = -8\pi T^{\phi}$, this becomes

$$T^{\phi} = -\eta_1\left(1 - \eta_{12}\phi^2\right)^{-1}\Sigma. \tag{9}$$

From (7) this last result immediately yields a form of the scalar field equation which is suitable for 3+1 calculations:

$$\Box\phi = \eta_{12}\left(1 - \eta_{12}\phi^2\right)^{-1}\Sigma\phi. \tag{10}$$

Note that for the minimally coupled ($\xi = 0$) and conformally coupled ($\xi = 1/6$) cases we have $T^{\phi} = -\Sigma$ and $T^{\phi} = 0$, respectively, and that for both these cases, the scalar field equation is just $\Box\phi = 0$. Also observe that the forms of equations (6) and (10) suggest that for a general $\xi > 0$, the theory could become pathalogical should $|\phi|$ attain a value of $1/\sqrt{8\pi\xi}$.

From this point on, I will restrict attention to the case of spherical symmetry. I adopt a specific coordinate system, (r, t, θ, φ), which has been widely used in spherically-symmetric work in numerical relativity (see, for example, Shapiro & Teukolsky 1986, Choptuik 1991) and which can be considered a generalization of the usual Schwarzschild coordinates to the case of time-dependent spacetimes. The line element is given by

$$ds^2 = -\alpha^2 (r, t) \, dt^2 + a^2 (r, t) \, dr^2 + r^2 \, d\Omega^2, \tag{11}$$

where α is the *lapse* function, a is the radial metric function and $d\Omega^2$ is the usual metric on the 2-sphere. Note that the radial coordinate, r, is *areal*—the proper area of a constant t and r surface is $4\pi r^2$. Corresponding to the single, non-trivial 3-metric component, a, is a single non-trivial component, $K^r{}_r$, of the extrinsic curvature tensor $K^i{}_j$, viz.

$$K^i{}_j = \text{diag}\left(K^r{}_r (r, t), 0, 0\right). \tag{12}$$

Thus, the extrinsic curvature tensor satisfies the so-called *polar* condition (Bardeen & Piran 1983),

$$K^r{}_r = K^i{}_i \equiv \text{Tr} K \equiv K, \tag{13}$$

which, when combined with an areal spatial coordinate in spherical symmetry, essentially fixes the coordinate system. In particular, the radial component, β, of the shift vector $\beta^i = (\beta, 0, 0)$, which would generically appear in a $dr \, dt$ term in the line element, identically vanishes in this system (that is, the spacetime trajectories of observers with constant spatial coordinates are *normal* to the $t = $ constant hypersurfaces).

The choice of *polar/radial* coordinates, as defined above, results in a particularly simple set of Einstein equations. In addition, because the coordinate choice eliminates two of the four basic dynamical variables (two 3-metric components and two extrinsic curvature components), it is possible to construct what Piran (1980) has termed a *fully constrained* evolution scheme for the combined gravito-scalar system. In such a scheme, the constraint equations, rather than evolution equations, are used to update appropriate geometric variables. Empirically, numerical implementations of these methods have generally been found to have superior accuracy and stability properties (at a given resolution) than their *free evolution* counterparts, and, largely for these reasons, I have adopted a fully-constrained approach in the work described here.

The Hamiltonian constraint is

$$^{(3)}R - K^i{}_j K^j{}_i + K^2 = 16\pi\rho = 16\pi\alpha^2 T^{tt}, \tag{14}$$

where $^{(3)}R$ is the curvature scalar of the hypersurface and ρ is the energy density of the scalar field as measured by "Eulerian" observers (I have dropped the superscript ϕ

on the stress-tensor from this and following equations). Although a direct calculation of the explicit form of this equation is not too difficult, given the simple form of the 3-metric in the current case, it is instructive to consider an alternative "derivation" which begins by rewriting the line element (11) as

$$ds^2 = -\alpha^2\,(r,t)\,dt^2 + \left(1 - \frac{2m\,(r,t)}{r}\right)^{-1} dr^2 + r^2\,d\Omega^2 \tag{15}$$

where the *mass-aspect* function, $m(r,t) = r(1 - a(r,t)^{-2})/2$, measures (in a region of vacuum) the total mass within the origin-centred sphere of radius r at time t. Then, it can be shown that (14) is equivalent to

$$\frac{dm}{dr} = 4\pi r^2 \rho, \tag{16}$$

which may be written as a first-order, non-linear, ordinary differential equation (ODE) for the radial metric function, a:

$$\frac{a'}{a} + \frac{a^2 - 1}{2r} - 4\pi r a^2 \rho = 0. \tag{17}$$

(' denotes differentiation with respect to r.) This equation is integrated from $r = 0$ outward, with an "initial condition", $a(0,t) = 1$, dictated by the demand of elementary flatness at the origin. The momentum constraint is

$$K_i{}^j{}_{|j} - K_{|i} = 8\pi j_i = 8\pi\alpha T^t{}_i, \tag{18}$$

where j^i ($j_i = g_{ik}j^k$) is the (Eulerian) momentum density and the bar subscript denotes covariant differentiation with respect to the 3-geometry. A straightforward calculation yields

$$K_i{}^j{}_{|j} - K_{|i} = K^r{}_r{}' + \frac{2}{r}K^r{}_r - K^r{}_r{}' = \frac{2}{r}K^r{}_r, \tag{19}$$

so,

$$K^r{}_r = 4\pi r j_r = 4\pi r\,\alpha\,T^t{}_r. \tag{20}$$

Thus, the momentum constraint is *algebraic* in this case.

As discussed above, I have implemented a fully constrained evolution scheme in my work; however, in order to write the various components of $T_{\mu\nu}$ in proper 3+1 form, it is necessary to use the evolution equation for a:

$$\dot{a} = -\alpha a\,K^r{}_r = -4\pi r\,\alpha^2 a\,T^t{}_r, \tag{21}$$

where the overdot denotes differentiation with respect to t. The lapse function, α, is governed by an equation which follows from the polar slicing condition, $K = K^r{}_r$.

Since the general extrinsic curvature tensor in spherical symmetry has the form $\mathtt{diag}\,(K^r{}_r, K^\theta{}_\theta, K^\theta{}_\theta)$, the polar condition implies that $K^\theta{}_\theta = 0$ for all t. The Einstein evolution equation for $K^\theta{}_\theta$ then reads:

$$\dot{K}^\theta{}_\theta = -\alpha^{|\theta}{}_{|\theta} + \alpha \left(R^\theta{}_\theta + 4\pi\,(S - \rho) - 8\pi S^\theta{}_\theta \right) = 0, \tag{22}$$

where S^i_j is the stress 3-tensor $(S_{ij} = T_{ij})$. Calculation then yields

$$\frac{\alpha'}{\alpha} - \frac{a'}{a} + \frac{1 - a^2}{r} + 4\pi r a^2 \left(\rho - S^r{}_r \right) = 0, \tag{23}$$

which is to be interpreted as a first-order ODE for α.

Considering now the scalar field equation of motion, I have found it useful to introduce two sets of auxiliary variables:

$$\Phi \equiv \phi' \qquad \Pi \equiv \frac{a}{\alpha}\,\dot{\phi}, \tag{24}$$

$$X \equiv \sqrt{2\pi}\,\frac{r\Phi}{a} \qquad Y \equiv \sqrt{2\pi}\,\frac{r\Pi}{a}. \tag{25}$$

Numerically, I adopt Φ and Π as a fundamental pair of dynamical variables for evolving the scalar field. However, as will be mentioned in Section 4, the dynamics of the scalar field is actually better described using the pair X and Y. I note that for the minimally coupled case we have the simple relationship:

$$\frac{dm}{dr} = X^2 + Y^2. \tag{26}$$

The equation of motion for Φ follows immediately from the definitions (24):

$$\dot{\Phi} = \left(\frac{\alpha}{a}\Pi \right)'. \tag{27}$$

Defining

$$\Gamma \equiv \left(-\det g_{\mu\nu} \right)^{\frac{1}{2}} = r^2 \alpha a, \tag{28}$$

and

$$s_2\,(\phi) \equiv \eta_{12}\,(1 - \eta_{12}\phi^2)^{-1}\,\phi, \tag{29}$$

equation (10) may be written as

$$\Box\phi = \Gamma^{-1} \left(\Gamma g^{\mu\nu}\phi_{,\mu} \right)_{,\nu} = s_2\Sigma, \tag{30}$$

or

$$\left(r^2\alpha a\,(-\alpha^{-2})\,\phi_{,t} \right)_{,t} + \left(r^2\alpha a\,(a^{-2})\,\phi_{,r} \right)_{,r} = \Gamma s_2\Sigma, \tag{31}$$

which yields an evolution equation for Π

$$\dot{\Pi} = \frac{1}{r^2}\left(r^2\frac{\alpha}{a}\Phi\right)' - \alpha a s_2 \Sigma = \frac{\alpha}{a}\frac{1}{r^2}\left(r^2\Phi\right)' + \frac{\alpha}{a}\Phi\left(\frac{\alpha'}{\alpha} - \frac{a'}{a}\right) - \alpha a\, s_2\Sigma. \qquad (32)$$

Equations (17), (23), (27) and (32) constitute the basic set of partial and ordinary differential equations which I solve numerically using finite difference techniques. These equations are not, however, in their final form, since the various parts of the stress energy tensor (ρ, j_i and S_{ij}) need to be explicitly expressed in terms of the scalar field variables. As previously noted, the non-minimal coupling of ϕ to the gravitational field complicates the task of casting the equations into true 3+1 form — that is, so that the right hand sides of all equations are evaluable solely in terms of quantities defined on a *single* hypersurface. Although space restrictions prohibit a complete treatment, I will briefly explain why it is desirable to have such a set of equations and then sketch how the calculation to place the equations in 3+1 form proceeds.

First consider the minimally coupled case; then equations (17), (23), (27) and (32) reduce to

$$\dot{\Phi} = \left(\frac{\alpha}{a}\Pi\right)',$$

$$\dot{\Pi} = \frac{1}{r^2}\left(r^2\frac{\alpha}{a}\Phi\right)',$$

$$\frac{\alpha'}{\alpha} - \frac{a'}{a} + \frac{1-a^2}{r} = 0,$$

$$\frac{a'}{a} + \frac{a^2-1}{2r} - 2\pi r\left(\Phi^2 + \Pi^2\right) = 0.$$

The structure of this system made it straightforward (Choptuik 1986) to devise a second-order accurate ($O(h^2)$ truncation error both in space *and* in time on a mesh with basic discretization scale, h) finite-difference analogue of the equations. Denoting the matter variables by $T \equiv \{\Phi, \Pi\}$ and the geometric variables by $G \equiv \{\alpha, a\}$, the difference scheme may be schematically written as

$$T^{n+1} := \mathbf{U}[T^n, T^{n-1}, G^n],$$
$$G^{n+1} := \mathbf{C}[T^{n+1}, G^{n+1}].$$

Thus, to advance the solution from time t^n to t^{n+1}, the scalar field variables are first updated using differenced versions of their equations of motion, and then the values of geometric variables at the advanced time are computed from the differenced Hamiltonian constraint and slicing condition. Moreover the latter stage can be further decomposed into a solution of the constraint equation for a, followed by a solution of the slicing equation for α. The net result then, is a scheme which is simultaneously $O(h^2)$ accurate and, apart from the non-linearity of the Hamiltonian constraint, completely explicit.

In order to preserve these desirable numerical properties of second-order accuracy and explicitness when treating the *non-minimally* coupled field, it is necessary to systematically eliminate certain time-derivatives (using the equations of motion themselves) which appear in the equations for the geometric variables. For example, consider the Hamiltonian constraint (equation (17)), which is to be solved for a:

$$\frac{a'}{a} + \frac{a^2 - 1}{2r} - 4\pi r a^2 \rho = 0.$$

Defining $\zeta(\phi) \equiv (1 - \eta_2 \phi^2)^{-1}$ and $h(\phi) \equiv \eta_{12}(1 - \eta_{12} \phi^2)^{-1} \phi^2 = s_2 \phi$, we find

$$\rho = \alpha^2 T^{tt} = \alpha^{-2} T_{tt} = \zeta\left(\rho^0 + 2\xi\left(-\Sigma - \alpha^{-2}\dot{\phi}^2 - \phi\Box\phi - \phi\phi_{;\mu\nu} n^\mu n^\nu\right)\right)$$

$$= \zeta\left(\rho^0 + 2\xi\left(-\Sigma(1 + h) - \left(\frac{\Pi}{a}\right)^2 - \frac{\phi}{\alpha}\left(\frac{\Pi}{a}\right)_{,t} + \frac{\phi\Phi}{a^2}\frac{\alpha'}{\alpha}\right)\right),$$

where $n^\mu = (\alpha^{-1}, 0, 0, 0)$ is the future-directed unit normal to the hypersurfaces. To get the Hamiltonian constraint into the desired form, we must eliminate the $(\Phi/a)_{,t}$ term using the equations of motion. For example, we have (equation (21)),

$$\dot{a} = -\alpha a K^r{}_r = -4\pi r \alpha^2 a T^t{}_r,$$

where the stress tensor component, $T^t{}_r$, is given by

$$T^t{}_r = \zeta\left(T^{0t}{}_r + 2\xi\left(-\phi^{;t}\phi_{;r} - \phi\phi^{;t}{}_{;r}\right)\right).$$

Again, we find that the term $-\phi\phi^{;t}{}_{;r}$ itself contains a term involving $\dot{a}$; the "coefficient" of this term must be isolated before $\dot{a}$ can be expressed *explicitly* in terms of quantities defined on a *single* hypersurface. Following this general strategy of eliminating time derivatives from the Hamiltonian constraint and slicing condition, we eventually get equations of the form:

$$\frac{\alpha'}{\alpha} - \frac{a'}{a} + C_{02}(T) + a^2 C_{03}(T) = 0, \tag{33}$$

$$C_{11}(T)\frac{a'}{a} + C_{12}(T) + a^2 C_{13}(T) = 0, \tag{34}$$

where the various "coefficient functions", $C_{ij}(T)$ depend only on the scalar field variables. As with the minimally-coupled system, these equations are then discretized and solved in a very straightforward fashion using $O(h^2)$ difference techniques on *uniform* grids (mesh spacings constant in each of the coordinate directions).

3 SUMMARY OF THE NUMERICAL ALGORITHM

The program which I have constructed to solve the differential equations governing the coupled scalar-gravitational dynamics is based on a general algorithm for hyperbolic equations described in Berger & Oliger (1984). The key feature of the Berger-Oliger algorithm is that it is *adaptive*; the local scale of discretization, h is allowed to vary from place to place in the computational domain in an attempt to keep the *local truncation error* below some user-specified threshold. Here, I will only present a brief outline of the method for problems in one space dimension and some of the extensions which I have made to it for my own work. The interested reader is referred to the above reference, as well as to Berger (1986) for more details, particularly for discussions of the application of the technique to problems in more than one spatial dimension.

The Berger-Oliger algorithm achieves adaptivity using the mechanism of *local, uniform, nested refinement*. This is a technique which has long been advocated by Brandt (Brandt 1977) as providing the basis of a general strategy for developing PDE solvers having near-optimal efficiency. In this approach, the computational domain (the set of all grid points at which finite differenced versions of the differential equations are defined) consists of a hierarchical *set* of grids, each of which has constant mesh spacings in both the spatial and temporal coordinate directions. For rectangular, simply connected domains such as those considered here, the hierarchy is rooted at a single, *base grid*, denoted G_0^0. Then, at any given integration time, the computational domain is a set $\{G_\ell^i\}$, where the subscript ℓ labels the *level* of refinement, $\ell = 1 \ldots L$, for some a *priori* specified maximum level of refinement, L, and the superscript i labels a particular grid on a given level. Each level of refinement is characterized by a *single* discretization scale, h_ℓ, with the various h_ℓ's typically satisfying

$$h_\ell = \rho\, h_{\ell+1} \qquad l = 0 \cdots L - 1 \tag{35}$$

for some (generally fixed) integer-valued *refinement ratio*, ρ. Here, it should be stressed that the local refinement is performed both in space *and in time*. In particular, for the calculations described below, the spatial mesh size, Δr_ℓ of any grid can always be identified with h_ℓ, while the temporal mesh spacing, Δt_ℓ is always some fixed fraction, λ, $\lambda < 1$ of Δr_ℓ. For hyperbolic equations which are finite differenced to the same order in space and time this is a natural way to proceed since spatial and temporal variations will usually be of comparable magnitude. In addition to the relation (35) between the different discretization scales, each grid, G_ℓ^i must be properly contained within the domain of some *parental* grid $G_{\ell-1}^{i'}$, (nesting property) and be positioned so that its outer boundaries lie along parental grid lines (alignment). Figure 1 depicts a simple example of the type of grid structure admitted by the algorithm with $\rho = 2$, $L = 2$.

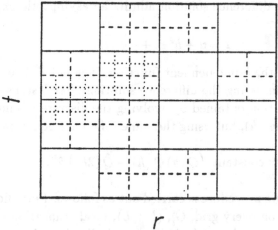

Figure 1. An example of the grid structure admitted by the Berger-Oliger mesh-refinement algorithm. The different line styles show grid lines on three distinct levels of discretization. Note that the local refinement is in both space and time.

As can be seen from figure 1, the spatial distribution of grid points is *not* static in general, and, of course, the process of dynamically regridding the domain results in the *adaptivity* of the Berger-Oliger algorithm. Berger and Oliger investigated several different regridding strategies and eventually settled on a technique which, as mentioned above, attempts to keep the local truncation error in the solution below some *a priori* specified threshold. I have also adopted this approach which, in fact, is based on principles basically identical to those employed in many modern, "automatic" ODE solvers. The key idea, which can be traced back to Richardson's early work (Richardson 1910) on finite-difference methods, is easily understood by considering a simple evolution equation, ($u_t = u_x$, for example) for a scalar unknown, u, which, for the sake of presentation, I will assume has been differenced to $O(h^2)$ using a two-level, explicit difference scheme. Thus the discrete unknown, $\hat{u}$ satisfies

$$\hat{u}^{n+1} = \hat{Q}(h)\hat{u}^n, \tag{36}$$

where $\hat{Q}(h)$ is the one-step "update operator" at mesh scale h. The local truncation error, τ^n, associated with $\hat{Q}(h)$ is defined by

$$\tau^n \equiv u^{n+1} - \hat{Q}(h)u^n, \tag{37}$$

and, given that the difference scheme is second order accurate, will have the form

$$\tau^n = h^3 t_2 + O(h^4), \tag{38}$$

where t_2 is a smooth (provided u is smooth), h-independent function, depending both on the difference scheme and the unknown u. Now, following Richardson, if we assume that (locally) the difference solution admits an asymptotic expansion of the form

$$\hat{u} = u + h^2 \, e_2 + \cdots, \tag{39}$$

where e_2 is another smooth, h-independent function, it is not difficult to show that we can estimate τ^n by comparing the difference solution after 2 steps on the h mesh to *another* difference solution obtained by evolving the discrete unknowns 1 step on a *coarsened* mesh (spacing $2h$), but using the same difference equations. Specifically:

$$\tau^n \approx \text{constant} \, \left(\hat{Q}\,(h)\,\hat{Q}\,(h) - \hat{Q}\,(2h) \right) \hat{u}^n. \tag{40}$$

Briefly then, the regridding procedure is carried out as follows. Periodically (typically every 4 time steps), and on every grid, G_l^i, $(l < L)$, local truncation error estimates are generated using the method sketched above and all grid points where the error estimate exceeds the threshold are flagged. From these flagged points, regions requiring refinement are identified and then covered with new grids $G_{l+1}^{i'}$ (see Berger and Oliger (1984) for details regarding grid-generation strategies).

A pseudo-code description of the overall flow of the Berger-Oliger algorithm is given in figure 2. The basic task of the pseudo-routine **time_step**[l] is to advance the level-l difference equations a single time-step. The routine first checks to see if it has been called at a regridding time; if it has, then the **regrid** routine is invoked to perform the truncation-error estimation and regridding procedure, not only at level l, but on all finer extant levels as well. In fact, the regridding process *starts* at the finest level and works "upwards"; this ensures that all grids are properly nested at all times. This part of the code also involves *initialization* of the discrete unknowns on newly created grids. In general, this involves a combination of the transfer of "old" level l values (on those parts of the domain which were previously covered by a level l grid) and *interpolation* of level $l - 1$ quantities (on newly refined regions). In fact, the regridding operation at level l always involves a *complete* rebuilding of the overall grid hierarchy (tree) at level l and finer, but in practice, the overhead incurred is typically a small percentage of the total computational cost.

Following any necessary regridding, the **time_step** routine applies the basic differenced evolution equations on all grids at level l. Then, if l is *not* the finest level, the routine recursively invokes itself to effect ρ time-steps at level $l + 1$. Note that the structure of **time_step** guarantees that coarse-level difference equations are advanced *before* fine-level ones. This enables boundary values for level l grids that have one or more boundaries in the *interior* of the numerical domain to be determined from *interpolation* of suitable values from the parent grid. When **time_step** has finished taking the ρ fine time-steps (end of the **FOR** loop), the level $l + 1$ grids have been advanced to the same time as the level l grids. At this point, another routine, **inject**, is called to overwrite parental grid values with any corresponding values from contained refinements. This ensures that, at all times, each grid contains the *best* available approximations to the continuum unknowns; if this were not done, (large) errors in coarse grid unknowns in refined regions would propagate to regions not requiring refinement with the potential of eventually corrupting the entire solution.

> **ROUTINE** time_step[l]
>> **IF** regridding time **AND** $l < L$ **THEN**
>>> **regrid** on *all* levels $l' \geq l$
>>
>> **END IF**
>> **FOR** $i = \cdots$ (all grids on level l)
>>> o advance solution on G^l_i (basic difference equations)
>>> o if necessary, compute boundary values from interpolation
>>>   in some G^{l-1}_j
>>
>> **END FOR**
>> **IF** $G^{l+1}_i \neq \emptyset$ **THEN**
>>> **FOR** $k = 1 \cdots \rho$
>>>> time_step[$l + 1$]
>>>
>>> **END FOR**
>>
>> **END IF**
>> **IF** $l > 0$ **AND** time-aligned with $l - 1$ **THEN**
>>> **inject** level l values into parental $l - 1$ grids
>>
>> **END IF**
>
> **END ROUTINE**

Figure 2. Pseudo-code description of the Berger/Oliger algorithm.

Although my implementation of the adaptive-mesh algorithm follows the description in Berger & Oliger (1984) quite closely, I have made several modifications and extensions to the method which have proven quite useful in obtaining the results described in the next section. For example, the original algorithm was designed for strictly *hyperbolic* systems and the need to solve "spatially non-local" equations such as the Hamiltonian constraint and slicing condition introduces complications, both conceptually and in practice. Although I have been able to incorporate the non-hyperbolic equations into my scheme without too much difficulty, I wish to emphasize that the treatment of such equations within the context of this sort of mesh-refinement algorithm still needs careful study, particularly for problems in two or three spatial dimensions. In order to investigate the accuracy and convergence properties of the solutions generated by the algorithm, I have added an alternate regridding module which, at regridding times, determines where refinements should be placed from a previously generated *script* rather than from dynamically produced local truncation error estimates. This allows me to perform standard convergence tests by running calculations with coarse mesh spacings of $1/2\,h_0, 1/4\,h_0, \ldots$ which read a script generated by a "normal" run executed with a coarse resolution of h_0. Finally, in order to create the initial $(t = 0)$ grid structure, the algorithm must ideally be able to evaluate the values of the dynamical unknowns to arbitrary accuracy at any location in the spatial domain. In many cases, this may be trivially accomplished simply by giving some specific "analytic" form for the initial data. However, for the purpose of investigating the dynamics of *arbitrary* initial data for the collapsing scalar field, I have found it useful to implement a routine which will fit *any* grid function, $f(r_j), j = 0, 1, \ldots, N$, using Chebyshev polynomials on a variable number of intervals $[r_i \cdots r_{i'}]$ to some specified tolerance. The routine then produces a callable subprogram (basically consisting of a database of intervals and coefficients, and a call to a general Chebyshev evaluator) which can be used, for example, in the refinement algorithm's initial data sequence.

4 PARAMETER SPACE SURVEYS AND "CRITICAL" BEHAVIOUR

In this section I summarize some key results concerning the strong-field phenomenology of the collapsing scalar field in spherical symmetry. My studies have focussed on the systematic evolution of sets of smooth initial data characterized by a single parameter, p, such that variation of p generates a family of spacetimes which "interpolate" between geometries which are (arbitrarily) flat and geometries which contain black holes. For any such interpolating family, there will be a *critical* value, $p^\star$, at which a black hole first appears, and in the parameter-space regime $p \approx p^\star$, the *strong-field* behaviour of the system is characterized by a number of intriguing and unexpected features. What emerges is a surprisingly simple picture of the strong-field dynamics of the coupled system. In particular, there is strong evidence for a type of *universality*, in that the near-critical, strong-field evolution of any family of initial data appears to be describable in terms of a *unique* trajectory in a suitably defined phase space.

A typical 1-parameter family of solutions may be generated by specifying an initial profile, $\phi(r, 0)$, for the scalar field, such as the following "Gaussian" form:

$$\phi(r, 0) = \phi_0\, r^3\, \exp\left(-\left((r - r_0)/\Delta\right)^q\right), \tag{41}$$

and demanding that the scalar radiation be purely ingoing at the initial time. For the time being, ϕ_0 is to be considered the variable parameter, while r_0, Δ and q will have fixed values for any particular family. (Alternatively, one can specify $\Phi(r, 0)$

and $\Pi(r, 0)$ or, for the minimally coupled case, $X(r, 0)$ and $Y(r, 0)$, and I have found it convenient at times to adopt one of these approaches.) The $t = 0$ values of the geometric variables, a and α, can then be computed from the Hamiltonian constraint and slicing condition to complete the process of determining initial data.

Associated with an initially-imploding pulse (or more correctly, shell, since the solutions are spherically symmetric) of scalar radiation, are *two* length scales. The first is simply the physical (radial) extent, L, of the pulse (a few $\times$ Δ for the Gaussian data defined above); the second is the gravitational (Schwarzschild) radius, R_S,

$$R_S = 2M_\infty = \lim_{r \to \infty} m(r, 0),\tag{42}$$

of the configuration. When $L \gg R_S$ ($\phi_0 \to 0$, for example), weak-field evolution results; departures from Minkowski spacetime are small and, for any value of the coupling constant, ξ, the scalar field dynamics is well-approximated by a (linear) solution of the usual wave equation in spherical symmetry, $(r\phi)_{tt} = (r\phi)_{rr}$. The imploding wave packet simply "self-reflects" through $r = 0$, then disperses to infinity. At the other extreme, ($\phi_0 \approx \phi_0^*$) the strong-field regime is roughly defined by the condition, $L \approx R_S$; here, the evolution is decidely *non-linear*, the scalar field has significant self-gravitation and black hole formation becomes possible. At this point I should remark that it is well known (Bardeen & Piran 1983) that the polar/radial coordinate system used here can *not* penetrate event horizons. However, in these coordinates, the signature of black hole formation in an actual numerical evolution is very clear. For example, the quantity $2m(r, t)/r$ will rapidly asymptote to a value of 1 at some radius r_{BH} from which an approximate mass of the hole can be immediately computed.

For any given family of initial data with parameter p, it is a straightforward matter to determine (approximately) the critical value, p^* which marks the transition from those spacetimes where all of the scalar radiation escapes to infinity to those which contain black holes. Assuming that the non-linearity of the interaction increases with increasing p (as is the case for the Gaussian data parameterized by ϕ_0), we can simply view the entire numerical evolution of given initial data as a single "function call" which returns a value indicating whether a black hole forms (implying $p > p^*$) or not ($p < p^*$). Given two initial values, p_{LO} and p_{HI}, known to bracket p^*, this then forms the basis for a simple binary (bisection) search which allows p^* to be determined to some specified precision. I have performed many such searches using a wide variety of initial profiles and parametrizations; in *all* of the cases I have examined, the quantity π, defined as

$$\pi \equiv \ln |p - p^*|,\tag{43}$$

is a natural choice for discussing "variations in parameter space". That is, the detailed behaviour of any of the dynamical variables exhibits an essentially linear dependence on π and hence an *exponential* dependence on the initial conditions. In addition, the nature of this extreme sensitivity to the initial data is such that structure on increasingly smaller spatio-temporal scales appears as p approaches p^*.

These and other features of the strong-field $(2m/r \approx 1)$ evolution of the scalar model are perhaps best understood at this time in terms of the following conjecture: for each value of ξ, there exists a *unique* (up to trivial rescalings) sequence of configurations of dynamic variables, such as X and Y which constitutes an "attractor" in the strong-field regime. (Here a "trivial rescaling" is a transformation $r \rightarrow cr$, $t \rightarrow ct$ for some arbitrary constant; X and Y (see equation (21)) are form invariant under such transformations and it is for this reason that they, rather than Φ and Π, are the more relevant quantities in this discussion.) In slightly more precise terms, I claim that for each ξ, there is a unique sequence, Z^*_ξ:

$$Z^*_\xi \equiv [X, Y]^*, \tag{44}$$

which is naturally expressed as a function of *logarithmic, physical* independent variables (coordinates), ρ and τ:

$$\rho \sim \ln(\text{proper (areal) radius}), \tag{45}$$

$$\tau \sim \ln(\text{proper time of a central } (r = 0) \text{ observer,}), \tag{46}$$

and which has an *"echoing"* property:

$$Z^*_\xi(\rho - \Delta_\xi, \tau - \Delta_\xi) \simeq Z^*_\xi(\rho, \tau), \tag{47}$$

where Δ_ξ is a constant which has (at most) a weak ξ-dependence for the cases I have been able to examine. Figure 3 illustrates examples of this "echoing" phenomena for near-critical evolutions using initial field profiles of the form (41) and coupling constant values of $\xi = 0$ and $\xi = 1/2$. In these graphs, the curves marked with open squares are profiles of the dynamical variable X plotted against $\rho \equiv \ln r$, on that portion of the domain where the gravitational self-interaction of the field is strong, and at some otherwise arbitrary time during the interaction. The curves marked with filled circles are profiles from the *same* evolutions, but at later times and on *smaller spatial scales*. If these configurations were *precisely* critical, the echoing property implies that if we continued examining the X profiles at logarithmic proper time intervals Δ_ξ, and on smaller and smaller spatial scales, we would find that the profiles would continue to be coincident. For such a case, the field would oscillate more and more rapidly near $r = 0$, but the entire process would "converge" at some finite central proper time, T^*. I have, in fact, *estimated* T^* for these calculations and then used this value to define $\tau \equiv \ln(T^* - T)$. Also, the quantity Δ_ρ quoted in each plot represents a value of Δ_ξ computed from the spatial rescaling factor which gives

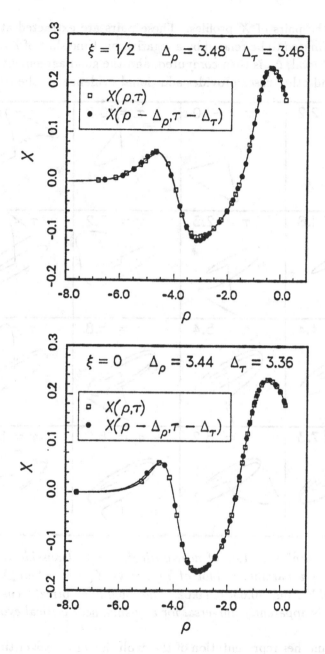

Figure 3. Plots of the dynamical variable X vs ρ ≡ ln r extracted from near-critical calculations (|p − p*|/p < 10⁻¹³), illustrating the "echoing" behaviour discussed in text. Each plot shows two separate "snapshots" of X from a single evolution which are very nearly coincident after spatial rescaling of the later waveform by a factor of about 30.

a best overlap for the pairs of X profiles. These pairs are extracted at integration times which are *estimated* to be related by a logarithmic time shift of Δ_ξ. The actual logarithmic time interval, Δ_τ is then *computed*, and the good agreement between Δ_ρ and Δ_τ for these and other cases provides additional evidence for the conjecture.

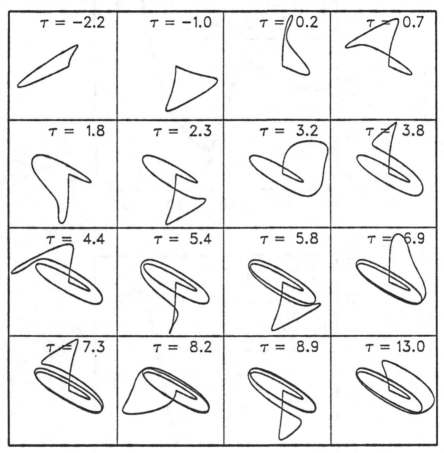

Figure 4. *"Phase-space" evolution of near-critical, $\xi = 0$, Gaussian ($q = 2$) initial data. Each frame is a parametric plot of $Y(r,\tau)$ vs $X(r,\tau)$, where X varies from -0.41 to $+0.43$ and Y from -0.63 to $+0.62$. The roughly elliptical locus about which the curve "wraps" is apparently universal for any such near-critical evolution.*

Figure 4 provides another representation of the evolution of a near-critical configuration for the case of minimal coupling. Each frame in this sequence is a parametric plot of $Y(r,\tau)$ vs $X(r,\tau)$, and thus constitutes a sort of "phase-space" portrait; the point near the centre of each frame where two segments meet corresponds to both $r = 0$ and $r = \infty$. The extremal values achieved by X and Y (quoted in the figure caption) in this, and other evolutions, are apparently functions only of ξ, as is the locus about which the X-Y curve "wraps" during the evolution. It is in this sense that the strong-field evolution appears to be *universal*; *all* of the families I have investigated exhibit the same wrapping behaviour, about the same locus, when examined in sufficient detail. In this regard I should note that at the conference I presented some prelimi-nary results which showed two configurations generated from non-Gaussian data and characterized by parameter values p_0 and p_1 satisfying $|p_0 - p_1|/p_0 < 1.0 \times 10^{-13}$. I remarked that while there was no sign of echoing behaviour in either calculation, the p_0 evolution did not form a black hole while the p_1 calculation did, and stated that this seemed to provide evidence that the echoing phenomena was *not* completely generic. However, using the Chebyshev fitting routine described at the end of the previous section, I was subsequently able to generate a reparametrized family of so-lutions (parameter P) such that new computations with $P = 0$ and $P = 1$ generated (to high accuracy) the same spacetimes as the p_0 and p_1 computations. Then a search for a critical value P^* revealed the usual echoing behaviour; the original initial data simply had so much detail that machine precision had been exhausted merely getting close to the critical value. As an indication of the performance of the adaptive algo-rithm, the runs to compute P^* used a base grid with about 600 points in the radial direction and then as many as 9 additional levels of $5:1$ refinement. Thus, it would have required about 10^9 grid points had the domain been covered by a single uniform mesh at the finest level; in practice the total number of grid points at any given time was always less than 2500.

One of the immediate corollaries of the echoing conjecture is that there will be *no* "mass-gap" for general 1-parameter interpolating families of solutions of the model–that is, black-hole formation generically "turns on" at *infinitesimal* mass, rather than at finite mass. Goldwirth & Piran (1987) (see also Goldwirth et al 1989) investigated this matter to a certain extent and appeared to find a mass gap and, in fact, a rather substantial mass (measured as a fraction of M_∞) for the smallest black hole in a family of Gaussian initial data. However, the basic scale of discretization they used was extremely coarse compared to that which is required to resolve behaviour such as that shown in figure 4, (where each wrap occurs on a scale some 30 times smaller than the previous), so their results are not really in contradiction with this work.

The results I have described so far concern solutions which do not contain black holes. Equally intriguing features appear in 1-parameter sequences of solutions which *do* contain black holes. In particular, I have found that, at least for sufficiently "compact" pulses of radiation, the masses of black holes which in this model are remarkably well fit by a *universal* power law:

$$M_{BH} \simeq c_f \, |p - p^*|^{\gamma_\xi} \,, \tag{48}$$

where c_f is a family-dependent constant, but γ_ξ is a *family-independent* exponent, which is only weakly dependent on ξ, and has a numerical value suspiciously close to e^{-1}. (I have investigated various families using coupling constant values in the range $-1/2 \leq \xi \leq 1/2$, various numerical and coordinate difficulties currently preclude study for significantly larger values of $|\xi|$). Figure 5 shows some data illustrating this scaling behaviour. The figure incorporates results from three distinct families of solutions generated from Gaussian initial profiles (equation (41), $q = 2$) and with variable parameters ϕ_0, Δ, and r_0 respectively. The masses of the black holes which form in the various calculations span a range of $\mu \equiv M_{BH}/M_\infty$, from less than 0.01 to greater than 0.90.

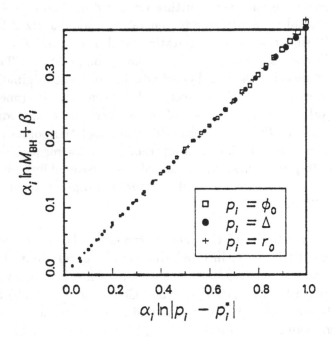

Figure 5. *Illustration of the conjectured black hole mass-scaling law (equation (48)) using computed data from three 1-parameter variations of Gaussian initial data. The diameter of each symbol is proportional to* $\ln \mu$ *where* μ *is the fraction of the total mass,* M_∞, *of the spacetime which ends up in the black hole. Each data set has been individually scaled* (α_i) *and shifted* (β_i) *to produce a uniform range of the abscissa and a common origin. The maximum ordinate value of the framed area is* e^{-1}.

5 CONCLUDING REMARKS

The computations underlying the results summarized in the previous section essentially exhaust the mesh-refinement algorithm's ability to supply *quantitative* information concerning the strong-field behaviour of the scalar model. Although the algorithm is capable of *resolving* the phenomena, the *accuracy* of the results is limited by 1) the use of second-order difference techniques, and 2) the degradation of the solution which the regridding process inevitably induces in the course of a long integration. At the current time, I feel that the various "pure numbers", such as Δ_ξ and γ_ξ, which appear in the strong-field limit are known to a few percent. It would be very interesting to compute these quantities more accurately, as well as to further investigate just how exact the scaling relations such as (47) and (48) are. It is unlikely that any simple modification of the current scheme will lead to a substantial increase in accuracy. One possible route to increased precision would be to use the results from the current algorithm to define an analytic specification of a "remeshing" function which could then be incorporated into a higher accuracy scheme.

It is essential to note, however, that an algorithm requires *proper resolution*–supplied, for example, through an "Eulerian" technique such as the Berger/Oliger algorithm, or a "Lagrangian" one such as the *a posteriori* specified coordinate transformation suggested above–before *any* higher order or higher accuracy scheme (including spectral methods) will be effective. Related to this observation is the view that the key issue in many, if not most, numerical solutions of PDEs is that of obtaining (locally) adequate resolution. It seems clear that for many of the most interesting outstanding problems in numerical relativity, including any realistic black hole calculations, some form of grid adaptation will be very valuable. The coding of multi-level algorithms for numerical relativity problems in two or three spatial dimensions will not be an easy task, but the benefits are clear. Such codes can truly abide by Brandt's Golden Rule of Numerical Analysis–*the amount of computational work expended in solving a model should be proportional to the amount of real physical changes in the computed system* (Brandt 1982), and will allow us to take full advantage of the continuing exponentiation of hardware resources.

ACKNOWLEDGEMENTS

The research was supported by the Natural Sciences and Engineering Research Council of Canada, by NSF grant PHY-8806567 and by Texas Advanced Research Project grant TARP-085. All numerical computations described here were performed using the Cray Y-MP8/864 at the Center for High Performance Computing, University of Texas with Cray time supported by a Cray University Research grant to Richard Matzner.

REFERENCES

Bardeen, J. M. and Piran, T. (1983). *Phys. Rep.*, **96**, 205.

Berger, M. J. and Oliger, J. (1984). *J. Comp. Phys*, **53**, 484.

Berger, M. J. (1986). *SIAM J. Sci. Stat. Comput.*, **7**, 904.

Brandt, A. (1977). *Math. Comput.*, **31**, 333.

Brandt, A. (1982). In *Lecture Notes in Mathematics: Multi grid Methods*, Vol. 960, W. Hackbusch and U. Trottenberg (eds.), Springer Verlag, New York.

Choptuik, M. W. (1986). Ph.D Thesis, University of British Columbia.

Choptuik, M. W. (1989). In *Frontiers in Numerical Relativity*, C. R. Evans, L. S. Finn and D. W. Hobill (eds.), Cambridge University Press, Cambridge.

Choptuik, M. W. (1991). *Phys. Rev. D*, **44**, 3124.

Christodoulou, D. (1986a). *Commun. Math. Phys.*, **105**, 337.

Christodoulou, D. (1986b). *Commun. Math. Phys.*, **106**, 587.

Christodoulou, D. (1987a). *Commun. Math. Phys.*, **109**, 591.

Christodoulou, D. (1987b). *Commun. Math. Phys.*, **109**, 613.

Goldwirth, D. and Piran, T. (1987). *Phys. Rev. D.*, **36**, 3575.

Goldwirth, D., Ori, A. and Piran, T. (1989). In *Frontiers in Numerical Relativity*, C. R. Evans, L. S. Finn and D. W. Hobill (eds.), Cambridge University Press, Cambridge.

Gómez, R. and Winicour, J. (1989). In *Frontiers in Numerical Relativity*, C. R. Evans, L. S. Finn and D. W. Hobill (eds.), Cambridge University Press, Cambridge.

Gómez, R., Isaacson, R. A. and Winicour, J. (1992). *J. Comp. Phys*, **98**, 11.

Gómez, R. and Winicour, J. (1992). *J. Math. Phys*, in press.

Misner, C, W., Thorne, K. S. and Wheeler, J. A. (1973). *Gravitation*, Freeman, San Francisco.

Piran, T. (1980). *J. Comp. Phys.*, **35**, 254.

Richardson, L. F. (1910). *Phil. Trans. Roy. Soc.*, **210**, 307.

Shapiro, S. L. and Teukolsky, S. A. (1986). *Astrophys. J.*, **398**, 33.

GODUNOV-TYPE METHODS APPLIED TO GENERAL RELATIVISTIC STELLAR COLLAPSE

José Mª. Ibáñez, José Mª. Martí, Juan A. Miralles and J.V. Romero

Department of Theoretical Physics, University of Valencia, Valencia, Spain

Abstract. We have extended some *high-resolution shock-capturing methods*, designed recently to solve nonlinear hyperbolic systems of conservation laws, to the general-relativistic hydrodynamic system of equations and applied them to the study of the gravitational collapse of spherically symmetric configurations.

1 INTRODUCTION

Several topics are of current interest among astrophysicists working in the field of stellar collapse: (i) The equation of state for both subnuclear and supranuclear densities. Alongside the theoretical problems concerned here, there is also the technical problem of making both approaches consistent with each other, as well as sufficiently fast to compute in stellar collapse calculations (see Lattimer and Swesty, 1992). (ii) The coupling between neutrinos and matter in connection with the feasibility of the so-called *delayed mechanism*. (iii) The correct modelling of shocks in order to conserve total energy along the propagation of the shock formed in the collapse after bounce. In the last years, a part of our research has been addressed to this point.

In a previous paper (Martí *et al.*, 1990, in the next MIM90) we have focussed on the shock formation and propagation such as it appears in the standard scenario of the *prompt mechanism*. In MIM90 we have undertaken Newtonian stellar collapse calculations with two codes: (i) A standard finite-difference scheme which uses an artificial viscosity technique. (ii) A Godunov-type method which uses a linearized Riemann solver. The initial model and the equation of state was kept fixed in order to be able to compare both methods directly. Differences in the behaviour of the velocity field and the global energetics of the collapse were found. These differences can be so dramatic as to produce a factor of two in the minimum of the velocity, at the infall epoch, and a factor of four in the kinetic energy of the material ejected. These results correspond to the case in which the artificial viscosity is inactivated when one of the following two conditions are verified: $\frac{\partial v}{\partial m} > 0$ or $\frac{\partial \rho}{\partial t} < 0$, where ρ and v are, respectively, density and velocity of the matter and m and t, the included mass (Lagrangian coordinate) and the time. Consequences might be of interest not only in the field of the *prompt mechanism* of Type II Supernovae, but even in the correct estimation of the efficiency of the energy released in the form of gravitational radiation in non-spherical collapse. Indeed, as Bonazzola and Marck (1989, see also these proceedings) have emphasized, the gravitational power decreases with viscosity, ν, as $1/\nu^6$; hence, viscosity of a numerical code — explicit or intrinsic — may have dramatic effects.

2 LOCAL CHARACTERISTIC APPROACH FOR THE GENERAL RELATIVISTIC HYDRODYNAMICS

The equations of general-relativistic hydrodynamics are the expression of the local laws of conservation of baryon number density and energy-momentum in a space-time M, described by the four dimensional metric tensor $g_{\mu\nu}$. In our procedure, the metric is split into the objects α and γ_{ij}, giving a line element of the form

$$ds^2 = -\alpha^2 dt^2 + \gamma_{ij} dx^i dx^j$$

(latin indexes run from 1 to 3). We will consider the energy-momentum tensor $T_{\mu\nu}$ corresponding to a perfect fluid

$$T_{\mu\nu} = \rho h u_\mu u_\nu + p g_{\mu\nu}$$

(greek indexes run from 0 to 3), where $h = 1 + \epsilon + p/\rho$ the specific enthalpy, ϵ the internal specific energy, p the pressure and ρ the rest-mass density.

By defining the following set of variables

$$D \equiv \rho W$$

$$S \equiv \alpha T^{0r} = \rho h W^2 v$$

$$\tau \equiv \alpha^2 T^{00} = \rho h W^2 - p$$

the general-relativistic hydrodynamic equations in the one dimensional case (for example, spherical symmetry) can be written as a system of conservation laws in the sense of Lax (1972)

$$\frac{\partial \mathbf{u}}{\partial t} + \frac{\partial \mathbf{f}(\mathbf{u})}{\partial r} = \mathbf{s}(\mathbf{u}),$$

where

$$\mathbf{u} = (D, S, \tau)^T$$

is the 3-dimensional vector of unknowns which defines the state of the system. The fluxes are

$$\mathbf{f} = \frac{\alpha}{\sqrt{\gamma_{rr}}} \left(\frac{DS}{\tau + p}, \frac{S^2}{\tau + p} + p, S \right)^T,$$

and source terms are free of derivatives of hydrodynamic quantities and read

$$
\mathbf{s(u)} = \Bigg(-D\frac{\partial ln\sqrt{\gamma}}{\partial t} - \frac{\alpha}{\sqrt{\gamma_{rr}}}\frac{DS}{\tau + p}\frac{\partial ln\gamma}{\partial r},
$$
$$
-S\frac{\partial ln(\sqrt{\gamma\gamma_{rr}})}{\partial t} - \frac{\alpha}{\sqrt{\gamma_{rr}}}\Big(\frac{S^2}{\tau + p}\Big)\frac{\partial ln\sqrt{\gamma}}{\partial r} - \frac{\alpha}{\sqrt{\gamma_{rr}}}\tau\frac{\partial ln\alpha}{\partial r} - \frac{\alpha}{\sqrt{\gamma_{rr}}}p\frac{\partial ln\sqrt{\gamma_{rr}}}{\partial r},
$$
$$
-\tau\frac{\partial ln\sqrt{\gamma}}{\partial t} - \frac{\alpha}{\sqrt{\gamma_{rr}}}S\frac{\partial ln\alpha\sqrt{\gamma}}{\partial r} - \frac{S^2}{\tau + p}\frac{\partial ln\sqrt{\gamma_{rr}}}{\partial t} - p\frac{\partial ln\sqrt{\gamma}}{\partial t}\Bigg)^T.
$$

In the above expressions, the quantity γ is the determinant of the matrix γ_{ij}, v is defined by $v \equiv \sqrt{\gamma_{rr}}u^r/\alpha u^0$ (indexes 0 and r stands for the temporal and radial components, respectively) and represents the fluid velocity relative to an inertial observer at rest in the coordinate frame. The Lorentz-like factor is defined by $W \equiv \alpha u^0$ and satisfies the following relation

$$
W = \frac{1}{\sqrt{1 - v^2}}.
$$

As usual, the above system of equations is closed by means of an equation of state (EOS), which we will assume is written in the form

$$
p = p(\rho, \epsilon).
$$

From the *conserved* quantities $\aleph = \{D, S, \tau\}$ we must obtain the set of quantities $\wp = \{\rho, v, \epsilon\}$ at each time step, by solving an implicit equation in the pressure. In the Newtonian limit, the set of new variables $\aleph = \{D, S, \tau\}$ tends to the set $\{\rho, \rho v, \rho\epsilon + (1/2)\rho v^2\}$.

The hyperbolic character of the relativistic hydrodynamic system of equations (see, for example, Anile 1990) allows us, as in the Newtonian calculations, to use Godunov-type methods, supported on the resolution of local Riemann problems. In the present applications, we have made use of our version of MUSCL (Monotonic Upstream Schemes for Conservation Laws), based on a generalization of Roe's approximate Riemann solver extended to systems of equations by means of a *local characteristic approach* (see Martí *et al.*, 1991, in the next MIM91, or Martí, 1991, for details).

Crucial in our procedure is the knowledge of the spectral decomposition of the Jacobian matrix of the system at hand. The eigenvalues, that is, the characteristic speeds associated with the above system are

$$
\lambda^0 = \frac{\alpha}{\sqrt{\gamma_{rr}}}v,
$$

$$\lambda^{\pm} = \frac{\alpha}{\sqrt{\gamma_{rr}}}(v \pm c_s)/(1 \pm vc_s),$$

c_s being the local sound speed in the fluid frame, and the corresponding eigenvectors, $\mathbf{e}^A(A = 0, \pm)$, have the components

$$e_1^A = \frac{(\lambda^A - v) - v(\lambda^A p_S + p_\tau)}{(\lambda^A - v)hW - v(p_\tau - p_D)},$$

$$e_2^A = \lambda^A,$$

$$e_3^A = 1 - e_1^A,$$

where subindexes S, τ and D stand for the corresponding partial derivatives of pressure.

3 SOME PRELIMINARY RESULTS

We have tested our code to reproduce some of the stationary solutions of the *spherical accretion onto a black hole* in two cases: (i) *dust* accreting onto a Schwarzschild black hole (Hawley *et al.* , 1984) and (ii) an *ideal gas* accreting onto a Schwarzschild black hole (Michel, 1972).   A grid of 50 points spans the interval $1.05 \leq r/2M \leq 10.0$. Discrepancies with the exact solution amount to less than 2%.   Details of these numerical experiments can be found elsewhere (Martí, 1991).

The initial model we have taken in present application is a white dwarf-like configuration having a central density 2.5×10^{10} g/cm^3. This is an equilibrium model for a particular EOS (Chandrasekhar's EOS with coulombian corrections) corresponding to the maximum of the "mass-radius" curve (see Ibáñez, 1984).

The numerical grid has been built up in such a way that the radius of the initial model is partitioned into 200 zones distributed in geometric progression in order to have a finer resolution near the centre.

The EOS we have used is a γ-law such that γ varies with density according to Van Riper's prescription (Van Riper, 1978)

$$\gamma = \gamma_{min} + \eta(\log \rho - \log \rho_b)$$

with: $\eta = 0$ if $\rho < \rho_b$ and $\eta > 0$ otherwise.   The parameters γ_{min}, η and ρ_b are, typically: 4/3, 1, and 2.7×10^{14} gcm^{-3}, respectively, but we have also considered other values for γ_{min} and η .

We have used Schwarzschild-type coordinates (Bondi 1964, Gourgoulhon 1992) in terms of which the 3-metric reads

$$\gamma_{ij} = diag(X^2, r^2, r^2 sin^2\theta)$$

where
$$X = (1 - 2m/r)^{-1/2}, \quad m = m(r,t).$$

The source terms have the particular expression

$$s^1 = \alpha X D \left[4\pi r p v - \frac{v}{r}(2 - 5m/r)\right]$$

$$s^2 = -\alpha X \left[8\pi r p(\tau + D - Sv) + S\frac{v}{r}(2 - 5m/r) + (\tau + D - p)\frac{m}{r^2}\right]$$

$$s^3 = \alpha X \left[-4\pi r p v D - \frac{2S}{r}(1 - 2m/r) + \frac{Dv}{r}(2 - 5m/r)\right].$$

At each time, step functions m and α are integrated along the radius according to

$$\frac{\partial m}{\partial r} = 4\pi r^2(\tau + D),$$

$$\frac{\partial \ln\alpha}{\partial r} = X^2 \left(\frac{m}{r^2} + 4\pi r(p + Sv)\right).$$

Our preliminary results allow to be confident in the procedure followed for extending Godunov-type methods to the general-relativistic hydrodynamics. Let us point out the behaviour of the velocity field as it can be seen in figure 1. Shock is sharply solved in two zones and is free of spurious oscillations. The minimum of velocity at the infall epoch is about -0.40, that is $\approx 25\%$ greater than the value reported by MIM90, in the Newtonian case, for the same initial model, same EOS and a lagrangian version of our code which uses the same Riemann solver. The difference is entirely due to general-relativistic effects.

Figure 2 shows the geometrical quantity α, as a function of radius, in a nonexplosive collapse to a black hole (simulated by doing $\eta = 0$ in the above equation of state). Values of α at the center of the star lower than 10^{-8} have been reached before stopping the calculation.

At present, we are considering the extension to multidimensional problems of these numerical methods. The spectral decomposition of the jacobian matrixes in each spatial direction has been already obtained. A new modern high-resolution shock-capturing algorithm, the so-called PHM (Marquina, 1991) promises to be robust enough to attack the ultrarelativistic regime (Marquina *et al.* , 1992) and has the optimum accuracy — globally third order accuracy, both in space and time — for multidimensional applications.

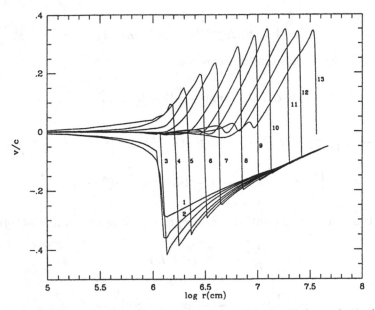

Figure 1. *Snapshots of the velocity (in units of the light velocity) versus radial coordinate (in logarithmic scale and units of cm.). Each curve is labeled by a number which establishes the temporal sequence into the interval* $90.17 \leq t(msec.) \leq 93.34$.

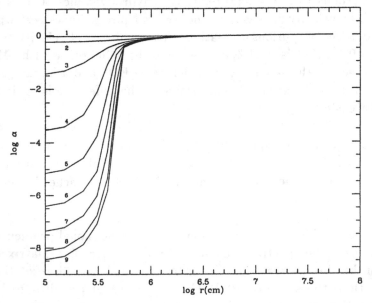

Figure 2. *Snapshots of the lapse versus radial coordinate (in logarithmic scale and units of cm.). Each curve is labeled by a number which establishes the temporal sequence into the interval* $88.96 \leq t(msec.) \leq 90.62$.

ACKNOWLEDGMENTS

This work has been supported by the Spanish DGICYT (reference numbers PB90-0516-C02-02 and PS90-0265). José Mª. Martí has benefited from a postdoctoral fellowship from the Ministerio de Educación y Ciencia (MEC) and Juan A. Miralles from a MEC/Fulbright grant.

REFERENCES

Anile, A. M. (1989). *Relativistic fluids and magneto-fluids*, Cambridge University Press, Cambridge.

Bonazzola, S. and Marck, J. A. (1989). In *Frontiers in Numerical Relativity*, C. R. Evans and L. S. Finn (eds.), Cambridge University Press, Cambridge, 239.

Bondi, H. (1964). *Proc. Roy. Soc. A*, **281**, 39.

Gourgoulhon, E. (1992). *Astron. and Astrophys.*, **252**, 651.

Hawley, J. F., Smarr, L. L. and Wilson, J. R. (1984). *Ap. J. Suppl.*, **55**, 211.

Lattimer, J. M. and Swesty, F. D. (1992). *Nucl. Phys.*, in press.

Lax, P. (1972). *Regional Conference Series Lectures in Applied Math.*, **11**, SIAM, Philadelphia.

Marquina, A. (1992). *SIAM J. Scient. Stat. Comp.*, submitted.

Marquina, A., Martí, J. Mª., Ibáñez, J. Mª., Miralles, J. A. and Donat, R. (1992). *Astron. and Astrophys.*, in press.

Martí, J. Mª., Ibáñez, J. Mª. and Miralles, J. A. (1990). *Astron. and Astrophys.*, **235**, 535 (in the text MIM90).

Martí, J. Mª. (1991). PhD. Thesis, University of Valencia.

Martí, J. Mª., Ibáñez, J. Mª. and Miralles, J. A. (1991). *Phys. Rev. D*, **43**, 3794 (in the text MIM91).

Michel, F. C. (1972). *Ap. Space Sci.*, **15**, 153.

Van Riper, K. A. (1978). *Ap. J.*, **221**, 304.

ASTROPHYSICAL SOURCES OF GRAVITATIONAL WAVES AND NEUTRINOS

Silvano Bonazzola, Eric Gourgoulhon, Pawel Haensel and Jean-Alain Marck

DARC, Meudon Observatory, Meudon, France

Abstract. Our project was inspired by the prospect that a new non-electromagnetic astronomy will develop by the end of the century. Projects like Virgo and Ligo will lead to detectors able to detect extra-Galactic and Galactic sources of gravitational radiation. New generation of neutrino detectors like Superkamiokande will be able to detect various Galactic neutrino sources. All these considerations motivated us to study in detail potential Galactic sources of bursts of the gravitational radiation and neutrinos. In this paper, our projects are described in some detail. The advantages and the drawbacks of the numerical technique used in our computer simulations (pseudospectral methods) are discussed. Possible applications of the numerical methods are illustrated by some examples of astrophysical interest: coalescence of two neutron stars, mini-collapse of a neutron star (phase transition) and formation of a black hole due to the collapse of a neutron star.

1 INTRODUCTION

The main idea, which motivated our project, is that massive stellar cores, involved in supernovae of type II, are not the only collapsing Galactic objects generating bursts of gravitational waves that could be detected by the next generation of gravitational wave detectors. It is quite likely that SNI and SNII are only an optically detectable subset of a larger class of collapse events, which are less spectacular (as far as electromagnetic radiation is concerned) but perhaps quite frequent, and which are able to radiate a conspicuous amount of gravitational radiation. These optically silent collapsing objects, in which the ejection of outer layers is aborted, are likely to be missed because of their low electromagnetic luminosity. There are at least three scenarios of an optically silent collapse:

a) core collapse at the end of the evolution of an ordinary massive star without ejection of outer layers, leaving at the end of the collapse a black hole, either directly or after the collapse of the freshly formed neutron star due to accretion of the infalling envelope.

b) collapse of a neutron star due to the matter accretion beyond the critical mass or due to the slowing down of rotation, resulting in a formation of a black hole or a neutron star in a more compact configuration (minicollapse due to a phase transition).

c) coalescence of two white dwarfs, without matter ejection, resulting in formation of a neutron star or a black hole.

All these catastrophic events occurring in our Galaxy or in the Magellanic Clouds could radiate enough gravitational energy to be detected by the next generation of gravitational radiation detectors (Virgo and Ligo experiments). Our project consists of studying exhaustively the neutrino burst, and gravitational radiation burst, accompanying these events.

The computation of the neutrino burst is important for two reasons. The first one is that already existing neutrino detectors seem to have detected until now only one burst: that from the Supernova 1987A. If the computation of the neutrino burst, accompanying an optically silent collapse, shows that the neutrino signal is strong enough to be detected by the present detectors, we can conclude that the Galactic events of such a type occur less frequently than once in 11 years (i.e. since the neutrino experiments became operational). The second reason is that new, much larger neutrino detectors, like Superkamiokande, will (hopefully) become operational at the begining of the next century. A simultaneous detection of the gravitational radiation burst and neutrino neutrino burst will be a major result of the new non-electromagnetic astronomy.

This paper is organised in the following way. In Section 2 we give a short description of the numerical technique used in our computer simulations (spectral methods), and we discuss the advantages of these methods and the way to overcome their drawbacks. Examples of applications, in which the particular advantages of the spectral methods are displayed, are presented in Section 3. Description of preliminary results for astrophysical scenarios of gravitational collapse is given in Sections 4 and 5. Finally, Section 6 contains our conclusions.

2 THE PSEUDOSPECTRAL METHOD

The techniques used in order to solve partial differential equations (PDE) systems in various geometries with pseudospectral methods are described in detail in Bonazzola and Marck (1989 and 1990). Consequently, we shall restrict our discussion to pointing out the main advantages and drawbacks of the pseudospectral methods, as well as possible way of overcoming the drawbacks of these methods.

Advantages
  - high precision
  - ability to handle the coordinates singularities and the boundary conditions
  - modularity
  - other

Drawbacks
  - severe Courant condition
  - difficulty in treating problems in which the solution has high contrast and the relative error has to be small
  - difficulty in handling discontinuities present in the solutions (as in most other numerical methods)

The first drawback is that the Courant condition is more severe than in the finite difference method. We therefore have to use implicit or semi-implicit schemes which implies that the pseudospectral methods are quite expensive to implement. We have overcome this difficulty by creating a library of subroutines able to treat implicitly the more common differential operators, like (∇, Δ), in simple geometries (cubic, cylindrical, toroidal, spherical and spherical shell) and all the geometries diffeomorphic to the above. The matrix of the linear system can be reduced easily to a well conditioned band-matrix (9 diagonals in the worst case). The Alternating Direction Method (ADM) is also used in multidimensional problems.

The physical problems in which the solution has high contrast and in which the relative error must stay locally bounded remain a challenge to spectral methods. Indeed, these methods are global and, consequently, the truncation error is spread out on all the grid points and the local relative error may become too large in the region, where the calculated value of the solution is small. We overcome this difficulty by using a multi-domain technique in which the solutions of the problem on two contiguous grids are matched. The ability of the spectral methods to treat exactly the boundary conditions turns out to be very useful in matching the two solutions. The matching of the solutions implies the solution of a linear system with a band-matrix. Note that the multi-domain technique one allows to switch continuously from a complete spectral method (when only one grid is present) to an ordinary finite difference scheme (where there are 3 degrees of freedom for each subdomain). Finally it seems to be interesting to mention that spectral methods and multi-domain technique are just some kind of wavelet expansion.

Discontinuities in the solution constitute a real problem for almost all numerical methods. The use of artificial viscosity should be considered as a last resort, very far from being a satisfactory way to solve this crucial problem (see Ibanez, *this volume*). One way to describe the discontinuities (for instance a shock) is to use a two grid technique, the boundaries of the grids being at the position of the shock, and to match the solutions with the appropriated jump conditions (e.g. the Rankine Hugoniot conditions in the case of a shock, see Bonazzola and Marck 1991).

If the geometry of the problem makes it too difficult to implement the above technique we are forced to use an appropriate "artificial viscosity". However, with this technique, the spectral methods lose their main advantage, namely the exponential decrease of the error with increasing number of the spatial grid points.

3 SOME EXAMPLES

Partial results obtained from studies which are still in progress are good examples of the applications of the techniques described above. In figure 1 we show the equilibrium configuration of a presupernova star, obtained with a relaxation method and using six spatial domains (grids). The density varies by about a factor of 10^{18} between the center and the surface of the star.

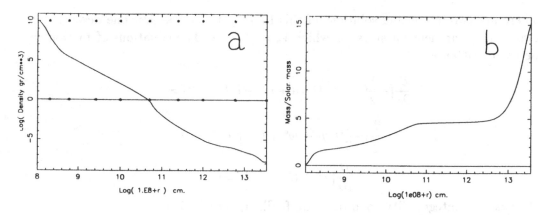

Figure 1. Density profile (fig. 1a) and mass profile (fig. 1b) of a presupernova.

A second example shows the way used to solve the elliptical partial differential equations in a domain diffeomorphic to a sphere, in order to find the steady state configuration and the velocity field of a close binary system of two neutron stars.

In a co-rotating reference frame, the equations of motion for one of the stars read

$$\frac{\partial V_x^a}{\partial t} - 2\Omega V_y^a - \Omega^2(x+d) - V^{ia}\partial_i V_x^a = -\frac{\partial}{\partial x}(w^a - U^a - U^{ba})$$

$$\frac{\partial V_y^a}{\partial t} - 2\Omega V_x^a - \Omega^2 y - V^{ia}\partial_i V_y^a = -\frac{\partial}{\partial y}(w^a - U^a - U^{ba})$$

$$\frac{\partial w^a}{\partial t} = -V^{ia}\partial_i w^a - w^a \frac{\partial \log w^a}{\partial \log \rho^a}\partial_i V^{ia}$$

$$\Delta U^a = -4\pi G \rho^a$$

where the indices a and b label the stars, and Ω is the orbital angular frequency, d the distance between the two centers of mass,

$$w = \int \frac{dP}{\rho}$$

is the enthalpy per unit mass, U^a is the gravitational self potential of the star a and U^{ba} is the potential of the gravitational field of the star b acting on star a.

A steady-state condition implies

$$\frac{\partial V_i^a}{\partial t} = 0 \quad \text{and} \quad \frac{\partial w^a}{\partial t} = 0.$$

There are two interesting solutions of this PDE system. The first one describes the two co-rotating neutron stars for which $V_x = V_y = 0$. The equations of motion can be then written as

$$\frac{\partial}{\partial x}[-\frac{1}{2}\Omega^2 x^2 - \Omega^2 x d + w^a + U^a + U^{ba}] = 0 \;,$$

$$\frac{\partial}{\partial y}[-\frac{1}{2}\Omega^2 y^2 + w^a + U^a + U^{ba}] = 0 \;,$$

$$\frac{\partial}{\partial z}[w^a + U^a + U^{ba}] = 0 \;.$$

Because the integrability conditions are fulfilled, one obtains

$$-\frac{1}{2}\Omega^2(x^2 + y^2 - x d) + w^a + U^a + U^{ba} = 0 \;.$$

The second solution is the case for which the neutron stars are not spinning in the Galilean rest frame. In the rotating frame tied to the binary system (see figure 2), the matter is counter-rotating. The components of matter velocity, with respect to this frame, satisfy

$$V_x = \Omega y + \frac{\partial \phi}{\partial x} \;,$$

$$V_y = -\Omega x + \frac{\partial \phi}{\partial y} \;,$$

$$V_z = \frac{\partial \phi}{\partial z} \,.$$

where $\nabla \phi$ is a potential velocity field introduced in order to satisfy the boundary conditions. The first integral of the equations of motion becomes

$$(\nabla \phi)^2 + 2\Omega y \frac{\partial \phi}{\partial x} - 2\Omega x \frac{\partial \phi}{\partial y} + w^a + U^a + + U^b + U^{ba} = \text{const.}$$

The velocity potential ϕ is determined using the equation of motion involving the enthalpy w. We have

$$\frac{\partial w}{\partial t} = 0 = \mathbf{v} \cdot \nabla w - w(\frac{\partial w}{\partial \log \rho})\Delta \phi$$

with the boundary conditions $\mathbf{v} \cdot \nabla w = 0$ on the surface of the star (which is defined by $w = 0$).

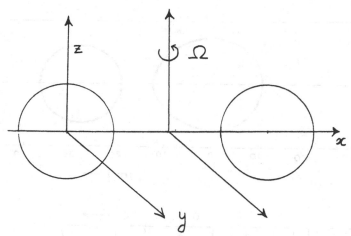

Figure 2. *Schematic representation of the coordinate system used to compute the steady state configuration of a binary system.*

The solution of the system of the PDE proceeds then in the following way (relaxation method):

1) guess an initial distribution for ρ^1 and ρ^2
2) compute on each grid the potentials U^a, U^b, U^{ab}, U^{ba}
3) compute w^a from the first integrals of motion
4) determine the surface of the stars by means of $w = 0$
5) solve the equation for ϕ (see below)
6) compute ρ^1, ρ^2, by means of the EOS
7) return to 1) with the new values of the densities ρ^1 and ρ^2.

The solution for ϕ in the stellar interior is obtained in the following way. Let $r = 1 + f(\theta)$ be the equation of the surface of the star. Consider the coordinate transformation

$$r = r' + R(r')f(\theta')$$

$$\theta = \theta' + G(r')\Psi(\theta')$$

where $0 \leq r' \leq 1$ is the new radial coordinate. Here, R and G are arbitrary functions, which vanish at the origin $r' = 0$, and which satisfy the boundary condition at $r' = 1$

$$R'(1) = 0 \; ; \; G(1) = 0 \; ; \; G'(1) = 1 \; .$$

It is easy to see that $r' = 1$ is the equation of the surface of the star and that the new system of coordinates is orthogonal on the stellar surface. The new equation for ϕ contains cross terms $\partial^2\phi/\partial r\partial\theta$ which are computed with the value of ϕ calculated at the previous iteration. An illustrative solution is presented in figure 3.

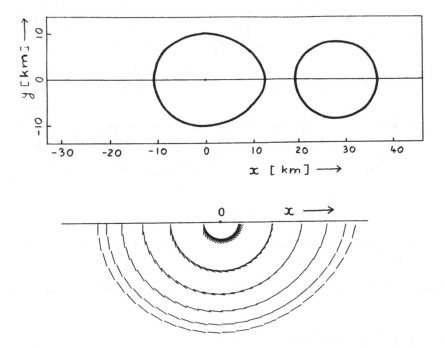

Figure 3.   *Velocity field (which is normalised) in the interior of the less massive neutron star (bottom) and the shape of the surfaces of the stars (top) just before they coalesce. The masses are $1.35M_\odot$ and $1.45M_\odot$.*

4 MINICOLLAPSE OF A NEUTRON STAR

An example of application of the multigrid technique is the minicollapse problem. If a phase transition in the superdense neutron star core occurs, the star will undergo a transition into a new more compact equilibrium configuration. We call this sudden transition a minicollapse of the neutron star. The possible mechanism for the first order phase transition is assumed to be theoretically predicted *pion condensation* in dense nucleon matter (Migdal *et al.* 1979, Ramaty *et al.* 1980, Haensel and Schaeffer *et al.* 1982, Haensel *et al.* 1990). Our ultimate goal is to compute the neutrino emission accompanying the minicollape of a non-rotating neutron star. This work is still in progress. In this paper we shall restrict ourselves to the description of the hydrodynamics of minicollapse.

We describe the metastable state of neutron star by a polytropic equation of state (EOS) $P = \kappa \rho^\gamma$, $\gamma = 2$. Once the phase transition starts, the EOS softens in the metastable core region,

$$P = \kappa \rho^\gamma , \quad \rho < \rho_c$$

$$P = \kappa \rho_c^\gamma , \quad \rho > \rho_c$$

where ρ_c is a critical constant density.

We assume that the phase transition occurs in a characteristic time which is much shorter than the dynamical time. This approximation is surely bad because it means a strongly supersonic propagation of the nucleation of the condensed phase. Of course, a less crude approximation is needed in order to obtain more physically meaningful results. In view of this, what we describe below should be considered just as a good example of application of the multigrid technique.

The initial condition is a metastable equilibrium configuration computed with a two grid code. The first grid spans the density range $\rho > \rho_c$, while the second one refers to $\rho < \rho_c$. The two grid code evolves this initial configuration in time, allowing for a very rapid phase transition in the core with $\rho > \rho_c$. The equations of motion are

$$\partial_t v = -v \partial_r v + \partial_r U + \nu \left(\partial_{rr} v + \frac{2}{r} \partial_r v - 2\frac{v}{r} \right) ,$$

$$\partial_t \rho = -v \partial_r \rho - \rho \left(\partial_r v + 2\frac{v}{r} \right) ,$$

$$\partial_{rr} U + \frac{2}{r} \partial_r U = -4\pi G \rho ,$$

for the inner grid, and

$$\partial_t v = -v \partial_r v - \partial_r w + \partial_r U ,$$

$$\partial_t \rho = -v\partial_r \rho - \rho(\partial_r v + 2\frac{v}{r}) \,,$$

$$\partial_{rr} U + \frac{2}{r}\partial_r U = -4\pi G\rho \,,$$

$$w = \frac{\gamma}{\gamma - 1}\kappa\rho^{\gamma-1} \,,$$

for the outer grid.

Note that the sound velocity vanishes on the inner grid. We have introduced the bulk viscosity, ν, because entropy is generated during the phase transition. On the contrary, the viscosity is neglected on the outer grid.

We need the matching conditions between the two grids. We shall discuss in more detail the analytical properties of the above system of equations. For simplicity, we shall assume that the gravitational potential U is frozen when the phase transition starts. At the time $t = 0$, we put $v = 0$ and

$$\partial_t v = \partial_r U \quad \text{and} \quad \partial_t \rho = 0 \,.$$

Some time later, the velocity $v(r)$ will reach some negative value

$$v(r) \sim \partial_r U dt \leq 0 \,.$$

We see that when $v(r) \neq 0$, the system of equations for the inner grid is hyperbolic with two ingoing characteristics. (Note that the equation for v is decoupled from the equation for ρ). Therefore, this system requires two boundary conditions: one for v and one for ρ, at the interface between the grids (edge of the condensing core). The boundary condition for ρ is $\rho = \rho_c$ (by definition of the edge of the inner grid).

The system of equations on the outer grid is also hyperbolic but with only one ingoing characteristics at the matching point. The boundary condition at the surface of the star R is $P(R) = P_0$ or, equivalently, $\rho(R) = \rho_0$. A second boundary condition has to be given at the matching point. As a result of the above discussion, we have three degrees of freedom v^+, v^- and the velocity of the grid. Mass conservation law and momentum conservation law yield two of the three boundary conditions required. At the discontinuity, we have

$$\rho^+ v^+ = \rho^- v^-$$

and

$$P^+ + \rho^+ v^{+2} = P^- + \rho^- v^{-2} + W^-$$

where W^- is the viscous stress in the inner region.

Moreover, $\rho^- = \rho_c$, $P^- = P_c = P^+$ and $\rho^+ = \rho_c$, by definition of the edge of the inner grid. Consequently, we have $\rho^- = \rho^+ = \rho_c$, $v^+ = v^-$ and $W^- = \nu\partial_r v^- = 0$. The first two relations are a consequence of the mass conservation and the third one is a consequence of the momentum conservation.

Figure 4 shows the velocity and density profiles in the interior of collapsing star at different times. We can see the sonic cone. Outside this cone, the matter "does not know" yet that the phase transition occurs. Notice the rarefaction wave which is propagating outward (fig. 4b). When the central density becomes larger then a given value, the condensation is completed, and the equation of state becomes stiff again. Consequently, the collapsing core bounces. A numerical study of this second, post-bounce stage of neutron star mini-collapse is still in progress.

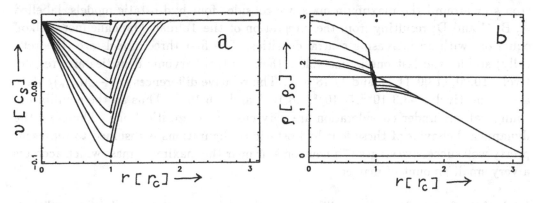

Figure 4. Velocity and density profiles of a minicollapsing neutron star. The radial coordinate is given in the units of the inner core radius. On the outer grid c_s is the local sound velocity, while on the inner grid it is $c_s = c_s(\rho_c)$. Density unit $\rho_0 = 1.67\ 10^{14}\ g\ cm^{-3}$

5 COLLAPSE OF A NEUTRON STAR INTO A BLACK HOLE

In this section we consider the fate of a neutron star which has the maximum allowable mass, M_{max}, and which accretes a small amount of matter. We studied this problem in the non-rotating case, using a complete general relativistic code (Gourgoulhon 1991, Gourgoulhon *et al.* 1991).

This problem is a good example of application of spectral methods when high precision is required. The equilibrium configuration is computed using a general relativistic code in Schwarzschild coordinates by means of a relaxation technique. This equilibrium configuration, combined with a prescription for its pertubation, constitutes an initial data for the time evolution hydro-code.

We use an equation of state for cold catalyzed matter obtained by matching smoothly the EOS of Baym, Pethick, and Sutherland (BPS, 1971) and that of Baym, Bethe, and Pethick (BBP, 1971) with the EOS Diaz II of Diaz-Alonzo (1985). This EOS is described in detail in Gourgoulhon (1991). The resulting EOS gives $M_{max} = 1.924 M_\odot$ for a central baryonic density of $n_{crit} = 11.16 n_{nuc}$ (where $n_{nuc} \equiv 10^{38} \; fm^{-3}$). The corresponding radius is $10.678 km$; it is only 1.87 times larger than the Schwarzschild radius, which means that such an object is highly relativistic.

(i) Evolution toward instability

It is well known that the maximum on the mass versus central density curve corresponds to a loss of stability. Equilibria for which the central density is less than the critical one [$n_{centr} < n_{crit}$, with $M(n_{crit}) = M_{max}$] are stable while those with $n_{centr} > n_{crit}$ are unstable. To simulate the stability change in an accreting neutron star around the maximum mass, we consider four hydrostatic models, labelled A, B, C and D, resulting from the integration of the Tolman–Oppenheimer–Volkoff equation, with an increasing central densities. The first three models are (theoretically) stable, the last one is unstable : their central baryonic densities are, respectively, $10.00, 11.00, 11.12$ and $11.18 \; n_{nuc}$. The relative differences $(M - M_{max})/M_{max}$ are, respectively, $-3.3 \; 10^{-3}, 5. \; 10^{-5}, 3.8 \; 10^{-6}$ and $3.6 \; 10^{-6}$. Thus all the equilibrium configurations under consideration are very close to the critical one. Therefore the dynamical behavior of these four hydrostatic configurations is assumed to represent pretty well successive stages of a neutron star near the maximum mass which accretes a very small amount of matter.

Each of the four hydrostatic equilibrium configurations was taken as initial conditions for the time evolution code. The velocity field was taken to be zero at the initial instant. The analysis of the characteristic directions of the hydrodynamical system shows that there is one ingoing characteristic, so that one boundary condition is required at this point. There is no boundary condition to impose at $r = 0$ but rather, due to the spherical-like coordinates used, regularity conditions. These are automatically satisfied, thanks to a suitable choice of the Chebyshev expansion basis (Galerkin procedure). We choose a constant surface pressure boundary condition. In the ideal case (no numerical errors), since we start from an equilibrium configuration, no subsequent evolution should be seen. Actually, the calculated source term for velocity is not exactly zero, because of round-off errors as well as discretisation errors (which are due to the finite number of Chebyshev polynomials used). So, some evolution in time appears. To be significant, this evolution should be computed over many hydrodynamical times. If the equilibrium is stable, the evolution should be limited to very small amplitude radial oscillations (sound waves). As we will show, when the equilibrium is *unstable* the round–off errors in the code are sufficient to trigger the dynamical instability.

The time evolution of the stellar radius over many dozens of hydrodynamical times is presented in figures 5a, 5b and 5c for the three stable configurations A, B, and C. The major feature is the conservation of equilibrium, modulo radial oscillations whose relative amplitude varies from 2.10^{-6} to 5.10^{-5}. Different modes of these oscillations clearly appear. The longest period corresponds to the fundamental mode. As it can be seen by examination of the velocity profile (or from the fact that, for this mode, the central density oscillates with the opposite phase to the stellar radius), there is no node between the star center and its surface. Thus the fundamental mode is a "breathing" mode: the star oscillates as a whole. The period of the fundamental mode is $T_0 = 1.08ms$, $2.95ms$ and $6.40ms$ for the configurations A, B and C, respectively.

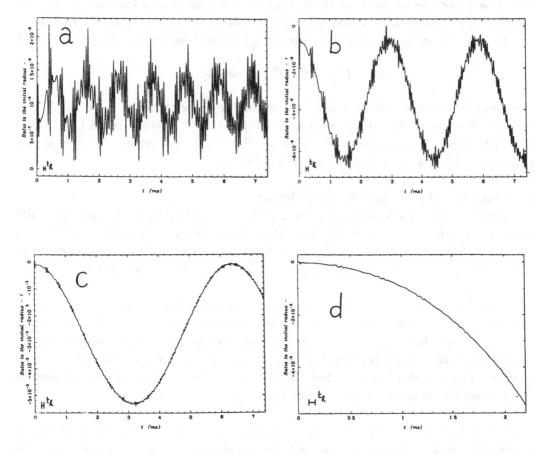

Figure 5

So, it increases when approaching the instability threshold. As it can be seen on figure 5, the relative amplitude of pulsations increases, too. The numerical values are : $1.0 \ 10^{-6}, 6.0 \ 10^{-6}$ and $5.4 \ 10^{-5}$, respectively. On the contrary, the periods and the amplitudes of the higher modes do not change: the periods are around the hydrodynamical time scale and the maximum amplitudes are, in the three cases considered, about $1.5 \ 10^{-6}$.

The time evolution of the star radius in case D, the unstable one, is presented on figure 5d. The higher order modes of radial pulsations are still present and have roughly the same period and amplitude. However, the component of the time evolution which, in the preceding cases, would have correspond to the fundamental mode, has a nonperiodic character: it develops into a collapse which will be analyzed in the following section.

This characteristic behavior of the fundamental mode is well known. A relation between the fundamental period T_0 and the difference between the star radius $R_\star$ and the critical radius (corresponding to M_{max}) is provided by Harrison *et al.* (1965)

$$(R_\star - R_{crit})T_0{}^2 = const \ .$$

This relation holds to within a few percent for the three stable cases A, B, C. The relative differences $(R_\star - R_{crit})$ are respectively $2.7 \ 10^{-2}, 3.7 \ 10^{-3}$ and $8.1 \ 10^{-4}$. We conclude that the fundamental mode of oscillations generated by the code is not spurious and has a theoretically expected behavior.

(ii) Collapse of the unstable equilibrium

For the case D, the initial configuration is beyond the instability threshold, and radial perturbations trigger collapse. The time evolution of the radial velocity field between times $t = 7.267 \ ms$ and $t = 7.377 \ ms$ is shown on figure 6a. The time evolution of the radial profile of the quantities $\sqrt{g_{11}}, \sqrt{-g_{00}}$ are also shown in figure 6.

The calculation stops at $t = t_f = 7.404ms$ because of the singularity in the metric coefficients and resulting very steep gradients in coordinates dependent quantities like g_{11}. Nevertheless, the evolution has gone sufficiently far in time so that no significative later evolution would be achieved, from the point of view of a distant observer ("frozen star" state). Indeed, at $t = t_f$, the star has almost entered its Schwarzschild radius: $R_\star(t_f)/R_s = 1.005$ and $v[R_\star(t_f)] = -0.99924c$.

The total baryon number and mass-energy of collapsing star evolves as shown in figure 6d. The total baryon number is conserved during the early collapse stage with a relative accuracy of $2. \ 10^{-8}$. The relative error remains under $4. \ 10^{-6}$ during the violent collapse stage and increases up to $5. \ 10^{-5}$ in the final sharp gradients phase.

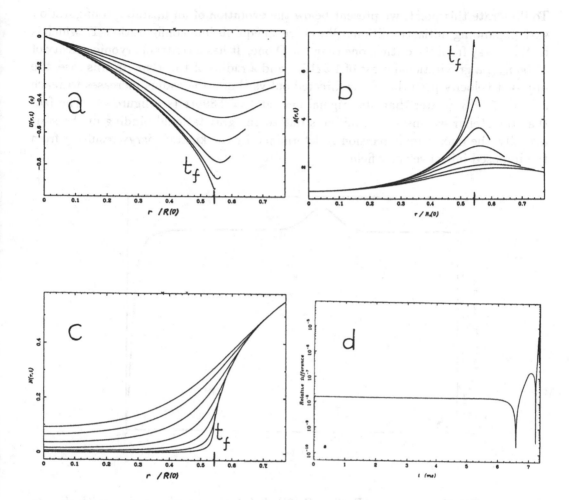

Figure 6. Matter velocity, U, (a); redshift parameter $A = \sqrt{g_{11}}$ (b); lapse function $N = \sqrt{-g_{00}}$ (c), versus radial Schwarzschild coordinate r. Curves corresponding to the last computed configuration are labeled t_f. The relative error in the calculated total baryon number of collapsing star (d).

One might ask the question : why did the unstable equilibrium configuration D collapse instead of exploding ? This is so because the initial numerical round-off errors turned out to generate a globally negative velocity field. Were they positive, the outcome would have been a mini-explosion.

To illustrate this point, we present below the evolution of an unstable configuration with expanding initial conditions. We chose an equilibrium configuration, E, which is further away from the critical one than the D one. It has a central baryonic density of $12.00n_{nuc}$, a gravitational mass of $1.921M_\odot$ and a radius of $10.44km$. In this case, the imposed velocity perturbation is directed outward and the radius increases to reach a value 5.5 % greater than its original one, at $t = 2.89ms$ (see figure 7). The fact that the stellar expansion is limited is due to the gravitational binding of the star: actually, the maximum expansion is determined by the kinetic energy resulting from the imposed initial velocity field.

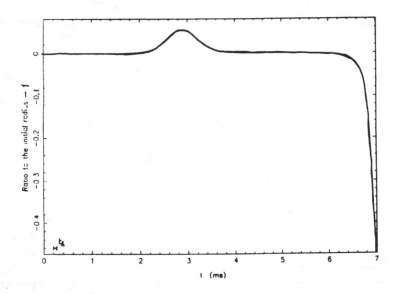

Figure 7. *Relative variation* $[R_\star(t) - R_\star(0)]/R_\star(0)$ *of the stellar radius with time for the equilibrium model E.*

Consequently, the subsequent contraction phase brings the star back to its initial state, but with the inward directed velocity field. This inward velocity field is sufficient to trigger the instability, but this time towards the center. More precisely, when the star comes back to its initial state, radial pulsations are being excited. The star seems "to hesitate" between re-expanding or collapsing, and, in this phase, its evolution is extremely sensitive to numerical round-off errors. Indeed , a calculation with a time step δt greater than the one used for figure 6, gives a re-expanding phase instead of collapse, and starts, in this manner, an infinite sequence of large amplitude oscillations.

The collapse phase is similar to that of configuration D. However, its timescale is shorter. This is a general finding, confirmed by other calculations: the further the unstable equilibrium is from the critical one, the shorter the collapse time.

Finally, we computed the neutrino burst accompanying collapse into a black hole. Figure 8 shows the electron antineutrino light curve and the neutrino mean energy. We see that the total energy of the neutrino burst is about 100 times lower than that in a typical SNII. This would mean that the neutrino burst accompanying a collapse of a neutron star in our Galaxy could (in principle) be detectable (with the largest present day neutrino detectors). On the other hand, the neutrino mean energy is about 4 times less than in the SNII case and below the threshold of the largest present day detectors with high time resolution (Kamiokande, IMB). Therefore, these detectors could not detect such a burst.

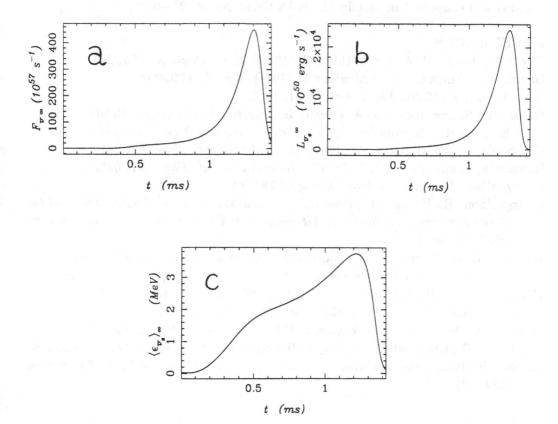

Figure 8. *Electron antineutrino number emission rate (a), energy luminosity (b) and mean energy (c), as measured by an observer at infinity*

6 CONCLUSIONS

We presented some applications of the pseudospectral methods for solving systems of partial differential equations. These methods seem well adapted to perform simulations of various astrophysical phenomena. Their drawbacks can be easily eliminated. These methods are also well suited to handle discontinuities in the solutions (for example acoustic shocks, see Marck and Bonazzola, *this volume*). The main results of astrophysical interest is that Galactic neutron stars collapsing into a black hole do not yield detectable neutrino burst (at least, with the present day large, high time resolution detectors). On the contrary, we estimate that such a phenomenon could be detected in the future, with the next generation of gravitational waves detectors. For the time being, however, we have no observational upper limit on the frequency of the gravitational collapse of neutron stars in our Galaxy.

P. Haensel (on leave of absence from N. Copernicus Astronomical Center, Warsaw, Poland) was supported in part by the KBN Grant No. 2-1244-91-01.

REFERENCES

Baym, G., Bethe, H. A. and Pethick, C. (1971). *Nucl. Phys. A*, **175**, 225.

Baym, G., Pethick C. and Sutherland P. (1971). *Ap. J.*, **170**, 299.

Diaz-Alonzo, J. (1985). *Phys. Rev. D*, **31**, 1315.

Bonazzola, S. and Marck, J. A. (1989). In *Frontiers in Numerical Relativity*, C. R. Evans and L. S. Finn (eds.), Cambridge University Press, Cambridge, 239.

Bonazzola, S. and Marck, J. A. (1990). *Journal Comput. Phys.*, **87**, 201.

Bonazzola, S. and Marck, J. A. (1991). *Journal Comput. Phys.*, **97**, 535.

Gourgoulhon, E. (1991). *Astron. Astroph.*, **252**, 651.

Gourgoulhon, E., Haensel, P., Bonazzola, S. and Marck, J. A. (1991). *Proc. of the Tenth Seminar on Relativistic Astrophysics and Gravitation*, to be published by World Scientific.

Harrison, B. K., Thorne, K. S., Wakano, M. and Wheeler, J. A. (1965). *Gravitation Theory and Gravitational Collapse*, University of Chicago Press, Chicago.

Haensel, P., Zdunik, J. L. and Schaeffer, P. (1986). *Astron. Astroph.*, **160**, 25.

Haensel, P. and Schaeffer, P. (1982). *Nucl. Phys. A*, **381**, 529.

Haensel, P., Denissov, A. and Popov, S. (1990). *Astron. Astroph.*, **240**, 18.

Migdal, A. B., Chernoutsan, A. I. and Mishustin, I. (1979). *Phys. Lett. B*, **83**, 158.

Ramaty, R., Bonazzola, S., Cline, T. L., Kazanas, D. and Meszaros, P. (1980). *Nature*, **287**, 122.

GRAVITATIONAL RADIATION FROM 3-D GRAVITATIONAL STELLAR CORE COLLAPSE

Jean-Alain Marck and Silvano Bonazzola

DARC, Meudon Observatory, Meudon, France

Abstract. Gravitational radiation from the first phase of the gravitational collapse of a stellar core, i.e. the dynamical phase which precedes the formation of a shock and a bounce, is studied by means of a 3-D pseudo-spectral self-gravitating hydro code. It is shown that the efficiency of this process is very low (of the order of a few percent) and insensitive to the equation of state and to whether the initial configuration is axisymmetric, with an initial quadrupole of rotational or tidal origin, or fully asymmetric. An attempt to treat shock waves in asymmetric situations is described and preliminary results obtained from stellar core bounce are presented.

1 INTRODUCTION

The gravitational collapse of a stellar core is one of the sources of gravitational radiation which is likely to be detected by the next generation of interferometric gravitational wave detectors (e.g. *VIRGO* and *LIGO* projects). However, we need an accurate prediction of the gravitational wave form for a wide range of collapse models in order to interpret the results of the gravitational wave observations.

During the last decade, various attempts have been made to predict the efficiency of this process and to predict wave forms. Most of these papers are based on numerical simulations. Some of them take account of the microphysics and some of them were performed in the framework of General Relativity. However, most of these preliminary works assumed axisymmetry (a complete review of this field can be found in Finn (1989)). Now, it is well known that the efficiency (e.g. in the energetic sense) will be zero for any spherically symmetric system (Birkhoff's theorem 1923) and the previous numerical simulations show that the energy efficiency in axisymmetric configurations is very low and cannot exceed an order of magnitude given by 10^{-7}. The fact that the efficiency is so much smaller than would be expected on purely dimensional grounds raises the following questions. Is this low value of efficiency attributable to the restriction of axisymmetry, in other words, is it possible to increase dramatically the gravitational efficiency of the collapse when dealing with fully 3–D collapse models ? Can the equation of state have a significant influence on the final outcome ?

The work reported in this paper is based on results obtained by a Newtonian pseudo-spectral 3–D hydro code (Bonazzola and Marck 1990 and Bonazzola et al., this volume) in order to evaluate the importance of the deviation from axisymmetry on the efficiency of the gravitational radiation process during the infall epoch. We note that most of the previous works, for which axisymmetry was assumed, give quantitative results in term of an *energetic* efficiency defined by the ratio of the amount of radiated energy to the total energy of the core. However, as pointed out by Thorne (1987), an interferometric gravitational wave detector does not measure directly the energy radiated but measures the fluctuation of the metric tensor which, in the linear approximation of the General Relativity, is given by

$$h_{ij} = \frac{2G}{c^4} P_{ij}{}^{kl} \ddot{Q}_{kl}$$

where D is the distance of the source, $\mathbf{P}$ is the projection tensor and Q_{ij} is the trace-free part of the quadrupole momentum of the source. Consequently, we have focused attention on the wave amplitude rather than the amount of radiated energy during the collapse and we give quantitative results for h in terms on an efficiency ϵ defined by means of h. (Note that, in the particular case for which the anisotropy of the core is axisymmetric due to rotational effects, this efficiency ϵ is closely related to the energetic efficiency of the process as described in details in Bonazzola and Marck 1992).

Section 2 of this paper gives a clear answer to the above questions as far as no bounce is concerned. Hence, defining the efficiency ϵ by means of

$$h = 1.4\ 10^{-19} \delta\epsilon \frac{1 Mpc}{D} \frac{M_\star}{1.5 M_\odot}$$

where h is the metric fluctuation measured by a gravitational dectector at a distance D from the source and δ is a dimensionless parameter characterizing the anisotropy of the core. It is shown that

- ϵ is insensitive to the equation of state,
- ϵ is insensitive to the value of the anisotropy parameter δ or to whether the anisotropy results from axisymmetric *or* non-axisymmetric effects,
- ϵ is very low, of the order of a few percent.

The above results hold only if no bounce is assumed. The gravitational energy emitted during the bounce should be more sensitive to the free parameters defining the initial core and, of course, to the equation of state. However, dealing with shock waves is (numerically speaking) a difficult task if one wants to give qualitative results which are not affected by the numerical tricks used to treat shock formation and propagation.

Most of the multi-dimensional codes use an artificial viscosity (which may be explicitly included in the numerical code or not) in order to smooth out the solutions. The main effect of this trick is to change the time-scale of the process. Now, if this artificial change does not have significant effects on the solution when only hydrodynamics is concerned, it appears that, in the study of gravitational collapse, such a technique may lead to dramatic errors in the final results (see e.g. Martí et al., (1990) and Ibáñez et al., this volume).

In section 3, we present a way of dealing with multi-dimensional shock formation and shock propagation in the framework of spectral methods. This technique is applied to the resolution of the 1-D Euler equations with periodic boundary conditions and to the simulation of a highly anisotropic core bounce.

2 THE INFALL

The gravitational collapse of a stellar core starts as a result either of photodissociation of iron nuclei or by electron capture (the so-called β–process). During this phase, the pressure decreases and is no longer able to balance gravity. The infall phase stops when the matter density reaches nuclear density (about $1.66 \ 10^{14} g.cm^{-3}$) and is followed by a bounce with matter ejection and/or the formation of a compact remnant (a black hole or a neutron star) with oscillations of the core (Saenz and Shapiro 1978).

The behaviour of the matter at and above the nuclear density is uncertain. However, during the infall phase, the inner core of the star can be modelled as a polytrope with index of order of 1.29. Moreover, as discussed in the next section, shock formation and propagation is hard to follow with the usual numerical codes. Consequently, we focus our attention on the first stage of the collapse and give the maximum value of the gravitational wave amplitude just before the core bounces, that is, when the central density reaches nuclear density.

The results presented in this paper were computed in the Newtonian approximation. In this approximation, the fluctuations of the background space-time, the quantity dectected by the interferometers, are given by (see e.g. M.T.W. 1973, Blanchet et al. 1990)

$$h_{ij} = \frac{2G}{c^4 D} P_{ij}{}^{kl} \ddot{Q}_{kl} \ ,$$

where D is the distance of the source and Q is the trace-free part of the quadrupole momentum of the core. However, especially when dealing with asymmetric calculations, it is more judicious to characterize the gravitational radiation process by its average value h, defined as

$$h = \int h_{ij} h^{ij} d\Omega \ ,$$

which is closely related to the polarization modes h_+ and $h_\times$ of the wave. It turns out that h can be expressed in term of the second time derivatives of Q as

$$h = \frac{2G}{c^4 D} \sqrt{\frac{8\pi}{3}} \ddot{Q}_{ij} \ddot{Q}^{ij} \ .$$

From the previous definition, we introduce the efficiency factor ϵ (not to be confused with the energetic efficiency of the process) by means of

$$h = 1.4 \ 10^{-19} \delta \epsilon \left(\frac{1 Mpc}{D}\right) \left(\frac{M_\star}{1.5 M_\odot}\right) \ ,$$

where δ is the dimensionless parameter characterizing the anisotropy of the core when the central density reaches nuclear density, i.e.

$$\delta = \frac{\sqrt{Q_{ij} Q^{ij}}}{I} \ ; \quad I = \int \rho r^2 d^3 x \ .$$

We have performed numerical simulations of the evolution of a rotating stellar core embedded in an external tidal potential. The initial conditions, which may play an important role in the final results (see e.g. Finn 1989) are steady state equilibrium configurations which are computed by relaxation as described in Bonazzola and Marck 1989 and Bonazzola et al., this volume. Several models have been examined for various values of the angular momentum $\Omega \in [0, \Omega_{max}]$ and of the external tidal potential $C \in [0, C_{max}]$ where Ω_{max} and C_{max} are the maximum values for which equilibrium exists. The collapse is then started by means of a decrease of the pressure. The evolution is then computed by numerical integration of the following system of equations

$$\frac{\partial \rho}{\partial t} = -\text{div}(\rho \mathbf{v}) \ ,$$

$$\frac{\partial \mu}{\partial t} = -\mathbf{v} \nabla \mu - (\gamma - 1) \text{div} \mathbf{v} \ ,$$

$$\frac{\partial \mathbf{v}}{\partial t} = -\mathbf{v} \cdot \nabla \mathbf{v} - \nabla \mu - \nabla U - \nabla C + \nu \Delta \mathbf{v} \ ,$$

$$\Delta U = -4\pi G \rho \ ,$$

with appropriate boundary conditions, where μ is the heat function defined by

$$\mu = \int \frac{dP}{\rho} \ .$$

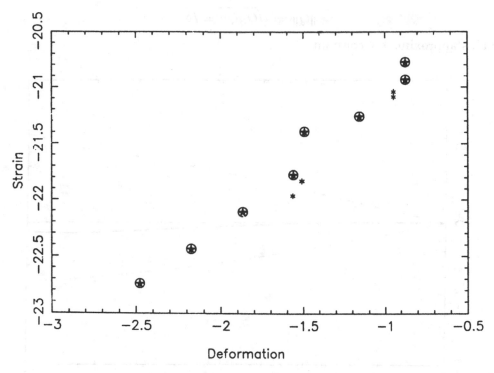

Figure 1. Plot of the average gravitational wave amplitude when the central density reaches nuclear density, as a function of the deformation parameter δ (log − log scales). The star symbol denotes purely axisymmetric calculations.

From the results plotted in figure 1, it appears that the efficiency ϵ is roughly constant, of the order of a few percent, and does not depend whether the anisotropy has an axisymmetric or triaxial origin. The only way we found of slightly increasing the amount of energy radiated for a given value of the anisotropy parameter δ is by altering the distribution of initial angular momentum (which confirms previous results obtained in the axisymmetric case by Müller 1984). In order to save computational time, these simulations involved a compression factor (the ratio of the initial and final central densities) which was rather low. For some of these calculations, we increased the compression factor (leading to very time consuming calculations) but, even for large compression factors, the efficiency ϵ still appeared to be roughly constant.

This rather low value of the efficiency results from a subtle compensation between the variations of the moment of inertia I and the deformation parameter δ during the infall stage, even for fully triaxial deformations. As shown in figure 2, the parameter δ increases during the collapse while the moment of inertia decreases in such a way that the product

$$\|Q\| := \sqrt{Q_{ij}Q^{ij}} = I\delta$$

remains approximately constant.

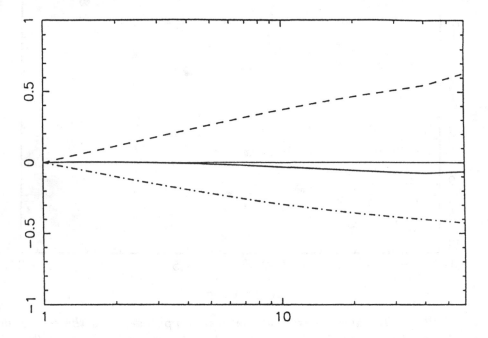

Figure 2. The relative variations of the quadrupole momentum $\|Q\|$ (solid line), the average eccentricity δ (dashed line) and the moment of inertia I (dot-dashed line) versus the compression factor.

The approximate constancy of the the norm $\|Q\|$ can obviously be interpreted as meaning that the time scale characterising its rate of variation is very long compared to the relevant free fall time scale, and it is hard to see how individual quadrupole components could have a more rapid variation than that of the norm if they are also constrained to axisymmetry, as will be the case if the only source of asymmetry is the initial rotation of the core. This does not by itself absolutely exclude the possibility of rapid variations that cancel in $\|Q\|$ but not in $\|\ddot{Q}\|$ for suitable triaxial configurations, but no significant effect of this sort has turned up in the simulations we have made so far, based on the assumption that the only deviation from axisymmetry is the modest effect of an external tidal field as would be the case in a binary system.

3 NUMERICAL TREATMENT OF SHOCKS

The handling of shock waves is one of the most difficult tasks in numerical hydro-dynamics, the troubles becoming even worse in numerical relativity. Basically, the problem arises because almost all the numerical codes compute spatial derivatives in order to solve the partial differential equations which is incompatible with the fact that, in cases of shock waves, the quantities are no more derivable. Various methods have been developed to overcome this difficulty. The main idea is to introduce some kind of artificial viscosity which acts as a local modification of the equation of state or to use intrinsic viscosity by means of a diffusive numerical method. However, as it has been pointed out by Martí et al. (1990) and Ibáñez et al. (this volume) the use of such tricks may lead to catastrophic errors in the final results of gravitational problems, even in the simple 1–D case.

Apart from some particular techniques of shock tracking and shock capture (see e.g. Bonazzola and Marck 1991 for applications to the problems for which we are concerned), which are useful only in special cases, there have been some attempts to handle shock formation and shock propagation in general situations, using [pseudo-] spectral methods.

The first approach is to add a dissipative operator to the source terms of some of the evolution equations of the system. The most natural operator is the Laplacian, which is added to the velocity equation. This technique (the so-called *natural viscosity*) has the advantage of smoothing out the desired solution adequately while it preserves the physical properties of the solutions. However, in this case, the main advantage of spectral methods (that the global error on the solution decreases exponentially with the number of degrees of freedom) is lost. Another approach has recently been developed by Maday et al. (1989). These authors suggest adding a dissipative operator which acts as an ordinary Laplacian, but only to the higher frequencies of the source terms in the evolution equations : the so-called vanishing viscosity. Typically, the $N/3$ highest frequencies of the solution are used in practice. The resulting solution is smooth enough to be handled by the code and spectral convergence is ensured. This technique leads to an oscillating numerical solution from which the physical solution can be extracted, when needed, by means of convolutions which are computed for each grid point. It is obvious (in view of the time computation) that the physical solution cannot be computed at every time step. The last remark shows that this technique can be applied only with extreme care in problems where non-linearities are dominant. A way of taking advantage of the beautiful properties of the vanishing viscosity, while keeping the computation time of highly non linear terms reasonable, is under development.

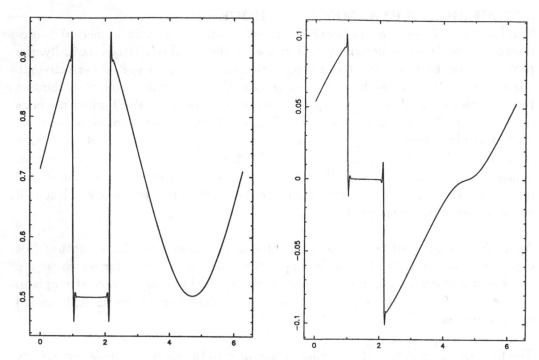

Figure 3. Plot of the density and velocity profiles when two shocks are present.

The second approach involves periodically filtering the solution. Several filters have been suggest by different authors. The most common ones can be found in McDonald (1989) and in references quoted therein. However, even if this technique leads to very accurate results when dealing with pure hydrodynamics, experience shows that, when dealing with gravitational problems, the accuracy of the solution is far from that one can expect from a spectral code.

In view of the above remarks, we tested a numerical technique which is intermediate between these two approaches. The idea is to replace the time derivatives in the evolution operators as

$$\frac{\partial}{\partial t} \longrightarrow \frac{\partial}{\partial t} + \nu \Delta^2$$

where Δ^2 is a spatial differential operator which acts as a squared Laplacian.

Such a technique has already been applied with success to hydrodynamics problems with large Reynolds number. It turns out that rewriting the evolution operators in this way is equivalent to seeking a solution for which the second spatial derivative is bounded. Application of this technique to a 1–D hydrodynamical problem with periodic boundary conditions is presented in figure 3. The initial conditions describe a smooth non-uniform initial density profile associated with a vanishing velocity field. The initial gradient of the pressure leads to an acceleration of the matter and eventually, because of the plane geometry, to the formation of shocks. Figure 3 is a snapshot of the density and velocity profiles at an instant where two shocks are present. One can show that this solution is of spectral accuracy except in a neighbourhood of the shock where the accuracy of the solution is of finite order (note the undershooting and overshooting on both sides of the shocks).

The technique described above has been applied to a highly anisotropic stellar core collapse. This calculation, which at present is no more than an illustration of the method, has been performed with the following equation of state

$$P := P(\rho) = \kappa_1 \rho^{\gamma_1} + \kappa_2 \rho^{\gamma_2} \ ,$$

with $\gamma_1 = 1.29$ and $\gamma_2 = 3$. The numerical code is the 3–D numerical code described in BM90 running on a moving grid with adaptive time step where we included the operator

$$\sqrt{(1 - x^2)} \Delta \{ \sqrt{(1 - x^2)} \Delta \{ \cdot \} \} \ ; \quad x \in [-1, +1]$$

which is treated in an implicit way. This simulation describes the collapse of a rapidly rotating stellar core in a strong external tidal potential. At the end of the dynamical phase, when the central density reaches nuclear density, the core starts to bounce in the $x - y$ direction while the matter is still falling in the z direction. Then, when the central density is high enough, a bounce occurs in the z direction. The dephasing of these two process leads to a second collapse and a second bounce in the $x - y$ direction which make the core oscillate. The second time derivatives of the components of the quadrupole moment (from which the above scenario can be deduced) are presented in figure 4. Comparison with results obtained from an axisymmetric calculation for which the initial deformation parameter δ had the same value, show that the efficiency of the bounce is about twice as large in the 3–D case as in the 2–D case.

4 CONCLUSION

We have shown in this paper that the gravitational efficiency of the infall stage of a stellar core collapse is rather small, roughly constant of a few percent and, moreover, that this efficiency does not depend on the way in which spherical symmetry is broken or on the equation of state used during the first epoch of the collapse. We showed that this surprising result, which holds when most of the gravitational energy is involved, results from a subtle compensation between the variation of the moment of inertia I and the deformation parameter δ measuring the anisotropy of the core.

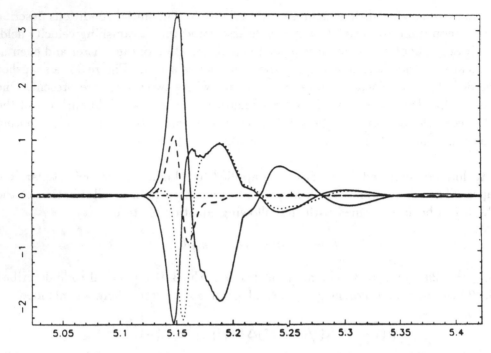

Figure 4.   The second time derivatives of the trace-free parts of the quadrupole moment. Note that two of these components vanish because of the planar symmetry assumed in this calculation.

The numerical values of the *averaged wave amplitude h* which we obtained in terms of the deformation parameter δ gives us a rule of thumb

$$h = \frac{10^{-21}}{D}\,\delta$$

where D is the distance to the source measured in Mpc; note that δ cannot exceeds 0.1. We emphasise that the efficiency law derived above is only valid for the infall epoch of the collapse, before a bounce occurs. However, one cannot expect a much higher emission rate during the bounce because most of the energy of the core is lost to neutrino emission.

The bounce phase, a crucial step of the collapse, must be handled carefully. It is well known that a strong shock is formed during this phase and handling it is a real problem. In our opinion, this important problem is often overlooked in the literature not only in axisymmetric numerical simulations but also in the simpler spherically symmetric cases. The astonishing result of Ibáñez et al. (this volume) supports this view. Bearing this problem in mind, we have performed numerous tests to show that our results on the infall phase are not altered by numerical viscocity.

Work is in progress in order to control the effects of the artificial viscocity used in the post-bounce phase. Despite the fact that this work is not yet finished, we are quite confident of the result suggesting an extra factor of two between the values of the efficiency in axisymmetric and full 3–D cases.

ACKNOWLEDGMENTS

All the numerical simulations providing the physical results presented in this paper were peformed on a *Silicon Graphics* 4D–320 GTI computer. We are grateful for the support of the S.M.P. and I.N.S.U. departments of the C.N.R.S.

REFERENCES

Birkhoff, G. (1923). *Modern Physics*, Harvard University Press, Cambridge, MA.

Blanchet, L., Damour, T. and Schäfer G. (1990). *M.N.R.A.S.*, **242**, 249.

Bonazzola, S. and Marck, J. A. (1989). In *Frontiers in Numerical Relativity*, C. R. Evans and L. S. Finn (eds.), Cambridge University Press, Cambridge, 239.

Bonazzola, S. and Marck, J. A. (1990). *Journal Comput. Phys.*, **87**, 201.

Bonazzola, S. and Marck, J. A. (1991). *Journal Comput. Phys.*, **97**, 535.

Bonazzola, S. and Marck, J. A. (1992). Submitted to *Astron. and Astrophys.*

Bonazzola, S., Gourgoulhon, E., Haensel, P. and Marck, J. A. (1992). This volume.

Finn, L. S. (1989). In *Frontiers in Numerical Relativity*, C. R. Evans and L. S. Finn (eds.), Cambridge University Press, Cambridge, 126.

Ibáñez, J.M., Martí, J. M., Miralles, J. A. and Romero, J. V. (1992). This volume.

Maday, Y. and Tadmor, E. (1989). *S.I.A.M. Jour. of Numer. Anal.*, **26**, 854.

Martí, J. M., Ibáñez, J. M. and Miralles, J. A. (1990). *Astron. Astrophys.*, **235**, 535.

McDonald, B. E. (1989). *Journal Comput. Phys.*, **82**, 413.

Misner, C., Thorne, K. and Wheeler, J. A. (1973). *Gravitation*, Freeman, San Francisco.

Müller, E. (1984). In *Problems of Collapse and Numerical Relativity*, D. Bancel and M. Signore (eds.), NATO ASI Series, vol. 134, 271.

Saenz, R. and Shapiro, S. (1978). *Ap. J.*, **221**, 286.

Thorne, K. (1987). In *300 Years of Gravitation*, S. W. Hawking and W. Israel (eds.), Cambridge University Press, Cambridge.

A VACUUM FULLY RELATIVISTIC 3D NUMERICAL CODE

C. Bona and J. Massó

Department of Physics, University of the Balearic Iles, Palma de Mallorca, Spain

Abstract. A numerical code for the vacuum Einstein Field Equations is presented. The full three-dimensional case is considered. The evolution equations are written as an hyperbolic system of balance laws and the second order finite differencing Mac-Cormack and Lax-Wendroff methods are used in an straightforward way. The code is tested by reproducing an analytical solution which can simulate inhomogeneous pancake collapse with periodic boundary conditions.

1 INTRODUCTION

In building a Numerical Relativity code to follow the evolution of the space-time, one must deal with a number of problems:

1. Coding Einstein's Field Equations.

2. Evolving the matter fields.

3. Setting up realistic initial data.

4. Imposing physical boundary conditions.

We have started our way by addressing just the first point. This means that we have built a code to evolve vacuum space-times which admit periodic boundary conditions and we have tested the code by evolving known (analytical) solutions. There are two well known requirements that we have tried to fulfill:

1. Singularity Avoidance.

2. Hyperbolicity of the evolution system.

and these coordinate-dependent features have oriented our choice of the coordinate system, as it is explained in what follows.

2 THE HARMONIC SLICING

As a first condition, we have written the line element in the form

$$ds^2 = -\alpha^2 dt^2 + g_{ij} dx^i dx^j \tag{1}$$

and this means that the time lines are orthogonal to the t =constant slices so that the mixed space-time components of the line element vanish (zero shift vector). This simple choice ensures that the congruence of temporal lines will be as regular as the time slicing itself.

The time slicing is governed by the lapse function α in (1). We will follow the prescription [1] of imposing the following evolution equation for the lapse,

$$\partial_t(\alpha/\sqrt{g}) = 0 , \tag{2}$$

which leads to the algebraic condition

$$\alpha = C(x)\sqrt{g} , \tag{3}$$

where g stands for the determinant of the three-dimensional metric g_{ij} and $C(x)$ is an integration constant.

Some authors would use the term "algebraic gauge" to refer to this coordinate choice. We will avoid using this term because it may be a little too general. A large number of algebraic gauges have already been introduced in numerical relativity. For this reason we have used the more specific term "harmonic slicing" to emphasize that the time coordinate is actually harmonic in that gauge:

$$\Box x^0 \equiv -g^{bc}\Gamma^0_{bc} = 0. \tag{4}$$

Equation (3) shows that the time evolution will slow down whenever the space volume element $\sqrt{g}$ decreases and this is the key for showing the singularity avoidance properties of the harmonic slicing. Let us consider for instance the time line with local equation $x^i = x_0^i$ and let us use the proper time τ to label points along this line. Let us suppose now that the three-dimensional volume element $\sqrt{g}$ vanishes at a given value τ_0 of τ ("crushing" singularity). The three-dimensional metric g_{ij} stops being invertible there and the evolution algorithm crashes.

The coordinate time elapsed before the crash occurs is given by the improper integral

$$t = \int^{\tau_0} 1/\alpha(\tau, x_0^i) \ d\tau \tag{5}$$

and singularity avoidance means that this integral diverges so that the singularity is not reached in a finite number of time steps. Allowing for (3), it is easy to show

[2] that the harmonic slicing avoids crushing singularities provided that the (proper time) rate of decreasing of the volume element does not diverge.

Note that singularity avoidance does not mean absence of coordinate singularities, as it is shown by the following example:

$$ds^2 = e^{-2t}(-dt^2 + dx^2) + dy^2 + dz^2 \ , \tag{6}$$

which is a Kasner-like form of the Minkowski metric; a coordinate singularity appears at a finite value of the proper time which is avoided by the harmonic slicing (it does not appear at any finite value of the coordinate time). The unphysical slicing is produced by an unphysical choice of initial conditions (the time derivative of the vacuum metric is chosen to be spatially constant).

3 A HYPERBOLIC EVOLUTION SYSTEM

The vacuum Einstein Field Equations can be written [3]:

$$\frac{1}{2}\left(\Box g^{ab} + \partial^a \Gamma^b + \partial^b \Gamma^a\right) - \Gamma^{acd}\Gamma^b_{cd} = 0 \ , \tag{7}$$

and the evolution system is constructed from the spatial components of (7). In the harmonic slicing, it can be written as a first order system of balance laws [4] in the quantities

$$Q^{ij} \equiv \partial_t g^{ij} \ , \quad D^{ij}_k \equiv \partial_k g^{ij} \tag{8}$$

as follows

$$\partial_t \left[\frac{\sqrt{g}}{\alpha}Q^{ij}\right] - \partial_k \left[\alpha\sqrt{g}(D^{kij} + g^{ki}\Gamma^j + g^{kj}\Gamma^i)\right] =$$

$$\frac{\sqrt{g}}{\alpha}Q^{ik}Q^j_k - 2\alpha\sqrt{g}\left[\Gamma^{ikl}\Gamma^j_{kl} + L^i L^j - \Gamma^i\Gamma^j\right] \ , \tag{9a}$$

$$\partial_t \left[D^{ij}_k\right] - \partial_k \left[Q^{ij}\right] = 0 \ , \tag{9b}$$

where we have noted

$$L^i \equiv \partial^i \ln \alpha \ , \tag{10a}$$

$$\Gamma^i \equiv -\Box x^i = \frac{1}{2}g_{jk}D^{ijk} - D^{ki}_k - L^i \ . \tag{10b}$$

One could complete the system (9) with the coordinate condition (2) and its spatial derivatives

$$\partial_t L_i + \frac{1}{2}\partial_i Q^k_k = 0 \tag{11}$$

to evolve α and L^i, respectively. The complete system (9,2,11) has the balance law structure [5]:

$$\partial_t D(u) + \partial_k F^k(u) = S(u) \ , \tag{12}$$

where the densities D, fluxes F^k and sources S are vector valued functions of the set of variables

$$u = (\alpha, L^i, g^{ij}, Q^{ij}, D_k^{ij}) \,. \tag{13}$$

The form (12) of the evolution system is well known to those who numerically solve the Navier-Stokes equations for hydrodynamics and standard numerical methods for this kind of systems have been developed. However, the consistency and stability of all such methods have been established only for hyperbolic systems, that is, systems in which the characteristic matrix (the Jacobi matrix of the projection of the Fluxes along a given direction) has real eigenvalues and a full set of independent eigenvectors. Actually, the most powerful methods do use the spectral decomposition of the characteristic matrix to obtain a finite difference version of the equations. And the complete evolution system for (13) is not hyperbolic [6].

A hyperbolic first order system in $\partial_t Q^{ij}, \partial_t D^{kij}$ can be obtained [1] by taking the time derivative of (9) and noticing that the mixed space-time components of (7) (the "momentum constraints" in our gauge) can be written as:

$$\partial_t \Gamma^i = Q_k^k L^i - 2Q_j^i L^j + \Gamma_{jk}^i Q^{jk} \,, \tag{14}$$

so that it can be used to eliminate the Γ^i terms in the principal part of the derived system.

We shall follow a much simpler way by considering (9) as an evolution system for the quantities

$$u = (\alpha, \Gamma^i, g^{ij}, Q^{ij}, D_k^{ij}) \tag{15}$$

(note that Γ^i has replaced L^i in (13)) and completing it with (2) and (14) (instead of (11), which is no longer used) so that L^i is obtained algebraically from Γ^i by using (10). The complete system (9,2,14) in the quantities (15) is hyperbolic as we will show in what follows.

4 THE TRANSPORT TERMS

To study the hyperbolicity of a system of the form (12) we need to consider only the principal part (densities and fluxes). This is closely related to the "operator splitting" approach in numerical applications, in which the time evolution is decomposed into a combination of two kinds of processes, namely:

i) transport, governed by the flux-conservative law

$$\partial_t D(u) + \partial_k F^k(u) = 0 \,. \tag{16}$$

ii) evolution, driven by the source terms only.

The hyperbolicity of the system can ensure the stability of the numerical methods used in the transport step only. The stability of the whole scheme would require in addition that of the system of nonlinear ordinary differential equations given by the source terms. This is beyond the scope of this work, where we will be concerned with the transport step only.

Going back to our evolution system, let us note that the ten quantities α, Γ^i, g^{ij} are constant in the transport step (they do not have flux terms) and that the evolution of the remaining twenty-four quantities can be written then

$$\partial_t \left[\frac{\sqrt{g}}{\alpha} Q^{ij} \right] - \partial_k \left[\alpha \sqrt{g} (g^{kl} \lambda_l^{ij}) \right] = 0 , \tag{17a}$$

$$\partial_t [\lambda_k^{ij}] - \partial_k [Q^{ij}] = 0 , \tag{17b}$$

where we have noted

$$\lambda_k^{ij} \equiv D_k^{ij} + \delta_k^i \Gamma^j + \delta_k^j \Gamma^i . \tag{18}$$

The interest of this change of variables (18) resides in the fact that the resulting system (17) consists of six uncoupled identical subsystems of four linear equations in Q^{ij}, λ_k^{ij} (with i, j fixed). This is important if one is planning to apply either spectral or implicit finite difference methods in the transport step, because it greatly reduces the dimensions of the matrices involved.

But the main point is that every subsystem in (17) has the structure of the wave equation in first order form. This shows that the evolution system is hyperbolic with characteristic surfaces given by the light cones (plus the time lines if one takes into account the quantities which are constant in the transport step), providing a good starting point for applying standard existence and uniqueness theorems. This also means that the arsenal of numerical methods designed and tested for the three-dimensional wave equations is at our disposal: this is a fast way of coming in on three-dimensional Numerical Relativity because one does not need to build an *ad hoc* code from scratch!

5 A 3D NUMERICAL SIMULATION

We have developed a code that implements standard second order finite differencing methods (MacCormack and Lax-Wendroff) to the system of wave equations (17). We have checked the code by reproducing the following form of the metric (6):

$$ds^2 = e^{-2t} \left[-dt^2 + (1 + c_1 \sin x) dx^2 \right] + (1 + c_2 \sin y) dy^2 + (1 + c_3 \sin z) dz^2 ,$$

which simulates an "inhomogeneous" pancake collapse with periodic boundary conditions, the spacing vectors going from $-\pi$ to π.

We give in the Figure our results for an "inhomogeneous" case ($c_1 = 0.3, c_2 = 0.2, c_3 = 0.1$) using various grid sizes (N is the number of points, which has been taken the same in each direction) with a MacCormack method. The calculation proceeds up to a freezing point where the lapse function gets very close to zero and the numerical errors start growing very quickly. In the finer grid, the freezing point is reached when the difference between static and collapsing components is of 12 orders of magnitude, with a maximum relative error on the lapse less than a 1%. The results for the Lax-Wenfroff method are similar, but the freezing point is reached a bit earlier.

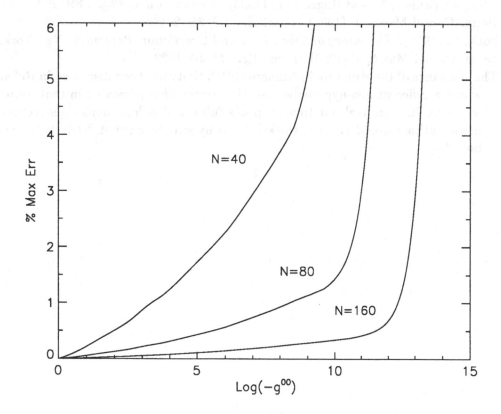

In the "homogeneous" case ($c_i = 0$) the PDE's are simply ODE's and we test the sources evolution, which can be followed up to 20 orders of magnitude (which is the limit set by the machine precision, considering that some square operations are involved) with a maximum error on the lapse function depending on the time step, typically less than a 1% for time steps equal to the ones used previously with the finer grid.

The code is written in the C language and uses dynamical memory allocation. The present performance (not optimized) is $7 \; 10^{-4}$ CPU seconds per zone-cycle in a 9 Mflops machine (the VAX-9000 at the Universitat de les Illes Balears).

ACKNOWLEDGMENTS

This work is supported by the Dirección General para la Investigación Científica y Técnica of Spain under project PB90-0516.

REFERENCES

[1] Choquet-Bruhat, Y. and Ruggeri, T. (1983). *Comm. Math. Phys.*, **89**, 269.

[2] Bona. C. and Massó, J. (1988). *Phys. Rev. D*, **38**, 2419.

[3] Fock, V. (1959). *The theory of Space Time and Gravitation*, Pergamon, New York.

[4] Bona, C. and Massó, J. (1989). *Phys. Rev. D*, **40**, 1022.

[5] The relevance of the structure in Numerical Relativity has been discussed in Ref.4, where a different non-hyperbolic evolution system was presented in that form.

[6] The eigenvalues are real but there is not a full set of independent eigenvectors. Some authors would say that the system is hyperbolic but not "strictly hyperbolic".

SOLUTION OF ELLIPTIC EQUATIONS IN NUMERICAL RELATIVITY USING MULTIQUADRICS

M. R. Dubal, S. R. Oliveira and R. A. Matzner

Center for Relativity, University of Texas at Austin, Austin, US

Abstract. We use the multiquadric approximation scheme for the solution of a three-dimensional elliptic partial differential equation occurring in $3 + 1$ numerical relativity. This equation describes two-black-hole initial data, which will be a starting point for time-evolution computations of interacting black holes and gravitational wave production.

1 INTRODUCTION

Adopting the Arnowitt-Deser-Misner (ADM) $3 + 1$ description of general relativity (1962) has, over the years, proved to be a fruitful approach for numerical relativity calculations. Using this description spacetime is constructed as a foliation of space-like hypersurfaces. This split into space plus time leads to a constrained system of equations so that initial data must be specified on a spatial hypersurface and evolved into the future. The specification of initial data necessarily involves the solution of elliptic partial differential equations; these being the Hamiltonian and momentum constraints. When combined with York's conformal approach (1979) the system of elliptic equations is well-posed for solution by numerical techniques.

Until a few years ago the standard approach adopted by numerical relativists for the construction of initial data consisted of finite-differencing the constraint equations and applying iterative techniques, such as simultaneous-over-relaxation, to the resulting matrix of algebraic equations. More recently direct matrix solvers such as conjugate gradient and its variations have been employed (Evans 1986; Oohara and Nakamura 1989; Laguna et al 1991). A sophisticated multilevel iterative scheme developed by Brandt (1977) has also been used in numerical relativity calculations, mainly by Choptuik (1982, 1986), Lanza (1986, 1987) and Cook (1990, 1991). This multigrid technique is currently the most efficient method for the solution of non-linear elliptic equations.

The approach to discretizing the partial differential equations has also changed in recent years; finite-differencing is no longer the automatically 'preferred' choice. Finite element and spectral methods can now compete on level terms with finite-difference schemes in numerical relativity calculations. The work described here investigates another method of obtaining a discrete representation of the constraint equations. This method is a type of global spectral technique known as the multiquadric (MQ) approximation scheme. Its major advantage is the ability to place grid points anywhere in the solution domain in any order. As shown here, for the construction of initial data of two colliding black holes this property can be very useful.

Next the MQ scheme is described within the wider context of radial basis functions. In order to demonstrate how it is utilized, in section 3, the solution of a model test problem is shown. An application of MQ to the construction of two-black-hole initial data is the subject of section 4. This involves the solution of a single elliptic equation (the Hamiltonian constraint). There are a number of numerical difficulties associated with the use of MQ, mainly due to the ill-conditioning of the resulting matrix of algebraic equations. These difficulties are discussed in section 5 along with suggestions for improvements and possible remedies. Some conclusions are drawn in section 6.

2 RADIAL BASIS FUNCTIONS AND MULTIQUADRICS

Using the radial basis function approximation a real valued function of d independent variables; $f(\mathbf{x})$; $\mathbf{x} \in \mathbf{R}^d$, is written as

$$f(\mathbf{x}) \approx \hat{f}(\mathbf{x}) = \sum_{j=1}^{N} a_j \phi(\|\mathbf{x} - \mathbf{x}_j\|),\tag{1}$$

where $\{\mathbf{x}_j : j = 1, 2, \ldots, N\}$ is a general set of data points in $\mathbf{R}^d$ and $\|\mathbf{x} - \mathbf{x}_j\|$ is the Euclidean vector norm. The coefficients, a_j, in the expansion are defined via the equations

$$\hat{f}(\mathbf{x}_j) = f(\mathbf{x}_j).\tag{2}$$

There are a number of possible choices for $\phi(\|\mathbf{x}-\mathbf{x}_j\|) = \phi(r_j)$; some of those currently under investigation by numerical analysts (see eg. Powell 1990) include:

$\phi(r_j) = r_j$	Linear	$\phi(r_j) = r_j^2 \ln r_j$	TPS
$\phi(r_j) = r_j^3$	Cubic	$\phi(r_j) = (r_j^2 + s^2)^{-1/2}$	RMQ
$\phi(r_j) = \exp(-r_j^2)$	Gaussian	$\phi(r_j) = (r_j^2 + s^2)^{1/2}$	MQ

where $r_j \geq 0$ and s is a non-zero, positive constant. In the above TPS denotes 'thin plate splines' and RMQ denotes reciprocal multiquadrics. All the basis functions, $\phi(r_j)$, are global in nature and therefore the number of operations required to evaluate the function value, $f(\mathbf{x})$, at any point scales as N.

In this paper we are concerned only with the application of the MQ scheme, however a property of all the basis functions shown above is that the $N \times N$ matrix

$$A_{ij} = \phi(\|\mathbf{x}_i - \mathbf{x}_j\|) \qquad i, j = 1, 2, \ldots, N\tag{3}$$

is non-singular for distinct data points (Micchelli 1985). A number of important developments in MQ theory and application have occurred since their introduction by Hardy in 1968 (see Hardy 1990). Franke (1982) applied a large number of global and local interpolants to fit a specified two-dimensional surface. In terms of accuracy MQ was found to outperform all those tested, particularly if the function to be fitted is steep. However, MQ performed poorly on flat and shallow surfaces.

A major difficulty with the application of MQ to real problems is in the choice of s^2. By experiment Tarwater (1985) discovered that the root-mean-square (RMS) error of any fit using MQ was a function of s^2. For increasing s^2 the RMS error was reduced to a minimum, however further increases in s^2 caused the error to grow rapidly afterwards. Formulae for estimating the 'optimal' value of s^2 have been given, usually as a function involving the number of data points, N, and the size and shape of the domain in which the points lie. Some specific examples are;

$$s^2 = (0.815d)^2, \tag{4}$$

given by Hardy (1971), where d is the average distance from each data point to a nearest neighbour; also

$$s^2 = \frac{1}{N(N-1)} \sum_{i=1}^{N} \sum_{j=1}^{N} [(x_i - x_j)^2 + (y_i - y_j)^2], \tag{5}$$

given by Schul'min and Mitel'man (1974) for two dimensional problems. Both expressions appear to be heuristic.

Kansa (1990a) has found that the accuracy of MQ can be increased substantially by making s^2 a function of the basis number j; specifically

$$s_j^2 = s_{min}^2 \left(\frac{s_{max}^2}{s_{min}^2} \right)^{(j-1)/(N-1)} \tag{6}$$

where now $s_{min/max}$ are input parameters which must be chosen in some way. The use of expression (6) produces a diverse range of basis function shapes which helps in approximating a wide variety of surfaces.

Recently Carlson and Foley (1991) have found that the 'optimal' value for s^2 is strongly dependent on the function values, $f(\mathbf{x}_j)$, and is quite independent of N and the locations of the data points; in direct contrast to previous beliefs. They choose s^2 according to how well a quadratic polynomial fits the data, using least squares to obtain a variance or average residual V. Then s^2 is given as

$$s^2 = 1/(1 + 120V)^2, \tag{7}$$

assuming that the domain of computation is scaled to unity in all dimensions.

Clearly the choice of s^2 is a difficulty in the practical use of MQ. Here we use Kansa's prescription (6) and choose $s_{min/max}$ by experimentation. This does introduce a dependence on the ordering of the points, and we choose this ordering such that j increases as the data points are taken further from the centre of the computational volume.

3 A MODEL TEST PROBLEM

Kansa (1990b) has pioneered the use of MQ for the solution of partial differential equations. In order to demonstrate this usage of the scheme a three-dimensional (3D) version of the test problem presented by Kansa will be described. This test problem consists of solving the model elliptic equation

$$\nabla^2 \phi(x, y, z) = (\lambda^2 + \mu^2 + \sigma^2) \exp(\lambda x + \mu y + \sigma z), \tag{8}$$

in the unit cube $0 \leq x, y, z \leq 1$, subject to the Dirichlet boundary conditions

$$\phi_{\text{Boundary}} = \exp(\lambda x + \mu y + \sigma z).$$

Using MO with an appended constant, ϕ is expanded as

$$\phi = a_1 + \sum_{j=2}^{N} a_j \tilde{g}_j \qquad \text{where} \qquad \tilde{g}_j = g_j - g_1 \tag{9}$$

and

$$g_j = [(x - x_j)^2 + (y - y_j)^2 + (z - z_j)^2 + s_j^2]^{1/2}, \tag{10}$$

with s_j^2 given by expression (6). In order to improve accuracy, for certain approximations polynomials can be appended to the right-hand side of (1). The resulting expansions occur naturally in some variational calculations (Powell 1990).

In this test we use $N = 406$ data points (x_j, y_j, z_j), some 206 of which lie on the six boundaries of the cube. The data points (boundary and interior) can be placed at any location within the solution domain; it is, however, essential that all regions of the domain be adequately covered, particularly those regions where the function values are expected to vary most rapidly.

Since the MQ basis function is infinitely differentiable it is a simple matter to substitute the expansion (9) into equation (8). The derivatives are approximated as, for example

$$\frac{\partial \phi}{\partial x} = \sum_{j=2}^{N} a_j \frac{\partial \tilde{g}_j}{\partial x} = \sum_{j=2}^{N} a_j \left[\frac{(x - x_j)}{g_j} - \frac{(x - x_1)}{g_1} \right]$$

and

$$\frac{\partial^2 \phi}{\partial x^2} = \sum_{j=2}^{N} a_j \frac{\partial^2 \tilde{g}_j}{\partial x^2} = \sum_{j=2}^{N} a_j \left[\frac{1}{g_j} - \frac{1}{g_1} - \frac{(x - x_j)^2}{g_j^3} + \frac{(x - x_1)^2}{g_1^3} \right].$$

The result is a system of N algebraic equations for the coefficients a_j, which can be written in the matrix form

$$\mathbf{Sa = b}, \tag{11}$$

with $\mathbf{a} = (a_1, a_2, \ldots, a_N)^T$ and where the entries of $\mathbf{S}$ and $\mathbf{b}$ are given by

$$\left. \begin{aligned} S_{i1} &= 1 \\ S_{ij} &= g_j - g_1 \\ b_i &= \exp(\lambda x_i + \mu y_i + \sigma z_i) \end{aligned} \right\} \quad \begin{aligned} &\text{if } (x_i, y_i, z_i) \text{ lies} \\ &\text{on a boundary} \\ &\text{of the cube} \end{aligned}$$

and

$$\left. \begin{aligned} S_{i1} &= 0 \\ S_{ij} &= 3 \left(\frac{1}{g_j} - \frac{1}{g_1} \right) - \left(\frac{r_{ij}^2}{g_j^3} - \frac{r_{i1}^2}{g_1^3} \right) \\ b_i &= (\lambda^2 + \mu^2 + \sigma^2) \exp(\lambda x_i + \mu y_i + \sigma z_i) \end{aligned} \right\} \quad \begin{aligned} &\text{if } (x_i, y_i, z_i) \text{ lies} \\ &\text{within the interior} \\ &\text{of the cube} \end{aligned}$$

where $r_{ij}^2 = (x_i - x_j)^2 + (y_i - y_j)^2 + (z_i - z_j)^2$. Kansa (1990a) and Dubal (1992), among others, have discussed the highly ill-conditioned nature of the system (11). Its solution is not amenable to straightforward LU decomposition techniques. However, as shown by Dubal, it can be solved successfully using the method of singular-value decomposition (SVD) (see eg. Stoer and Bulirsch 1980; or Press *et al* 1986) with single-precision FORTRAN arithmetic on the Cray Y-MP8/864. The routine SVD-CMP described in Press *et al* worked well for this problem.

Results from the solution of the test problem are presented in the following listing. We have taken $\lambda = 1$, $\mu = 2$ and $\sigma = -1$. Note that once the coefficients a_j have been determined it is possible to treat the expansion for ϕ as an analytic expression. Therefore in the listing output we have computed; (i) the function value, (ii) the sum of the first derivatives, and (iii) the sum of the second derivatives, at selected points (unrelated to the locations of the data points). The first three columns show the position the quantity is to be evaluated at; the fourth and fifth columns show respectively the analytic and MQ values of the quantity; the last column is a percentage relative error.

```
INPUT NUMBER OF BOUNDARY POINTS = 206
INPUT NUMBER OF INTERIOR POINTS = 200
INPUT SMIN, SMAX AND WCUT =  1.0 3.0 5.E-15
INPUT LAMBDA, MU AND SIGMA = 1.0 2.0 -1.0
CONDITION NUMBER OF MATRIX = 3.3805810739144E+17
```

SOLUTION

X	Y	Z	ANALYTIC	COMPUTED	% ERROR
0.00	0.00	0.00	1.00000000	0.99999873	0.00012733
0.00	0.00	0.50	0.60653066	0.60651324	0.00287199
0.00	0.00	1.00	0.36787944	0.36787949	0.00001282
0.00	0.50	0.00	2.71828183	2.71826971	0.00044567
0.00	0.50	0.50	1.64872127	1.64871938	0.00011476
0.00	0.50	1.00	1.00000000	0.99999232	0.00076808
0.00	1.00	0.00	7.38905610	7.38905627	0.00000236
0.00	1.00	0.50	4.48168907	4.48169620	0.00015901
0.00	1.00	1.00	2.71828183	2.71828196	0.00000494
0.50	0.00	0.00	1.64872127	1.64873297	0.00070950
0.50	0.00	0.50	1.00000000	0.99999923	0.00007667
0.50	0.00	1.00	0.60653066	0.60651393	0.00275898
0.50	0.50	0.00	4.48168907	4.48168374	0.00011895
0.50	0.50	0.50	2.71828183	2.71829096	0.00033604
0.50	0.50	1.00	1.64872127	1.64871762	0.00022141
0.50	1.00	0.00	12.18249396	12.18259949	0.00086626
0.50	1.00	0.50	7.38905610	7.38904921	0.00009323
0.50	1.00	1.00	4.48168907	4.48169755	0.00018927
1.00	0.00	0.00	2.71828183	2.71828158	0.00000931
1.00	0.00	0.50	1.64872127	1.64868502	0.00219893
1.00	0.00	1.00	1.00000000	1.00000019	0.00001870
1.00	0.50	0.00	7.38905610	7.38902382	0.00043686
1.00	0.50	0.50	4.48168907	4.48168651	0.00005711
1.00	0.50	1.00	2.71828183	2.71827556	0.00023078
1.00	1.00	0.00	20.08553692	20.08553658	0.00000172
1.00	1.00	0.50	12.18249396	12.18247671	0.00014162
1.00	1.00	1.00	7.38905610	7.38905635	0.00000337

AV. % ERR = 4.8058083186682E-4

SUM OF FIRST DERIVATIVES

X	Y	Z	ANALYTIC	COMPUTED	% ERROR
0.00	0.00	0.00	2.00000000	2.00120611	0.06030574
0.00	0.00	0.50	1.21306132	1.21362689	0.04662353
0.00	0.00	1.00	0.73575888	0.73660722	0.11530155
0.00	0.50	0.00	5.43656366	5.43731701	0.01385722
0.00	0.50	0.50	3.29744254	3.29792842	0.01473509
0.00	0.50	1.00	2.00000000	2.00093466	0.04673302
0.00	1.00	0.00	14.77811220	14.78308827	0.03367189
0.00	1.00	0.50	8.96337814	8.96333683	0.00046093
0.00	1.00	1.00	5.43656366	5.43704307	0.00881833
0.50	0.00	0.00	3.29744254	3.29710931	0.01010588
0.50	0.00	0.50	2.00000000	2.00008219	0.00410974
0.50	0.00	1.00	1.21306132	1.21266997	0.03226142
0.50	0.50	0.00	8.96337814	8.96373561	0.00398811
0.50	0.50	0.50	5.43656366	5.43653613	0.00050627
0.50	0.50	1.00	3.29744254	3.29738352	0.00179000
0.50	1.00	0.00	24.36498792	24.36649078	0.00616811
0.50	1.00	0.50	14.77811220	14.77811912	0.00004682
0.50	1.00	1.00	8.96337814	8.96331802	0.00067074
1.00	0.00	0.00	5.43656366	5.43782824	0.02326076
1.00	0.00	0.50	3.29744254	3.29712936	0.00949762
1.00	0.00	1.00	2.00000000	2.00071645	0.03582239
1.00	0.50	0.00	14.77811220	14.77789137	0.00149431
1.00	0.50	0.50	8.96337814	8.96333867	0.00044032
1.00	0.50	1.00	5.43656366	5.43618089	0.00704061
1.00	1.00	0.00	40.17107385	40.15741181	0.03400963
1.00	1.00	0.50	24.36498792	24.36445495	0.00218743
1.00	1.00	1.00	14.77811220	14.77757636	0.00362591

AV. % ERR = 1.9167902632611E-2

```
SUM OF SECOND DERIVATIVES
X    Y    Z          ANALYTIC          COMPUTED          % ERROR
0.00 0.00 0.00       6.00000000        5.95894998        0.68416695
0.00 0.00 0.50       3.63918396        3.63019463        0.24701483
0.00 0.00 1.00       2.20727665        2.13486536        3.28057150
0.00 0.50 0.00      16.30969097       16.29406455        0.09581064
0.00 0.50 0.50       9.89232762        9.89151102        0.00825489
0.00 0.50 1.00       6.00000000        5.96919784        0.51336934
0.00 1.00 0.00      44.33433659       44.26487206        0.15668337
0.00 1.00 0.50      26.89013442       26.88993451        0.00074344
0.00 1.00 1.00      16.30969097       16.26935065        0.24733959
0.50 0.00 0.00       9.89232762        9.89014331        0.02208094
0.50 0.00 0.50       6.00000000        5.99938798        0.01020034
0.50 0.00 1.00       3.63918396        3.63354801        0.15486856
0.50 0.50 0.00      26.89013442       26.89385527        0.01383724
0.50 0.50 0.50      16.30969097       16.31004855        0.00219244
0.50 0.50 1.00       9.89232762        9.89293543        0.00614416
0.50 1.00 0.00      73.09496376       73.01538481        0.10887064
0.50 1.00 0.50      44.33433659       44.33765626        0.00748780
0.50 1.00 1.00      26.89013442       26.87647635        0.05079211
1.00 0.00 0.00      16.30969097       16.22041936        0.54735316
1.00 0.00 0.50       9.89232762        9.93066654        0.38756209
1.00 0.00 1.00       6.00000000        6.04404079        0.73401313
1.00 0.50 0.00      44.33433659       44.38875803        0.12275235
1.00 0.50 0.50      26.89013442       26.88689396        0.01205075
1.00 0.50 1.00      16.30969097       16.29703893        0.07757379
1.00 1.00 0.00     120.51322154      120.22770507        0.23691713
1.00 1.00 0.50      73.09496376       73.08553302        0.01290204
1.00 1.00 1.00      44.33433659       44.31389070        0.04611752
AV. % ERR = 0.2884322504363
```

The values of $s_{min/max}$ were obtained by experimentation. Higher accuracy can be obtained by increasing the number of grid points, however the number of operations of the SVD algorithm scales as $O(N^3)$ so this can be expensive (Dubal 1992). The CPU time required to solve the problem was two seconds on the Cray Y-MP.

4 TWO-BLACK-HOLE INITIAL DATA IN THREE-DIMENSIONS

In this section the MQ scheme is applied to the construction of two-black-hole initial data in Cartesian, (x, y, z), coordinates. This data will be the starting point for time-evolution calculations of interacting binary black-holes.

Dubal (1992) has discussed the difficulties of using standard approaches for the construction of multi-dimensional black-hole initial data, particularly in accurately representing the inner boundaries (i.e. the holes). Various numerical experiments showed that 3D single-black-hole data can be produced to quite high accuracy using the MQ scheme. The ability to place grid points in arbitrary locations was crucial for the success of these calculations. Here we describe an extension of that work for two-black-hole initial data.

The single, non-linear elliptic equation to be solved is

$$\nabla^2 \Psi = -\frac{1}{8}\hat{K}^{ij}\hat{K}_{ij}\Psi^{-7},$$
(12)

where Ψ is a conformal factor and $\hat{K}^{ij}$ is the conformally transformed extrinsic curvature of the spatial hypersurface. Here lower case Latin letters, $i, j, \ldots$, indicate spatial tensor indices with values $1, 2, 3$. The Hamiltonian constraint equation (12) arises in the context of the ADM $3 + 1$ description of general relativity when combined with York's conformal treatment (1979). The hypersurface is taken to be maximal and the spacetime is vacuum and conformally flat. The momentum constraint equations for $\hat{K}^{ij}$ are linear and may be solved analytically for multiple-black-hole initial data (see eg. Kulkarni *et al* 1983; Bowen and York 1980). Encoded within the 'source' term, $H = \hat{K}^{ij}\hat{K}_{ij}$, is information on the masses, spins and linear momenta of the holes.

There are a number of boundary conditions which must be considered. Since the spacetime is asymptotically flat then $\Psi \to 1$ as $r \to \infty$ if r is taken to be a flat-space radial coordinate with its origin at the 'centre' of the gravitating system. For numerical calculations the outer boundary is at a finite radius, R, and the Robin boundary condition is used

$$n^i D_i \Psi + \frac{\Psi - 1}{r} = 0 \qquad \text{at } r = R, \tag{13}$$

where n^i is the unit normal to the spherical surface at $r = R$ and D_i is the covariant derivative compatible with the conformal metric. In addition there are two inner boundary conditions; one for each hole

$$n_{1,2}^i D_i \Psi + \frac{\Psi}{2a_{1,2}} = 0 \qquad \text{at } r_{1,2} = a_{1,2}. \tag{14}$$

If $(x_{1,2}, y_{1,2}, z_{1,2})$ are the positions of holes 1 and 2 then

$$r_{1,2}^2 = (x - x_{1,2})^2 + (y - y_{1,2})^2 + (z - z_{1,2})^2; \tag{15}$$

$a_{1,2}$ are the radii of the holes and $n_{1,2}^i$ are unit normals to their spherical surfaces.

Previous work (Dubal 1992) has shown that a straightforward application of MQ to equation (12) produces poor results, due to the asymptotically flat behaviour of Ψ. Based on that work for single-black-hole initial data we transform to a new variable, Φ, where

$$\Psi = 1 + \left[\left(\frac{a_1}{r_1} \right)^3 + \left(\frac{a_2}{r_2} \right)^3 + \left(\frac{a_1 a_2}{r_1 r_2} \right)^{3/2} \right] \Phi. \tag{16}$$

Then equation (12) is transformed to become

$$\left[\left(\frac{a_1}{r_1}\right)^3 + \left(\frac{a_2}{r_2}\right)^3 + \left(\frac{a_1 a_2}{r_1 r_2}\right)^{3/2}\right]\nabla^2\Phi$$

$$-6\left\{\frac{a_1^3}{r_1^5}\left[1 + \frac{1}{2}\left(\frac{r_1 a_2}{r_2 a_1}\right)^{3/2}\right]\left[(x - x_1)\frac{\partial\Phi}{\partial x} + (y - y_1)\frac{\partial\Phi}{\partial y} + (z - z_1)\frac{\partial\Phi}{\partial z}\right]\right.$$

$$\left.+\frac{a_2^3}{r_2^5}\left[1 + \frac{1}{2}\left(\frac{r_2 a_1}{r_1 a_2}\right)^{3/2}\right]\left[(x - x_2)\frac{\partial\Phi}{\partial x} + (y - y_2)\frac{\partial\Phi}{\partial y} + (z - z_2)\frac{\partial\Phi}{\partial z}\right]\right\}$$

$$+6\left[\frac{a_1^3}{r_1^5} + \frac{a_2^3}{r_2^5}\right]\Phi + \frac{3}{4}\left(\frac{a_1 a_2}{r_1 r_2}\right)^{3/2}\left[\frac{1}{r_1} + \frac{1}{r_2} + \frac{6D}{r_1^2 r_2^2}\right]\Phi$$

$$= -\frac{1}{8}H\left\{1 + \left[\left(\frac{a_1}{r_1}\right)^3 + \left(\frac{a_2}{r_2}\right)^3 + \left(\frac{a_1 a_2}{r_1 r_2}\right)^{3/2}\right]\Phi\right\}^{-7}, \qquad (17)$$

where $D = (x - x_1)(x - x_2) + (y - y_1)(y - y_2) + (z - z_1)(z - z_2)$. The boundary conditions also need to be transformed. At the outer boundary, expression (13) is written as

$$\left[\left(\frac{a_1}{r_1}\right)^3\left(1 - \frac{3D_1}{r_1^2}\right) + \left(\frac{a_2}{r_2}\right)^3\left(1 - \frac{3D_2}{r_2^2}\right) + \left(\frac{a_1 a_2}{r_1 r_2}\right)^{3/2}\left(1 - \frac{3D_1}{2r_1^2} - \frac{3D_2}{2r_2^2}\right)\right]\Phi$$

$$+\left[\left(\frac{a_1}{r_1}\right)^3 + \left(\frac{a_2}{r_2}\right)^3 + \left(\frac{a_1 a_2}{r_1 r_2}\right)^{3/2}\right]\left[x\frac{\partial\Phi}{\partial x} + y\frac{\partial\Phi}{\partial y} + z\frac{\partial\Phi}{\partial z}\right] = 0, \qquad (18)$$

at $x^2 + y^2 + z^2 = R^2$, where $D_{1,2} = x(x - x_{1,2}) + y(y - y_{1,2}) + z(z - z_{1,2})$. The inner boundary condition for hole 2 in terms of Φ is written as

$$\left[\left(\frac{a_1}{r_1}\right)^3 + \left(\frac{a_2}{r_2}\right)^3 + \left(\frac{a_1 a_2}{r_1 r_2}\right)^{3/2}\right]\left[(x - x_2)\frac{\partial\Phi}{\partial x} + (y - y_2)\frac{\partial\Phi}{\partial y} + (z - z_2)\frac{\partial\Phi}{\partial z}\right]$$

$$-\left[2\left(\frac{a_2}{r_2}\right)^3 + \left(\frac{a_2}{r_2}\right)^3\left(\frac{3D}{r_1^2} - \frac{1}{2}\right) + \left(\frac{a_1 a_2}{r_1 r_2}\right)^{3/2}\left(1 + \frac{3D}{2r_1^2}\right)\right]\Phi = -\frac{1}{2}, \qquad (19)$$

at $(x - x_2)^2 + (y - y_2)^2 + (z - z_2)^2 = a_2^2$. A similar expression can be written for the other hole.

Although the resulting system is more cumbersome than the original one, it is nevertheless still a simple procedure to solve it using the MQ scheme. As described in the previous section, an approximation for the function Φ is written as $\Phi = a_1 + \sum_{j=2}^{N} a_j \tilde{g}_j$ and substituted into equation (17) and the boundary conditions. A matrix system (11) is constructed in the same way, and this is solved, again using SVD. Note that the only bookkeeping required is to determine whether a data point lies in the interior of the domain, or on one of the three boundaries (and if so which one). Data points are clustered around the two black holes since Φ will have the greatest variation there.

We describe results for the construction of Misner type data (Misner 1960) in which the two holes are of equal radii and the source term, H, is zero (no spins or linear momenta). The initial hypersurface is then time-symmetric. Misner provides analytic expressions for Ψ and the ADM energy of the system, which can be used to test the results of our code. Cook (1990) has obtained a generalization of the Misner solution for the case when the hole radii are unequal. Since it is always possible to find a line of symmetry joining the centres of the black holes, all of these solutions are axisymmetric. In Cartesian coordinates, however, Ψ is three-dimensional and the condition of axisymmetry provides a further test for the code. Solutions with non-zero H will be presented elsewhere (Dubal *et al* 1992).

Table 1 below shows results from several runs to produce Misner data. The holes become increasingly close to each other (as measured by β, the separation of the hole centres in flat space). In all cases $N = 2042$, the hole radii are $a_1 = a_2 = 1$, and the outer boundary is at $R = 800$. The accuracy of the solutions is evaluated by comparing the computed ADM energy (see Dubal 1992, equation (18)) with the analytic one given by Misner (1960) and also Cook (1990). In most cases the relative error is $\leq 1\%$. A better estimate of the error is the percentage average solution error for Ψ, given by $100 \sum_{j=1}^{N} |\Psi_j^{analytic} - \Psi_j^{computed}|/(N\Psi_j^{analytic})$. This is shown in the last column of the table. Again errors of $< 1\%$ are found.

β	ADM Energy (computed)	ADM Energy (analytic)	Diff. (%)	Ψ Solution Av. (%) error
12	4.3642	4.3639	0.007	0.08
10	4.4390	4.4449	0.133	0.12
8	4.5511	4.5727	0.472	0.32
6	4.7481	4.8046	1.176	0.50
4	5.3452	5.3642	0.354	0.60

Table 1. *Construction of Misner Data and test of the code.*

As stated previously, the solution for Ψ may be treated as an analytic function once the coefficients, a_j, of the expansion have been found. This is clearly demonstrated in figure 1 which shows contour plots of the conformal factor, Ψ, for Misner data, with $\beta = 12$, at various scales of the grid. In all four cases the Ψ values have been evaluated on a 128×128 regular lattice to produce the contour plots. Even at high 'magnification' (see last box of figure 1) the holes appear perfectly spherical in the Cartesian coordinate system. For convenience of presentation both hole centres lie in the $y = 0$ plane.

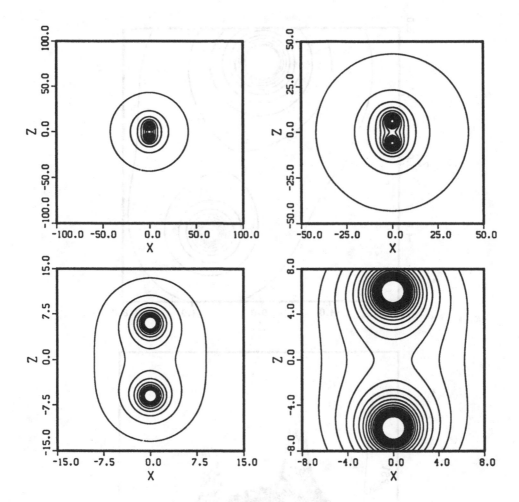

Figure 1. *Contour plots of* Ψ *for Misner data with* $\beta = 12$. *The outer boundary is at* $R = 800$. *A sequence of increasing 'magnification' is shown.*

In figure 2(a) we show a contour plot of Ψ for the case of two black holes with unequal radii ($a_1 = 1$ and $a_2 = 0.5$); again with $H = 0$. The centre of the second hole is offset from the x-axis, nevertheless the solution is axisymmetric to high accuracy about a line joining the hole centres. In order to present the contour plot the holes are again placed in the $y = 0$ plane. Figure 2(b) shows a surface plot of Ψ (interpolated onto a 64×64 regular lattice). The distribution of the data points used for the construction of this initial data is shown in figure 3. On the left-hand side all $N = 2042$ data points are shown in projection. The spherical outer boundary at $R = 800$ is clearly shown.

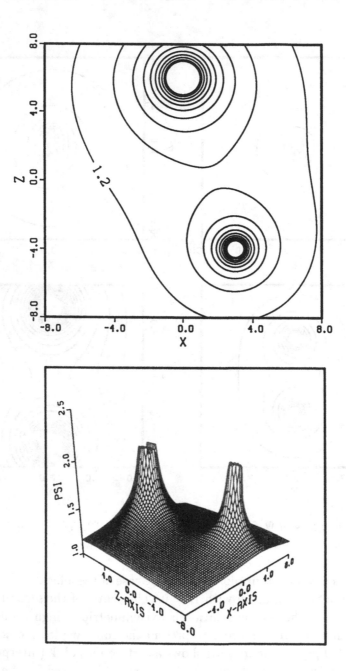

Figure 2. *Initial data for two black holes of unequal radii* ($a_1 = 1, a_2 = 0.5$), *with zero spin and linear momenta. (a) Contour plot of* Ψ. *(b) Surface plot of* Ψ.

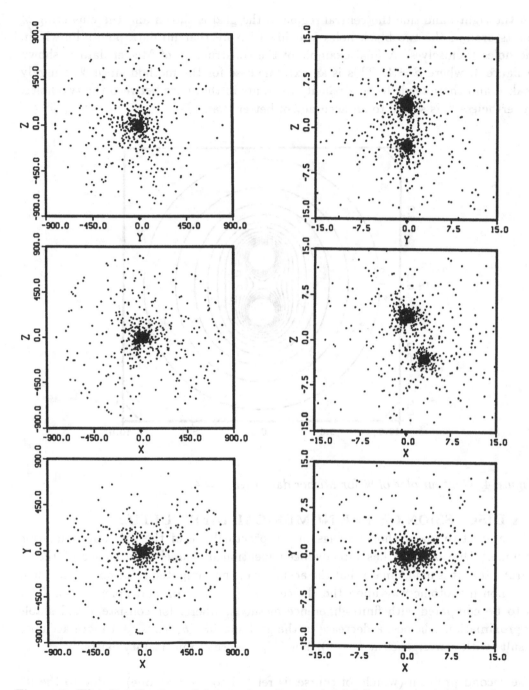

Figure 3. The distribution of data points for the initial data shown in figure 2.

On the right-hand side the central region of the grid is shown and the clustering of points around the two black holes is evident. Note that no data points lie within the holes themselves. A final example of the construction of Misner data is shown in figure 4, where $\beta = 4$. This is an extreme case for the method since Ψ is highly peaked and there tends to be a paucity of points in the space between the two holes. Nevertheless Ψ is solved to an accuracy of better than 1%.

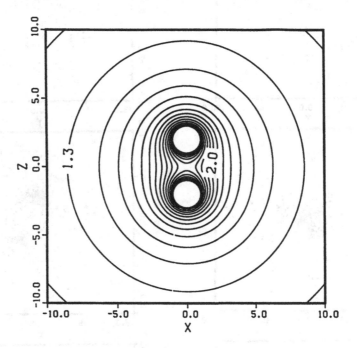

Figure 4. *Contour plot of Ψ for Misner data with $\beta = 4$.*

5 A DISCUSSION OF THE NUMERICAL DIFFICULTIES

There are two major shortcomings in the practical use of the MQ approximation scheme. One of these has already been mentioned; namely the choice of the s^2 parameter. It is unfortunate that the accuracy of the computed solution is dependent in a non-monotonic way upon the choice of s^2 (see Carlson and Foley 1991). This is to be compared with finite-difference methods, where, for consistent and stable approximation schemes, a decrease in the grid spacing, h, leads to a more accurate result (with convergence at $h \rightarrow 0$). For MQ a sense of uncertainty exists.

The second problem (which, of course, is related to the first one) is due to the ill-conditioned nature of the MQ matrix system. Standard iterative and direct methods will not solve the equations, due to round-off error. As in the problem of least-squares fitting, the SVD algorithm can solve the equations, but the operation count is high at $O(N^3)$. For many 3D problems N can be considerably smaller in the case of MQ than for finite-difference, for a given level of accuracy. However, it would be extremely useful to have a fast iterative method for the MQ system. Before any iterative methods could be used some preconditioning of the matrix would be necessary (generally to make it more diagonally dominant). Dyn *et al* (1986) describe preconditioning operators for systems with general scattered data points using various radial basis function interpolations (including MQ). The preconditioning matrices are obtained from a discrete representation of the Laplace operator on a 'nice' triangulation of the data points. This triangulation will introduce considerably more bookkeeping than the original MQ scheme (particularly in 3D), and, of course, it introduces locality. However, given the more complex scheme to produce a well-conditioned matrix, fast iterative schemes such as algebraic multigrid could be employed.

Another way to reduce the condition numbers of MQ matrices is to use domain decomposition (Kansa 1990a). Again this increases the complexity of the scheme by introducing more bookkeeping and locality.

6 CONCLUSIONS
It has been shown that the MQ approximation scheme can be used successfully for the solution of elliptic equations of the type occurring in $3 + 1$ numerical relativity. The ability to place data points in arbitrary locations with an arbitrary ordering enabled us to produce accurate two-black-hole initial data in 3D with relative ease.

A number of deficiencies of the MQ scheme have been mentioned, and it would be useful to remedy these before going onto more complex applications. At present little is known about the mathematical and numerical properties of the MQ basis function (and radial basis functions in general), however they are the subject of active research by numerical analysts (Powell 1990). Future investigations should provide us with an idea of why they work so well, and, more importantly, when and how to use them in an optimal way.

ACKNOWLEDGMENTS
All numerical computations in this paper were performed using the Cray Y-MP8/864 at the Center for High Performance Computing, University of Texas with Cray time supported by a Cray University Research grant to Richard Matzner. MRD acknowledges the receipt of a NATO/SERC postdoctoral fellowship. SRO was supported in part by CAPES-Brazilian Education Council, and Universidade de Brasilia, Brazil. This work was supported in part by NSF grant PHY-8806567 and by Texas Advanced Research Project TARP-085.

REFERENCES

Arnowitt, R., Deser, S. and Misner, C. W. (1962). In *Gravitation: An Introduction to Current Research*, L Witten (ed.), Wiley, New York.

Bowen, J. M. and York, J. W. (1980). *Phys. Rev. D*, **21**, 2047.

Brandt, A. (1977). *Math. Comput.*, **31**, 333.

Brandt, A. and Lanza, A. (1988). *Class. Quantum Grav.*, **5**, 713.

Carlson, R. E. and Foley, T. A. (1991). *Computers Math. Applic.*, **21**, 29.

Choptuik, M. (1982). M.Sc. Thesis, University of British Columbia.

Choptuik, M. and Unruh, W. G. (1986). *Gen. Rel. Grav.*, **18**, 813.

Cook, G. B. (1990). Ph.D. Thesis, University of North Carolina.

Cook, G. B. (1991). *Phys. Rev. D*, **44**, 2983.

Dubal, M. R. (1992). *Phys. Rev. D*, **45**, 1178.

Dubal, M. R., Oliveira, S. R. and Matzner, R. A. (1992). In preparation.

Dyn, N., Levin, D. and Rippa, S. (1986). *SIAM J. Sci. Stat. Comput.*, **7**, 639.

Evans, C. R. (1986). In *Dynamical Spacetimes and Numerical Relativity*, J. M. Centrella (ed.), Cambridge University Press, Cambridge, 3.

Franke, R. (1982). *Math. Comput.*, **38**, 181.

Hardy, R. L. (1971). *J. Geophys. Res.*, **76**, 1905.

Hardy, R. L. (1990). *Computers Math. Applic.*, **19**, 163.

Kansa, E. J. (1990a). *Computers Math. Applic.*, **19**, 127.

Kansa, E. J. (1990b). *Computers Math. Applic.*, **19**, 147.

Kulkarni, A., Shepley, L. C. and York, J. W. (1983). *Phys. Lett.*, **96A**, 228.

Laguna, P., Kurki-Suonio, H. and Matzner, R. A. (1991). *Phys. Rev. D*, **44**, 3077.

Lanza, A. (1986). Ph.D. Thesis, SISSA, Trieste.

Micchelli, C. A. (1986). *Constr. Approx.*, **2**, 11.

Misner, C. W. (1960). *Phys. Rev.*, **118**, 1110.

Oohara, K-I. and Nakamura, T. (1989). In *Frontiers in Numerical Relativity*, C. R. Evans, L. S. Finn and D. W. Hobill (eds.), Cambridge University Press, Cambridge, 74.

Powell, M. J. D. (1990). "The Theory of Radial Basis Function Approximation in 1990", DAMTP NA11 Report, University of Cambridge.

Press, W. H., Flannery, B. P., Teukolsky, S. A. and Vetterling, W. T. (1986). *Numerical Recipes*, Cambridge University Press, Cambridge.

Schul'min, M. V. and Mitel'man, Y. Y. (1974). *Geodesy Mapp. Photogram.*, **16**, 13.

Stoer, J. and Bulirsch, R. (1980). *Introduction to Numerical Analysis*, Springer-Verlag.

Tarwater, A. E. (1985). *A Parameter Study of Hardy's Multiquadric Method for Scattered Data Interpolation*, UCRL-54670.

York, J. W. (1979). In *Sources of Gravitational Radiation*, L. L. Smarr (ed.), Cambridge University Press, 83.

SELF–GRAVITATING THIN DISCS AROUND ROTATING BLACK HOLES

A. Lanza

SISSA, Trieste, Italy

Abstract. We have constructed sequences of equilibrium numerical models for self-gravitating thin discs around rotating black holes. The multigrid method has been used for solving numerically the stationary and axisymmetric Einstein equations describing the problem.

1 INTRODUCTION

We solved numerically Einstein's equations for equilibrium configurations made by self–gravitating thin discs around rapidly rotating black holes. These configurations may play an important role in modelling active galactic nuclei (AGN) since the self–gravity may induce the so–called "runaway" instability which may be connected with X–ray variability observed in AGNs (Abramowicz et al., 1980, see however Wilson, 1984). Also, such configurations are seen formed in numerical simulations of general relativistic collapse to a black hole (Nakamura, 1981, Stark and Piran, 1986, Nakamura et al., 1987) for some intial conditions.

We consider here only the case in which the disc is thin; for pressure dominated discs (thick discs) work is in progress (see Nishida, Eriguchi and Lanza, 1992). Self–gravitating discs and rings have been considered in the past by Bardeen and Wagoner (1971) (BW) without central body; Will (1974, 1975) has studied weakly self–gravitating rings around slowly rotating black holes.

In order to solve such a highly non–linear problem we employed the multigrid method (MG) as a strategy to solve the finite difference equations which derive from the discretization of Einstein's equations. MG methods make use of different grids with different mesh sizes h_i with $i = 1, ..., m$ where $i = m$ is the finest and $i = 1$ is the coarsest grid (usually $h_i/h_{i+1} = 2$). All of these grids usually extend over all the domain of integration, and finer grids may be added on partial regions if more resolution is needed. Multigrid is a tool to solve efficiently the problem, coarser levels are used to accelerate the convergence, whereas only the finest grid determines the accuracy of the numerical solution.

The reader is referred to Briggs (1987) for a tutorial introduction to the method and to Brandt (1977, 1984) and Stuben and Trottenberg (1982) for more advanced descriptions. Here we mention that with this method an accurate solution is obtained in a computational time which is proportional to the number of grid points of the finest grid used. The magnitude of the constant of proportionality tends to one when the problem to solve is multidimensional.

There are other applications of MG to general relativity, all of them concerning black hole initial data (Choptuik and Unruh, 1986, Lanza, 1986, 1987, Rauber, 1986, Cook, 1989, Lanza and Dubal, 1990). In all of these papers it is demonstrated that MG is an efficient and accurate solver in numerical relativity for solving a single elliptic problem, within the present application we demonstrated this for a coupled set.

2 THE PROBLEM

Any stationary and axisymmetric space–time can be described by the metric

$$ds^2 = -X^2 dt^2 + r^2 \sin^2 \theta \frac{B^2}{X^2} (d\phi - \omega dt)^2 + \frac{\psi^2}{X^2} (dr^2 + r^2 d\theta^2), \tag{1}$$

where the four metric functions B, ω, X and ψ are functions only of the spherical polar coordinates r and θ.

We here use Bardeen's formalism in which the field equations are written in the local orthonormal tetrad of the ZAMOs (zero angular momentum observers) (see Bardeen, 1973); the notation used here and the minor changes with respect to Bardeen's formalism are discussed in Lanza (1992a)

$$\partial_s \left(\frac{2}{h} \hat{r}^2 \sin^2 \theta B_{,s} \right) + \partial_\theta \left(\frac{2}{h} \hat{r}^2 \sin^2 \theta B_{,\theta} \right) = 0 \tag{2}$$

$$\partial_s \left(\hat{r}^3 \sin^3 \theta \frac{B^3}{X^4} \hat{\omega}_{,s} \right) + \partial_\theta \left(\hat{r}^3 \sin^3 \theta \frac{B^3}{X^4} \hat{\omega}_{,\theta} \right) = 0 \tag{3}$$

$$\partial_s \left(\frac{4}{h^2} \hat{r} \sin \theta \frac{B}{X} X_{,s} - \frac{1}{2} \hat{r}^3 \sin^3 \theta \frac{B^3}{X^4} \omega_H^2 \hat{\omega} \hat{\omega}_{,s} \right) +$$
$$\partial_\theta \left(\frac{4}{h^2} \hat{r} \sin \theta \frac{B}{X} X_{,\theta} - \frac{1}{2} \hat{r}^3 \sin^3 \theta \frac{B^3}{X^4} \omega_H^2 \hat{\omega} \hat{\omega}_{,\theta} \right) = 0 \tag{4}$$

$$\frac{\psi_{,\theta}}{\psi} = \left[(B_{,\theta} + B\cot\theta)^2 + (B_{,s} + B)^2 \right]^{-1}$$

$$\{(B_{,s} + B) \left[B_{,s}\cot\theta - \frac{1}{8} h^2\omega_H^2 \hat{r}^2 \sin^2\theta \frac{B^3}{X^4}\hat{\omega}_{,s}\hat{\omega}_{,\theta} + 2B\frac{X_{,s}X_{,\theta}}{X^2} + B_{,s\theta} \right] +$$

$$\frac{1}{2} (B_{,\theta} + B\cot\theta) \{ 2B_{,\theta}\cot\theta + B_{,\theta\theta} - B_{,ss} -$$

$$2B \left[\left(\frac{X_{,s}}{X}\right)^2 - \left(\frac{X_{,\theta}}{X}\right)^2 \right] + \frac{1}{8} h^2\omega_H^2 \hat{r}^2 \sin^2\theta \frac{B^3}{X^4} \left[(\hat{\omega}_{,s})^2 - (\hat{\omega}_{,\theta})^2 \right] \}\}. \tag{5}$$

It is convenient (see Lanza, 1992a) to use $s = \ln\hat{r} = \ln(2r/h)$ and $\hat{\omega} = \omega/\omega_H$ where $h/2$ is the coordinate radius of the horizon and ω_H is the angular velocity of the black hole. The use of a radial logarithm coordinate allows to take the advantage of getting automatically more resolution near the horizon, where the metric functions have most of their variation, but also it makes Einstein equations isotropic (same diffusion coefficients in the radial and angular derivatives terms). This allows to use point Gauss–Seidel relaxation instead of line relaxation which for the system we are dealing with is quite expensive in terms of computational time (see Brandt and Lanza, 1988, and Lanza, 1992a).

In the limit of an infinitesimally thin disc, the pressure is negligible compared with the total energy density (Wagoner 1965, Salpeter and Wagoner 1971) and the pressure gradient force is important only in the direction perpendicular to the plane of the disc. This assumption is equivalent to taking $p = 0$ and the perfect fluid stress energy tensor becomes

$$T_{ij} = \sigma u_i u_j, \tag{6}$$

where u_i is the four–velocity of the matter and σ is the surface energy density defined as

$$\sigma(r) = 2 \int_0^{\frac{\pi}{2}} e \, \frac{\psi^2}{X^2} r d\theta. \tag{7}$$

In the thin disc limit, Einstein's equations simplify considerably. They are essentially those for vacuum, and the matter source terms are introduced as boundary conditions. If the disc is located in the equatorial plane ($\theta = \pi/2$), by integrating the pressure–free Einstein equations (with the matter source terms being non–zero only in the disc (BW)) one gets

$$\frac{X_{,\theta}}{X} = \pi h \hat{r} \sigma \frac{1 + v^2}{1 - v^2} \tag{8}$$

$$\hat{\omega}_{,\theta} = -\frac{4\pi h \hat{r} \sigma}{\omega_H} \frac{\Omega - \omega_H \hat{\omega}}{1 - v^2}, \tag{9}$$

where Ω and v are the coordinate angular velocity and linear velocity in the *locally non–rotating frame* of a fluid element, and are given by

$$\Omega \equiv \frac{d\phi}{dt} = \frac{u^\phi}{u^t} \tag{10}$$

$$v = \frac{h}{2}\left(\Omega - \omega_H\hat{\omega}\right)\hat{r}\sin\theta\frac{B}{X^2}. \tag{12}$$

Equations (8) and (9) are the boundary conditions for equations (3) and (4) on the equatorial plane only for $\hat{r}_{in} \leq \hat{r} \leq \hat{r}_{out}$; outside the disc the equatorial reflection symmetry holds. The same symmetry holds for the metric function B for the whole range of r.

The boundary conditions on the axis of symmetry are

$$B_{,\theta} = \hat{\omega}_{,\theta} = X_{,\theta} = 0 \quad \psi = B \quad \text{at} \quad \theta = 0. \tag{12}$$

On the horizon we have

$$\begin{aligned} B &= 0 \\ X &= 0 \qquad\qquad\qquad\qquad \text{at} \quad \hat{r} = 1 \\ \hat{\omega} &= 1 \end{aligned} \tag{13}$$

plus the regularity condition for the ratios B/X and ψ/X. At infinity,

$$X = 1 - \frac{2M}{h\hat{r}} + O\left(\hat{r}^{-2}\right)$$

$$\hat{\omega} = \frac{16J}{\omega_H h^3\hat{r}^3} - \frac{96JM}{h^4\hat{r}^4\omega_H} + O\left(\hat{r}^{-5}\right) \tag{14}$$

$$B = 1 + O\left(\hat{r}^{-2}\right)$$

where

$$M = M_H + \frac{\pi}{2}h^2\int_{s_{in}}^{s_{out}}\hat{r}^2\sigma(s)ds\int_0^{\frac{\pi}{2}}\sin\theta d\theta\frac{B}{1-v^2}\left[1 + v^2 + h\hat{r}\sin\theta\omega_H\frac{\hat{\omega}B}{X^2}v\right] \tag{15}$$

$$J = J_H + \frac{\pi}{4}h^3\int_{s_{in}}^{s_{out}}\hat{r}^3\sigma(s)ds\int_0^{\frac{\pi}{2}}\sin^2\theta d\theta\frac{B^2}{X^2}\frac{v}{1-v^2} \tag{16}$$

are the total mass and total angular momentum, and

$$M_H = \frac{1}{4\pi}K_H A_H + 2\omega_H J_H \tag{17}$$

$$J_H = -\frac{h^3 \omega_H}{32} \int_0^{\frac{\pi}{2}} d\theta \sin^3 \theta \left[\hat{r}^3 \frac{B^3}{X^4} \hat{\omega}_{,s} \right]_{s=0} \tag{18}$$

are the mass and angular momentum of the black hole; $(K_H/8\pi)$ and A_H represent the "surface gravity" and area of the horizon and are given by

$$A_H = h^2 \pi \int_0^{\frac{\pi}{2}} d\theta \sin \theta \left[\hat{r}^2 \frac{B}{X} \frac{\psi}{X} \right]_{s=0} \tag{19}$$

$$K_H = \frac{2}{h} \left[\frac{1}{r} X_{,s} \frac{X}{\psi} \right]_{s=0}. \tag{20}$$

The mathematical description of the system is completed by writing the conservation laws

$$h_i^k \nabla_j T_k^j = 0 \tag{21}$$

where h_i^k is the projection tensor perpendicular to the direction of the four-velocity and ∇_j is the covariant derivative. For a pressure–free configuration (21) reduces to

$$\dot{u}_i = 0 \tag{22}$$

or

$$g_{tt,r} + 2\Omega g_{t\phi,r} + \Omega^2 g_{\phi\phi,r} = 0 \tag{23}$$

which by using (1) becomes

$$(-X^2)_{,s} + (\Omega - \omega_H \hat{\omega})^2 \left(\frac{4}{h^2} \hat{r}^2 \sin^2 \theta \frac{B^2}{X^2} \right)_{,s} - 2 (\Omega - \omega_H \hat{\omega}) \omega_H \hat{\omega}_{,s} \left(\frac{4}{h^2} \hat{r}^2 \sin^2 \theta \frac{B^2}{X^2} \right) = 0. \tag{24}$$

For a given surface density, this equation determines the angular velocity Ω of the fluid element which, in turn, gives the specific angular momentum (see Abramowicz et al., 1978)

$$\ell = -\frac{g_{t\phi} + \Omega g_{\phi\phi}}{g_{tt} + \Omega g_{t\phi}} \tag{25}$$

and the t–component of the four velocity

$$(u_t)^2 = \frac{g_{t\phi}^2 - g_{\phi\phi} g_{tt}}{g_{\phi\phi} + 2\ell g_{t\phi} + \ell^2 g_{tt}}. \tag{26}$$

Equations (2)-(5) and (24) together with the boundary conditions (8), (9), (12) and (14) form the complete set of equations. They are highly non–linear and a numerical approach is needed for solving them. Usually, they are linearised and discretised on a finite grid, the resulting approximated equations are then solved iteratively. Due to the high non–linearity, the iteration should be started with a good initial guess.

3 INITIAL APPROXIMATION

In order to start an iteration for solving the equations described above, we need an initial approximation for the metric functions which describe the space time and the distribution of matter. We here specify the surface density profile $\sigma(s)$ in equations (8)-(9) and calculate the corresponding rotation law from equation (24). The metric components are initially approximated by using the Kerr metric. These solutions are characterized by keeping constant J_H and A_H or ω_H and h.

The surface density profile has been determined using the theory of non self–gravitating thick discs around Kerr black holes (Abramowicz et al., 1978, Kozlowski et al. 1978). We calculate what the rest mass density would be for a non self–gravitating torus with constant angular momentum and then calculate the corresponding surface density integrating over the disc. The disc is then imagined to be squeezed onto the equatorial plane and this is taken as initial approximation.

For the case of constant specific angular momentum and a polytropic equation of state, the conservation laws for a torus orbiting around a black hole can be integrated to give, for a radiation dominated gas ($\gamma = 4/3$),

$$\hat{\rho} \equiv \frac{\rho}{\rho_c} = \left(\frac{u_{in} - u_t\, u_c}{u_{in} - u_c\, u_t} \right)^3 . \tag{27}$$

In the thin disc approximation, the surface density becomes

$$\sigma = \hat{r} \rho_c \int_0^{\frac{\pi}{2}} \hat{\rho}\, \frac{\psi^2}{X^2} d\theta. \tag{28}$$

The location of the outer edge can be obtained by using the fact that, for constant specific angular momentum, the conservation law also gives

$$u_t(\hat{r}_{in}, \theta) = u_t(\hat{r}_{out}, \theta), \tag{29}$$

which can be then solved for $\hat{r}_{out}$ on the equatorial plane ($\theta = \pi/2$), once an inner edge has been fixed.

One can then calculate the rest mass of the disc which in our dimensionless coordinates,

$$m = 4\pi \left(\frac{h}{2} \right) \int_{s_{in}}^{s_{out}} \hat{r}^2 \sigma(s) ds \int_0^{\frac{\pi}{2}} u^t B \sin \theta d\theta. \tag{30}$$

Besides the rest mass, the total gravitational mass and angular momentum can be calculated by using equations (15) and (16). These can then be used in the asymptotic boundary conditions.

Knowing an initial approximation the numerical iteration can be started. This is done by using the numerical code based on the multigrid method described in Lanza (1992a). At the end of each multigrid cycle, at each level, the angular velocity and the specific angular momentum are updated (using (24) and (26)) as well as the total gravitational mass and angular momentum and hence the boundary conditions at infinity. This is done a few times per level, keeping the rest mass of the disc constant until convergence is achieved.

4 NUMERICAL RESULTS AND DISCUSSION

Because of the highly non–linear mixing–up of the quantities describing the geometry and the matter for the two objects, we cannot clearly distinguish between the contributions to global quantities coming from the disc and from the hole. However, it is clear that as the rest mass of the disc is increased, its gravitational potential soon becomes non–negligible. A particle located between the hole and the disc will need less angular momentum to stay in a Keplerian orbit because it is helped by the gravitational attraction of the disc. This effect is shown in figure 1. This figure shows the distribution of Keplerian angular momentum for test particles in vacuum (solid line) and the modification resulting from self–gravity of the disc (dashed line). Beyond the center of the disc, close to the hole, the amount of angular momentum needed for a Keplerian orbit becomes smaller, whereas on the other side, it becomes greater because now the pull of the disc is in the same direction of that of the hole.

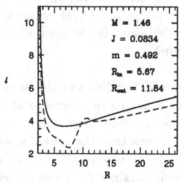

Figure 1. Keplerian specific angular momentum ℓ versus the proper distance R for a test particle in vacuum (solid line) and the corresponding modification by the self–gravity of the disc (dashed line).

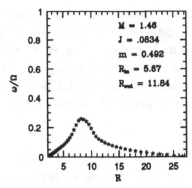

Figure 2. The ratio ω/Ω evaluated on the equatorial plane is plotted against the proper distance from the black hole. Note that the max frame dragging effect corresponds to the centre of the disc.

As already noted by Will (1974), the mixing–up of quantities related to each individual object makes it possible for the black hole to have zero angular momentum but non–zero angular velocity, or zero angular velocity and negative angular momentum. This is an effect of the dragging of the inertial frames by the external rotating disc. In order to investigate this, we have run a case with angular velocity $\omega_H = 0.00025$ and $h = 1$ (corresponding to an *initial* black hole with $a/m = 0.001$). In the case of such a slowly rotating black hole, the metric function $\hat{\omega}$ no longer has its maximum value at the horizon, but at the center of the disc and the maximum value increases with m. In figure 2 we plot the quantity $\hat{\omega}/\Omega$ versus the radial coordinate r in the equatorial plane. The figure indicates the location of the maximum of $\hat{\omega}$ (i.e. where frame dragging is a maximum).

Both the angular momentum of the disc and the angular momentum of the black hole contribute to the frame dragging in general, and in particular to ω_H. If ω_H is kept constant along the sequence, while the contribution to the dragging from matter in the disc is increased, it follows that the contribution from the black hole must decrease. Eventually this leads even to negative values of J_H (see figure 3). This is in agreement with Will's result obtained for a weakly self–gravitating ring of matter around a slowly rotating black hole.

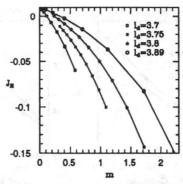

Figure 3. Angular momentum of the black hole J_H versus increasing rest mass m of the disc, for different values of ℓ_d. The parameters of the black hole are $\omega_H = 0.0025$ and $h = 1$.

The intrinsic geometry of the horizon is changed by the effect of the self–gravity of the disc. The horizon, a surface of constant time and $\hat{r} = 1$, is described by the 2–metric

$$ds_h^2 = \frac{h^2}{4}\left[\hat{r}^2\sin^2\theta\frac{B^2}{X^2}d\phi^2 + \hat{r}^2\frac{\psi^2}{X^2}d\theta^2\right]_{\hat{r}=1} \tag{31}$$

We can measure the deviation from sphericity of the horizon by taking the difference between the equatorial proper circumference ($\theta = 0$)

$$C_e = \pi h\hat{r}\frac{B}{X}, \tag{32}$$

and the polar proper circumference ($\theta = \pi/2$)

$$C_p = 2h\int_0^{\frac{\pi}{2}}\hat{r}\frac{\psi}{X}d\theta. \tag{33}$$

In figure 4, these two quantities are plotted against the rest mass of the disc for two extreme cases: when the black hole is slowly rotating, and when the black hole is rapidly rotating.

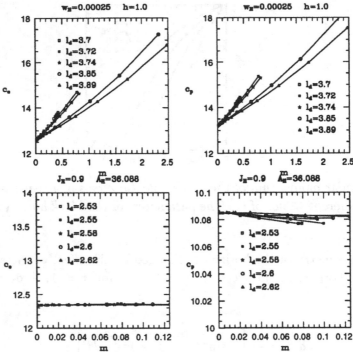

Figure 4. Proper equatorial circumference c_e *and proper polar circunference* c_p *versus* m *for a slowly rotating black hole case* ($\omega_H = 0.00025$, $h = 1$) *and for a rapidly rotating case* ($J_H = 0.9$, $A_H = 36.088$).

It is interesting to note that when the black hole is slowly rotating, $C_e < C_p$, whereas for rapidly rotating black holes $C_e > C_p$. As expected, the overall effect of increasing self-gravity is to enhance the deviation from sphericity.

In this paper we have presented a summary of results for equilibrium sequences of self-gravitating thin discs around rotating black holes. More detailed discussions of these models can be found in Lanza (1992b).

5 REFERENCES

Abramowicz, M. A., Calvani, M. and Nobili, L. 1980. *Nature*, **302**, 597.

Abramowicz, M. A., Jaroszynski, M. and Sikora, M. 1978. *Astr. Ap.*, **63**, 221.

Bardeen, J. 1973. In *Black Holes*, B. De Witt and C. De Witt (eds.), Gordon and Breach, New York.

Bardeen, J. and Wagoner, R. V. 1971. *Ap. J.*, **167**, 359.

Brandt, A. 1977. *Math. of Comp.*, **31**, 333.

Brandt, A. 1984. "Multigrid techniques: 1984 guide with applications to fluids dynamics", GMD-Studie n. 85, GMD-F1T Postfach 1240, D-5205, St. Augustin, Germany.

Brandt, A. and Lanza, A. 1988. *Class. Quant. Grav.*, **5**, 713.

Briggs, W. L. 1987. *A Multigrid Tutorial*, SIAM, Philadelphia.

Choptuik, M. and Unruh W. G. 1986. *Gen. Rel. Grav.*, **18**, 813.

Cook G., B. 1989. In *Frontiers in Numerical Relativity*, C.R. Evans, L. S. Finn and D. W. Hobill (eds.), Cambridge University Press, Cambridge, 222.

Kozlowski, M., Jaroszynski, M. and Abramowicz, M. A. 1978. *Astr. Ap.*, **63**, 209.

Lanza, A. 1986. Ph.D. Thesis, SISSA, Trieste.

Lanza, A. 1987. In *Proc. 7th Italian Conf. on General Relativity and Gravitational Physics*, U. Bruzzo et al. (eds.), World Scientific, Singapore, 67.

Lanza, A. 1992a. "Multigrid in general relativity: II. Kerr space-time", in press.

Lanza, A. 1992b. "Self–gravitating thin disks around rapidly rotating black holes", in press.

Lanza, A. and Dubal, M. R. 1990. *Comp. and Math. with Appl.*, **19**, 77.

Nakamura, T. 1981. *Progr. Theor. Phys.*, **65**, 1876.

Nakamura, T., Oohara, K. and Kojima, Y. 1987. *Prog. Theo. Phys. Suppl.*, **90**, 1.

Nishida, S., Eriguchi, Y. and Lanza, A. 1992. *Prog. Theor. Phys. Suppl.*, submitted.

Rauber, J.D. 1986. In *Dynamical Spacetimes and Numerical Relativity*, J. M. Centrella (ed.), Cambridge University Press, Cambridge, 304.

Salpeter, E. E. and Wagoner, R. V. 1971. *Ap. J.*, **164**, 557.

Stark, R. F. and Piran, T. 1986. In *Proc. of the XIV Yamada Conference on Gravitational Collapse and Relativity*, H. Sato and T. Nakamura (eds.), World Scientific, Singapore, 249.

Stüben, K. and Trottenberg, U. 1982. In *Multigrid Methods*, W. Hackbusch and U. Trottenberg (eds.), Lecture Notes in Math. 960, Springer-Verlag, 1.

Will, C. W. 1974. *Ap. J.*, **191**, 521.

Will, C. W. 1975. *Ap. J.*, **196**, 41.

Wagoner, R. V. 1965. *Phys. Rev. B*, **138**, 1583.

Wilson, D. B. 1984. *Nature*, **312**, 620.

AN ADI SCHEME FOR A BLACK HOLE PROBLEM

Gabrielle D. Allen and Bernard F. Schutz

University of Wales College of Cardiff, Cardiff, Wales, UK

Abstract. We outline an implicit finite differencing scheme for solving hyperbolic partial differential equations, and describe a 3-D black hole problem to which the scheme is applied.

1 INTRODUCTION

One issue in numerical relativity is how to solve the hyperbolic partial differential equations (PDEs) which appear as evolution equations for the geometric and matter variables. Typically a finite grid is introduced, and the PDEs are replaced by finite difference approximations, so that the geometric and matter variables are numerically calculated at the grid points. However, once we choose to use finite differencing we have to decide between an *explicit* or *implicit* scheme.

Traditionally explicit differencing schemes have been used, these are easy to derive and to implement in a code. They suffer, however, from a time step constraint needed to maintain stability. For the d-dimensional wave equation this constraint is, $\Delta t \leq \Delta/c\sqrt{d}$, where Δ is the grid spacing and c is the wave speed. Alternatively an implicit scheme is in general *unconditionally stable*, that is the size of the time step may be decided by accuracy considerations alone. Implicit schemes are usually avoided because they involve inverting huge matrices, however in Cardiff we have been developing *Alternating Direction Implicit* or *ADI* schemes. These ADI schemes can be constructed to remain as accurate as explicit or implicit schemes, as stable as implicit schemes, and yet as computationally tractable as explicit schemes.

2 AN ADI SCHEME FOR THE WAVE EQUATION

Consider the usual 3-D wave equation with sources

$$\frac{1}{c^2}\frac{\partial^2\Phi}{\partial t^2} - \nabla^2\Phi = 4\pi\rho(\mathbf{x}, t). \tag{1}$$

Defining

$$\Phi^n \equiv \Phi^n_{i,j,k} := \Phi(n\Delta t, i\Delta, j\Delta, k\Delta),$$
$$\delta_x^2\Phi^n := \Phi^n_{i+1,j,k} - 2\Phi^n_{i,j,k} + \Phi^n_{i-1,j,k}, \quad \text{etc.},$$

we can write one possible ADI scheme for (1), due to Lees (1962), as a system of three tridiagonal matrix equations, each of which must be solved over the grid to advance the solution by one timestep

$$(1 - \theta r^2 \delta_x^2)\Phi^{*n+1} = 2\Phi^n - \Phi^{n-1} + r^2(\delta_x^2 + \delta_y^2 + \delta_z^2)[(1 - 2\theta)\Phi^n + \theta\Phi^{n-1}]$$

$$+ r^2\theta(\delta_y^2 + \delta_z^2)\Phi^{n-1} + 4\pi c^2 \Delta t^2 \rho^n \tag{2a}$$

$$(1 - \theta r^2 \delta_y^2)\Phi^{**n+1} = \Phi^{*n+1} - \theta r^2 \delta_y^2 \Phi^{n-1} \tag{2b}$$

$$(1 - \theta r^2 \delta_z^2)\Phi^{n+1} = \Phi^{**n+1} - \theta r^2 \delta_z^2 \Phi^{n-1}. \tag{2c}$$

Here r is the *Courant parameter*, $r = c\Delta t/\Delta$ and θ is the *implicit weighting parameter*.

Eliminating the intermediate solutions Φ^{*n+1} and Φ^{**n+1} reduces (2) to the standard second order centred implicit approximation with the addition of an extra term of order $O(\Delta t^4)$, thus overall a second order scheme is retained. A stability analysis reveals that if $\theta \geq .25$ this scheme is unconditionally stable.

3 THE BLACK HOLE PROBLEM

To test our numerical schemes on a non-linear relativistic problem we are now applying them to a perturbed Schwarzschild black hole problem. Using a linearised formulation we find that the evolution equations resemble the 3-D wave equation making the application of the ADI scheme trivial. Also modelling the linearised problem we are free from the problems associated with evolving black holes. This problem has been investigated previously using Green-function techniques *e.g.* Oohara (1986).

3.1 The Evolution Equations

To derive the field equations for the perturbations we expand the Einstein field equations, $R_{\alpha\beta} - \frac{1}{2}g_{\alpha\beta}R = 8\pi T_{\alpha\beta}$, $(c = G = 1)$, about a given background metric $g_{\alpha\beta}^{(0)}$, and background matter distribution $T_{\alpha\beta}^{(0)}$. We thus write

$$g_{\alpha\beta} = g_{\alpha\beta}^{(0)} + h_{\alpha\beta}, \qquad T_{\alpha\beta} = T_{\alpha\beta}^{(0)} + \theta_{\alpha\beta}, \tag{3}$$

where $h_{\alpha\beta} \ll g_{\alpha\beta}^{(0)}$ and $\theta_{\alpha\beta}$ describes the matter causing the perturbations. We can then write the zeroth and first order field equations

$$R_{\alpha\beta}^{(0)} - \frac{1}{2}g_{\alpha\beta}^{(0)}R^{(0)} = 8\pi T_{\alpha\beta}^{(0)}, \tag{4}$$

$$-\frac{1}{2}\nabla^\epsilon \nabla_\epsilon h_{\alpha\beta} + \frac{1}{2}\nabla_\alpha \nabla_\gamma h^\gamma{}_\beta + \frac{1}{2}\nabla_\beta \nabla_\gamma h^\gamma{}_\alpha - \frac{1}{2}\nabla_\beta \nabla_\alpha h$$

$$+\frac{1}{2}g_{\alpha\beta}^{(0)}\nabla^\epsilon \nabla_\epsilon h - \frac{1}{2}g_{\alpha\beta}^{(0)}\nabla_\gamma \nabla_\delta h^{\gamma\delta} + \frac{1}{2}g_{\alpha\beta}^{(0)}R_{\gamma\delta}^{(0)}h^{\gamma\delta} - \frac{1}{2}R^{(0)}h_{\alpha\beta} \tag{5}$$

$$+\frac{1}{2}R_{\alpha\gamma}^{(0)}h^\gamma{}_\beta + \frac{1}{2}R_{\beta\gamma}^{(0)}h^\gamma{}_\alpha + R_{\gamma\alpha\beta\delta}^{(0)}h^{\gamma\delta} = 8\pi\theta_{\alpha\beta}.$$

Here the Riemann tensor, $R_{\alpha\beta\gamma\delta}^{(0)}$, and the covariant derivatives, ∇_α, are calculated with respect to the background metric, $g_{\alpha\beta}^{(0)}$, only.

We have to choose coordinate gauges for both the background spacetime and the perturbation field. Our ADI schemes have been formulated in Cartesian coordinates since we believe that for generality the coordinate system used should have no special association with the black hole(s). This has lead us to use quasi-Cartesian isotropic coordinates (t, x, y, z) for the background Schwarzschild spacetime, giving the background line element

$$ds^2 = -\left(\frac{2r - M}{2r + M}\right)^2 + \left(1 + \frac{M}{2r}\right)^4 (dx^2 + dy^2 + dz^2) \tag{6}$$

where $r = \sqrt{(x^2 + y^2 + z^2)}$ and M is the black hole mass. For the linearised equations we work in the *harmonic* or *de Donder* gauge which constrains $h_{\alpha\beta}$ via

$$\nabla_\alpha (h^{\alpha\beta} - \frac{1}{2} g^{(0)\alpha\beta} h) = 0. \tag{7}$$

This condition greatly simplifies the perturbation field equation (5), which reduces to a wave-like equation for each of the ten components $h_{\alpha\beta}$

$$\nabla^\gamma \nabla_\gamma \bar{h}_{\alpha\beta} = 2R^{(0)}_{\gamma\alpha\beta\delta}(\bar{h}^{\gamma\delta} - \frac{1}{2} g^{(0)\gamma\delta} \bar{h}) - 16\pi\theta_{\alpha\beta} \tag{8}$$

where $\bar{h}_{\alpha\beta}$ is the trace-reversed perturbation tensor, and the fact that the background spacetime is Ricci-flat, $R^{(0)}_{\alpha\beta} = 0$, has been used.

The evolution equations, (8), in the quasi-Cartesian isotropic background coordinates were expanded using Macsyma. The TEX output generated by Macsyma for the evolution equation of the $\bar{h}_{xy}$ component is shown in Figure 1.

3.2 The Constraint Equations
The perturbation tensor, $\bar{h}_{\alpha\beta}$, must satisfy constraint equations on the initial timeslice and subsequently on following timeslices. The constraint equations are the harmonic gauge condition and its first time derivative

$$\bar{h}^{\alpha\beta}_{;\beta} = 0, \qquad (\bar{h}^{\alpha\beta}_{;\beta})_{;t} = 0. \tag{9}$$

It can be shown that the harmonic gauge condition (9) is preserved by the evolution equations (5). Thus we only need solve (9) to obtain initial values for the simulation. From the constraint equations we obtain a set of elliptic equations. These are solved by assuming that initially all the spatial perturbations, $\bar{h}_{ij}$, are zero, leading to four coupled second order elliptic equations to be solved for $\bar{h}_{tt}, \bar{h}_{tx}, \bar{h}_{ty}, \bar{h}_{tz}$. The unphysical waves on the grid caused by setting $\bar{h}_{ij} = 0$ will 'wash' off the grid at the start of the simulation. The constraint equations have been solved using an SOR method.

$$\frac{\frac{d^2\bar{h}_{12}}{dt^2}\,(2\,r+M)^2}{2\,(2\,r-M)^2} - \frac{8\,\frac{d^2\bar{h}_{12}}{dz^2}\,r^4}{(2\,r+M)^4} - \frac{8\,\frac{d^2\bar{h}_{12}}{dy^2}\,r^4}{(2\,r+M)^4} - \frac{8\,\frac{d^2\bar{h}_{12}}{dx^2}\,r^4}{(2\,r+M)^4} =$$

$$-\frac{32\,\bar{h}_{12}\,M\,\left(24\,r^3\,z^2 - 40\,M\,r^2\,z^2 + 18\,M^2\,r\,z^2 - 3\,M^3\,z^2 - 8\,r^5 + 12\,M\,r^4 - 6\,M^2\,r^3 + 2\,M^3\,r^2\right)}{(2\,r-M)^2\,(2\,r+M)^6}$$

$$+\frac{32\,\bar{h}_{13}\,M\,\left(24\,r^3 - 40\,M\,r^2 + 18\,M^2\,r - 3\,M^3\right)\,y\,z}{(2\,r-M)^2\,(2\,r+M)^6} + \frac{32\,\frac{d\bar{h}_{23}}{dz}\,M\,r^2\,x}{(2\,r+M)^5}$$

$$+\frac{16\,\frac{d\bar{h}_{12}}{dz}\,M\,r^2\,(8\,r-3\,M)\,z}{(2\,r-M)\,(2\,r+M)^5} - \frac{32\,\frac{d\bar{h}_{23}}{dx}\,M\,r^2\,z}{(2\,r+M)^5} - \frac{32\,\frac{d\bar{h}_{13}}{dy}\,M\,r^2\,z}{(2\,r+M)^5} - \frac{4\,\bar{h}_{00}\,M\,\left(12\,r^2 - 8\,M\,r + 3\,M^2\right)\,x\,y}{r^3\,(2\,r-M)^4}$$

$$-\frac{32\,\bar{h}_{22}\,M^2\,\left(8\,r^2 - 4\,M\,r + M^2\right)\,x\,y}{(2\,r-M)^2\,(2\,r+M)^6} - \frac{32\,\bar{h}_{11}\,M^2\,\left(8\,r^2 - 4\,M\,r + M^2\right)\,x\,y}{(2\,r-M)^2\,(2\,r+M)^6} - \frac{64\,\bar{h}_{33}\,M\,\left(3\,r - M\right)\,x\,y}{(2\,r+M)^6}$$

$$+\frac{16\,\frac{d\bar{h}_{12}}{dy}\,M\,r^2\,(8\,r-3\,M)\,y}{(2\,r-M)\,(2\,r+M)^5} + \frac{4\,\frac{d\bar{h}_{01}}{dt}\,M\,(2\,r+M)\,y}{r\,(2\,r-M)^3} - \frac{32\,\frac{d\bar{h}_{22}}{dx}\,M\,r^2\,y}{(2\,r+M)^5} + \frac{32\,\frac{d\bar{h}_{13}}{dz}\,M\,r^2\,y}{(2\,r+M)^5}$$

$$+\frac{32\,\frac{d\bar{h}_{11}}{dx}\,M\,r^2\,y}{(2\,r+M)^5} + \frac{16\,\frac{d\bar{h}_{12}}{dx}\,M\,r^2\,(8\,r-3\,M)\,x}{(2\,r-M)\,(2\,r+M)^5} + \frac{4\,\frac{d\bar{h}_{02}}{dt}\,M\,(2\,r+M)\,x}{r\,(2\,r-M)^3}$$

$$+\frac{32\,\bar{h}_{23}\,M\,\left(24\,r^3 - 40\,M\,r^2 + 18\,M^2\,r - 3\,M^3\right)\,x\,z}{(2\,r-M)^2\,(2\,r+M)^6} + \frac{32\,\frac{d\bar{h}_{22}}{dy}\,M\,r^2\,x}{(2\,r+M)^5} - \frac{32\,\frac{d\bar{h}_{11}}{dy}\,M\,r^2\,x}{(-2\,r+M)^5} + 8\,\pi\,\theta_{12}$$

Figure 1. *The evolution equation for the $\bar{h}_{xy}$ component of the perturbation, calculated using Macsyma.*

3.3 The Matter Equations

The matter making up $\theta_{\alpha\beta}$ is assumed to follow geodesics of the background metric $g^{(0)}_{\alpha\beta}$. The matter is taken to be an extended point particle, with rest mass μ ,($\mu \ll M$), and proper volume V. The energy-momentum tensor is then defined as

$$\theta^{\alpha\beta} = \frac{\mu}{V}\frac{dx^\alpha}{dt}\frac{dx^\beta}{dt}\left(\frac{dt}{d\tau}\right)^2. \tag{10}$$

The geodesic equations are solved for the position and velocity of the particle using a second order leapfrog scheme.

4 IMPLEMENTING THE ADI SCHEME

The gauge constraints (9) were used to eliminate the first order time derivatives from the evolution equations (8), giving 10 second order hyperbolic equations of the form

$$\frac{\partial^2 \bar{h}_{\alpha\beta}}{\partial t^2} - 16r^4\frac{(2r-M)^2}{(2r+M)^6}\nabla^2\bar{h}_{\alpha\beta} = 16\pi\theta'_{\alpha\beta} \tag{11}$$

where $\theta'_{\alpha\beta}$ is now a 'source' term which includes terms in $\bar{h}_{\alpha\beta}$ and $\bar{h}_{\alpha\beta,i}$, as well as the matter source term $\theta_{\alpha\beta}$. Comparing (11) with the usual wave equation (1) it is seen that the ADI scheme (2) is now easy to apply. The new 'source' terms $\theta'_{\alpha\beta}$ were finite differenced and written into Fortran code using Macsyma.

Outgoing wave boundary conditions have been applied to the grid exterior. They fit in with the ADI scheme in a natural way, the condition on the boundaries at constant x is applied with the first equation of (2) *etc.* The boundary conditions used are perfectly absorbing for spherically symmetric waves. A static boundary condition has been applied at the event horizon, $r = M/2$.

5 CURRENT WORK AND CONCLUSION

We hope to soon demonstrate that it is possible to include ADI schemes in a black hole problem, and that using such schemes we will be able to take much longer timesteps than those possible with explicit schemes while retaining a second order computationally feasible stable scheme.

We will also be investigating this problem in a rotating frame, using new ADI schemes developed by Alcubierre (1992). These schemes use causal reconnection to allow grid velocities faster than the speed of light, whilst maintaining stability and second order convergence.

Ultimately we want to model a coalescing black hole problem where the black holes orbit a number of times on roughly circular paths before coalescing. In this regime, in a frame co-rotating with the black holes, ADI schemes would appear to be a necessity.

REFERENCES

Alcubierre, M. and Schutz, B.F. (1992). This volume.

Lees, M. (1962). *J. Soc. ind. appl. Math.*, **10**, 610.

Oohara, O. (1986). In *Dynamical Spacetimes and Numerical Relativity*, J. M. Centrella (ed.), Cambridge University Press, Cambridge, 365-378.

TIME-SYMMETRIC ADI AND CAUSAL RECONNECTION

Miguel Alcubierre and Bernard F. Schutz

Department of Physics and Astronomy, University of Wales College of Cardiff, Cardiff, Wales, UK

Abstract. We study the effect of a moving grid on the stability of the finite difference approximations to the wave equation. We introduce two techniques, which we call "causal reconnection" and "time-symmetric ADI" that together provide efficient, accurate and stable integration schemes for all grid velocities in any number of dimensions.

1 INTRODUCTION

In the numerical study of wave phenomena it is often necessary to use a reference frame that is moving with respect to the medium in which the waves propagate. In this paper, by studying the simple wave equation, we show that the consistent application to such a problem of two fundamental physical principles — causality and time-reversal-invariance — produces remarkably stable, efficient and accurate integration methods.

Our principal motivation for studying these techniques is the development of algorithms for the numerical simulation of moving, interacting black-holes. If we imagine a black hole moving "through" a finite difference grid then some requirements become clear. As the hole moves, grid points ahead of it will fall inside the horizon, while others will emerge on the other side. This requires grids that shift faster than light. Moreover, in situations when the dynamical time scale is large, one would like to be free of the Courant stability condition on time-steps, *i.e.* one wants to use implicit methods. Full implicit schemes require the inversion of huge sparse matrices. Alternating Direction Implicit (ADI) schemes reduce the computational burden by turning the integration into a succession of one-dimensional implicit integrations.

We will present here our most important results, leaving a more detailed derivation and analysis for a future paper (Alcubierre and Schutz (1992)).

2 THE WAVE EQUATION ON A MOVING GRID

We want to find a finite difference approximation to the wave equation using a grid that moves with an arbitrary non-uniform speed as seen in an inertial reference frame

(t, ξ^i). To represent this situation, we need to introduce a new coordinate system (t, x^i) that will be comoving with the grid

$$x^i = x^i(t, \xi^k). \tag{1}$$

In the new coordinates, the wave equation takes the form

$$(g^{ik} - \beta^i \beta^k) \frac{\partial^2 \phi}{\partial x^i \, \partial x^k} + \frac{2}{c} \beta^i \frac{\partial^2 \phi}{\partial x^i \, \partial t} - \Gamma^i \frac{\partial \phi}{\partial x^i} - \frac{1}{c^2} \frac{\partial^2 \phi}{\partial t^2} = 0 \tag{2}$$

where we have introduced the following quantities

- The spatial metric tensor g_{ij}.

- The shift vector β^i defined in the standard way (Misner, Thorne and Wheeler (1973))

$$x^i(\xi^j, t + dt) \approx x^i(\xi^j, t) - c\beta^i dt. \tag{3}$$

- The acceleration Γ^i defined in terms of the spatial Christoffel symbols

$$\Gamma^i := g^{kj} \Gamma^i_{kj} = -\frac{1}{\sqrt{g}} \left\{ \frac{\partial}{\partial t} \left(\sqrt{g} \beta^i \right) + \frac{\partial}{\partial x^j} \left[\sqrt{g} \left(g^{ij} - \beta^i \beta^j \right) \right] \right\}. \tag{4}$$

3 THE ONE DIMENSIONAL CASE

3.1 Finite Difference Approximation

In this case the metric, shift, and acceleration reduce to scalar functions

$$s^2(x, t) := g_{11}(x, t), \qquad \beta(x, t) := \beta^1(x, t), \qquad \Gamma(x, t) := \Gamma^1(x, t). \tag{5}$$

Using these expressions, (2) becomes

$$(\frac{1}{s^2} - \beta^2) \frac{\partial^2 \phi}{\partial x^2} + \frac{2\beta}{c} \frac{\partial^2 \phi}{\partial x \, \partial t} - \Gamma \frac{\partial \phi}{\partial x} - \frac{1}{c^2} \frac{\partial^2 \phi}{\partial t^2} = 0. \tag{6}$$

For the finite difference approximation to this equation we employ the usual notation

$$\phi_i^j := \phi(i \, \Delta x, j \, \Delta t).$$

We define the first and second centered spatial differences as

$$\delta_x \phi_i^j := \phi_{i+1}^j - \phi_{i-1}^j, \qquad \delta_x^2 \phi_i^j := \phi_{i+1}^j - 2\phi_i^j + \phi_{i-1}^j. \tag{8}$$

We can then write a second order accurate *implicit* finite difference approximation to (6) in the following way

$$\rho^2 \left(\frac{1}{s^2} - \beta^2 \right) \left[\frac{\theta}{2} \left\{ \delta_x^2 \phi_i^{j+1} + \delta_x^2 \phi_i^{j-1} \right\} + (1 - \theta) \left\{ \delta_x^2 \phi_i^j \right\} \right]$$

$$+ \frac{\rho\beta}{2} [\delta_x \phi_i^{j+1} - \delta_x \phi_i^{j-1}] - \frac{\rho (c\,\Delta t)}{2} \Gamma [\delta_x \phi_i^j] - [\phi_i^{j+1} - 2\phi_i^j + \phi_i^{j-1}] = 0 \qquad (9)$$

with θ an arbitrary parameter that gives the weight of the implicit terms and ρ the "Courant" parameter given by

$$\rho := (c\,\Delta t)/\Delta x. \qquad (10)$$

The above finite difference approximation will be implicit whenever the shift vector is different from zero, even when $\theta = 0$. Therefore the use of implicit approximations for the spatial derivatives does not add any extra numerical difficulty.

3.2 Local stability

It is well known (Richtmyer and Morton (1967)) that the implicit approximation to the wave equation can be made unconditionally stable when $\beta = \Gamma = 0$ and $s = 1$ by taking $\theta \geq 1/2$. We have been interested in studying under what conditions this property is preserved with a shifting grid. The existence of a shift introduces a major difficulty: the coefficients in the equation generally depend on both position and time. This means that the stability analysis must be *local*: we will only consider the stability of the difference equation obtained from (9) by, at each point (x, t), taking the coefficients to be constant, with their values at that point.

When all the parameters are free to take any value, the resulting stability condition is very complicated, and it is then difficult to find its consequences analytically. We have therefore studied this condition numerically, in order to find regions of the parameter space in which the finite difference scheme is stable.

We have searched through many values of the parameters, and our local stability analysis suggests that the finite difference scheme will be stable for all time steps if

$$\theta \geq 1/2, \qquad |s\beta| < 1, \qquad \Gamma \text{ irrelevant.} \qquad (11)$$

3.3 Causal reconnection of the computational molecules

The causal structure of a grid shifting faster than the wave speed is particularly clear in the original (ξ, t) coordinates. In Figure 1 we see how, for a very large shift, the individual grid points move outside the light-cone. It then seems plausible that the instability found in the previous section should arise because the causal structure in not represented properly any more. This suggests that we should not build the computational molecules from grid points with fixed index labels, but instead from those points that have the closest *causal* relationship (Figure 2).

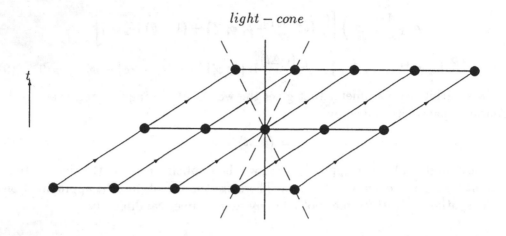

Figure 1: Grid moving faster than the waves.

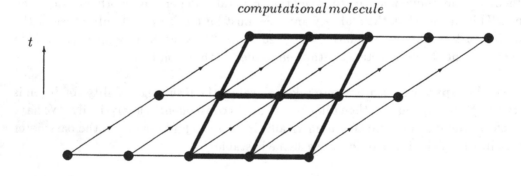

Figure 2: Causal computational molecule.

In Alcubierre and Schutz (1992) we introduce an algorithm to find the points that will form these *causal molecule*. Here we will assume that we have already found these points. The causal reconnection of the molecule is now implemented by introducing a new local coordinate system (x', t') adapted to the causal molecule. It can easily be seen that this new coordinate system (x', t') will move with respect to the old one (x, t) with a certain speed B at the intermediate time level, and with a constant acceleration A. In general the value of A and B will change from molecule to molecule, so the above change of coordinates must be repeated for each molecule.

In the primed coordinate system, the finite difference approximation will have the same form as before (equation (9)), except for the substitutions

$$\beta \longrightarrow \beta + \frac{B}{c}, \qquad \Gamma \longrightarrow \Gamma - \frac{A}{c^2}. \tag{12}$$

Causal reconnection can be implemented only when all the points in a given time level are inside the light-cone of some point in the previous time level. In the case when $\beta \neq \beta(x)$, this requirement takes the simple form

$$\Delta t \geq \frac{\max(s)\,\Delta x}{2\,c}. \tag{13}$$

This sets a *minimum* on Δt, and has a clear geometrical interpretation: the maximum proper distance between grid points ($\max(s)\,\Delta x$) must be smaller than the spread of the light-cone ($2c\,\Delta t$).

4 THE TWO-DIMENSIONAL CASE

4.1 Alternating Direction Implicit Methods
In matrix notation, the general form of the finite difference approximation to the wave equation in two dimensions is

$$\hat{Q}_2\,\phi^{j+1} = \hat{A}\,\phi^{j} + \hat{B}\,\phi^{j-1}, \tag{14}$$

with $\hat{Q}_2$, $\hat{A}$ and $\hat{B}$ spatial difference operators.

A straightforward generalization of the one-dimensional case will provide us with the most direct implicit approximation to the two-dimensional wave equation. We call this the "fully implicit" scheme. However, the numerical solution of this fully implicit scheme is considerably more time-consuming than in the one dimensional case. This is due to the fact that, if we have N grid points in each of n spatial directions, the matrix $\hat{Q}_n$ acting on ϕ^{j+1} will have N^n rows and columns. Most importantly, this matrix will *not* be tridiagonal. The matrix will still be sparse, but the number of operations involved in solving it may be very large indeed.

Alternating Direction Implicit (ADI) schemes reduce the numerical work involved in an n-dimensional problem by replacing the original large sparse matrix $\hat{Q}_n$ by one that can be factored into a product of tridiagonal matrices for each spatial direction. If we assume that we have the same number N of grid points in all directions, we will have to invert a series of N^{n-1} tridiagonal matrices of size $N \times N$ for each spatial dimension. This means that we will need only $O(nN^n)$ operations to solve the system. The reason that one can contemplate replacing the original operator $\hat{Q}_n$

with a different one is that the fully implicit scheme is only an approximation to the differential equation, so if we modify it by adding extra high-order terms that are of the same order as those neglected in the original approximation, the accuracy of the scheme will not be affected.

For our two-dimensional wave equation, the operator acting on ϕ^{j+1} turns out to be

$$\hat{Q}_2 := -1 + \frac{\rho}{2}\left(\beta^x\,\delta_x + \beta^y\,\delta_y\right)$$
$$+ \rho^2\frac{\theta}{2}\left\{\left[g^{xx} - (\beta^x)^2\right]\delta_x^2 + \left[g^{yy} - (\beta^y)^2\right]\delta_y^2\right\}. \tag{15}$$

We want to add high-order terms to this expression to transform it into

$$\hat{Q}_2' = \hat{Q}_x\hat{Q}_y$$
$$:= -\left\{1 - \frac{\rho\,\beta^x}{2}\,\delta_x - \rho^2\frac{\theta}{2}\left[g^{xx} - (\beta^x)^2\right]\delta_x^2\right\}$$
$$\times\left\{1 - \frac{\rho\,\beta^y}{2}\,\delta_y - \rho^2\frac{\theta}{2}\left[g^{yy} - (\beta^y)^2\right]\delta_y^2\right\}. \tag{16}$$

Let us define $\hat{S}$ to be the difference between these operators

$$\hat{S} := \hat{Q}_2' - \hat{Q}_2. \tag{17}$$

From this definition we find

$$\hat{S} = \hat{S}_{even} + \hat{S}_{odd}, \tag{18}$$

where we have separated terms linear and quadratic in β^i

$$\hat{S}_{even} = -\frac{\rho^2}{4}\,\beta^x\beta^y\,\delta_x\delta_y - \frac{\rho^4\theta^2}{4}\left(g^{xx} - (\beta^x)^2\right)\left(g^{yy} - (\beta^y)^2\right)\delta_x^2\delta_y^2,$$
$$\hat{S}_{odd} = -\frac{\rho^3\theta}{4}\left\{\beta^x\left(g^{yy} - (\beta^y)^2\right)\delta_x\delta_y^2 + \beta^y\left(g^{xx} - (\beta^x)^2\right)\delta_x^2\delta_y\right\}.$$

Now, we can't just add $\hat{S}\,(\phi^{j+1})$ to the finite difference approximation because in the limit when $\Delta x \to 0$ and $\Delta t \to 0$ we do not recover the original wave equation. There are many different ways to get around this problem

- Lees' first scheme. The most straightforward approach is that introduced by Lees in 1962 (Lees (1962), Fairweather and Mitchell (1965)) for the case of the ordinary wave equation on a fixed grid. In this method we add to the difference equation

$$\hat{S}\,\left(\phi^{j+1} - \phi^{j-1}\right). \tag{19}$$

It is clear that as $\Delta t \to 0$ the extra term will vanish, and we will recover the original differential equation. To find the accuracy of the scheme, we substitute the finite differences in the last expression for derivatives. We then see that the extra terms do not vanish as fast as the errors in the original approximation. Lees' first method is therefore only first-order accurate.

- Lees' second scheme. Another way to modify the equation is to add instead

$$\hat{S}\left(\phi^{j+1} - 2\phi^j + \phi^{j-1}\right).\tag{20}$$

Here again we recover the original differential equation in the limit $\Delta t \to 0$. It turns out that this method does not sacrifice accuracy: the introduced terms are of the same order as the original truncation error.

- Time-symmetric scheme. Both the original differential equation and the fully implicit scheme have the property of time-reversal invariance, that is, their form is preserved after the transformation

$$t \longrightarrow -t \qquad \beta^i \longrightarrow -\beta^i.\tag{21}$$

The operator $\hat{S}$ is not itself invariant: it contains linear and quadratic terms in β^i. Therefore, since Lees' first and second schemes both add terms in which $\hat{S}$ operates on an expression with a definite time-symmetry, neither scheme is time-reversal invariant. If we want to preserve the time-symmetry, we must allow the even and odd parts off $\hat{S}$ to act, respectively, on even and odd extra terms. That is, we add to the fully implicit scheme the term

$$\hat{S}_e\left(\phi^{j+1} - 2\phi^j + \phi^{j-1}\right) + \hat{S}_o\left(\phi^{j+1} - \phi^{j-1}\right),\tag{22}$$

Again, this scheme turns out to be just as accurate as the fully implicit method.

Whichever ADI method we choose, we will always produce an equation of the form

$$\hat{Q}'_2\,\phi^{j+1} = \hat{A}'\,\phi^j + \hat{B}'\,\phi^{j-1},\tag{23}$$

with $\hat{A}'$ and $\hat{B}'$ new spatial finite difference operators whose specific form will depend on the method chosen. From the the definition of $\hat{Q}'_2$ we see that the last equation can be decomposed into a system of two coupled equations in the following way

$$\left\{1 - \frac{\rho\,\beta^y}{2}\,\delta_y - \rho^2\frac{\theta}{2}\left[g^{yy} - (\beta^y)^2\right]\delta_y^2\right\}\phi^{j+1} := \phi^{*\,j+1},\tag{24}$$

$$\left\{1 - \frac{\rho\,\beta^x}{2}\,\delta_x - \rho^2\frac{\theta}{2}\left[g^{xx} - (\beta^x)^2\right]\delta_x^2\right\}\phi^{*\,j+1} = \hat{A}\,\phi^j + \hat{B}\,\phi^{j-1},\tag{25}$$

where the first equation defines the so-called *intermediate value* $\phi^{*\,j+1}$.

4.2 Local Stability

The different ADI schemes differ not only on their accuracy, but also in their stability properties. We have studied numerically the local stability condition for each of these schemes. The details of this stability analysis, together with plots in parameter space of the stability properties of the different schemes, can be found in Alcubierre and Schutz (1992). Here we will only summarize our most important results:

- All three methods become unconditionally unstable as soon as one of the components of the shift is larger than the wave speed.

- For $\theta < 1/2$ all three methods become unstable for at least some value of the Courant parameter ρ, regardless of the value of the shift vector considered.

- For $\theta \geq 1/2$, Lees' first method turns unstable for at least some value of ρ whenever the grid speed is not aligned with one of the coordinate axis. The instabilities for speeds below that of the waves are mild, but nevertheless significant as our numerical experiments show.

- For Lees' second method the situation is even worst: the instabilities grow faster than for the first scheme.

- For $\theta \geq 1/2$, the time-symmetric ADI seems to be unconditionally stable for all values of the shift up to the wave speed. This scheme is therefore superior to both of Lees' methods when we have a moving grid.

5 NUMERICAL EXAMPLES

We have tested the difference methods introduced in the last sections in a number of different situations. In the one-dimensional case, we have tested causal reconnection using a grid that oscillates harmonically. We find that the direct approach, without causal reconnection, goes unstable as soon as the maximum grid speed is larger than the wave speed. Causal reconnection, on the other hand, has allowed us to use grid speeds of up to 15 times the wave speed without any instabilities.

The different ADI schemes have been tested in a uniformly moving two-dimensional grid. In accordance with our stability analysis, both Lees' methods turn unstable for quite small grid velocities. Time-symmetric ADI, however, remains stable as long as the grid speed doesn't reach the wave speed. In Figure 3 we see one such calculation, where we compare the results of the evolution of a gaussian wave packet on a moving grid for both Lees' first method and time-symmetric ADI. The grid is moving slower than the waves, but nevertheless Lees' first method shows a very clear instability after only 30 time steps. The calculation using time-symmetric ADI remains stable.

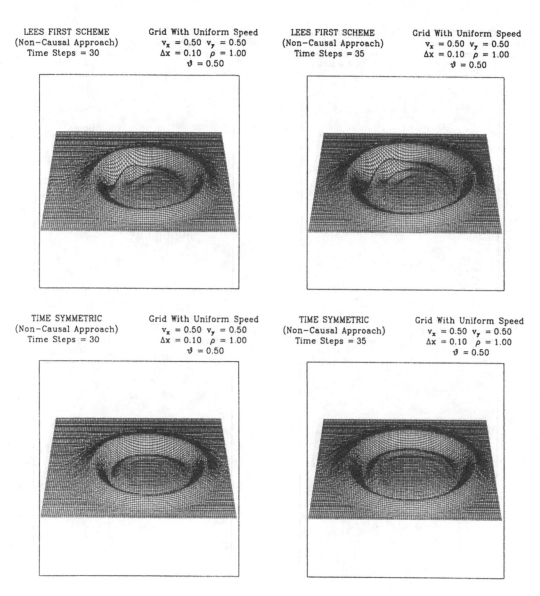

Figure 3. Uniformly moving grid: Lees' first method and time-symmetric ADI.

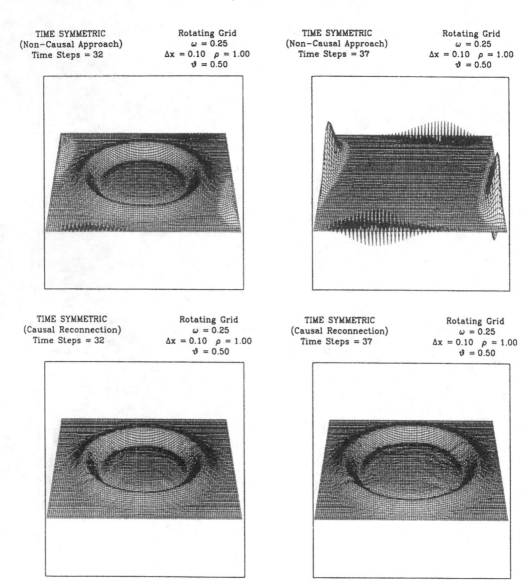

Figure 4. Rotating grid: non-causal method and causal reconnection.

Taking the time-symmetric ADI scheme as a starting point, we have tested causal reconnection in two dimensions using a rotating grid. We find that we can use grids whose edges move at many times the wave speed, without encountering any instabilities. In Figure 4 we show one example of a grid rotating in such a way that the speed at the edges is 1.25 of the wave speed, and we consider the evolution of a gaussian wave packet initially at rest at the center of the grid. When we don't reconnect the molecules, an instability forms after only 32 time steps, and grows so fast that 5 time steps later the original wave is no longer visible (the scale is set automatically to show the largest value). When we use causal reconnection, the instability is not present.

6 CONCLUSIONS

The wave equation is a prototype for more complex equations of mathematical physics. One would expect the instabilities we have found here to be *generic:* any numerical approximation to a hyperbolic system on a shifting grid should exhibit them. Only experience will show us just how well our cures for these generic instabilities transfer to more interesting equations. However, the instabilities we have described here are cured by the application of two clear physical principles, causality and time-reflection invariance. It seems clear that it would be asking for trouble *not* to incorporate these principles into the design of algorithms for the numerical integration of any fundamental physical equation. We are confident that causal reconnection and time-symmetric ADI will generalize easily to many problems in numerical relativity. These methods are stable, offer all the computational advantages of ADI schemes, and remain second order accurate.

REFERENCES

Alcubierre, M. and Schutz, B. F. (1992). "Time-Symmetric ADI and Causal Reconnection: Stable Numerical Techniques for Hyperbolic Systems on Moving Grids", to be published.

Misner, C. W., Thorne, K. S. and Wheeler, J. A. (1973). *Gravitation*, Freeman, San Francisco.

Richtmyer, R. D. and Morton, K. W. (1967). *Difference Methods for Initial-Value Problems*, 2nd. ed., Interscience.

Lees, M. (1962). *J. Soc. ind. appl. Math.*, **10**, 610.

Fairweather, G. and Mitchell, A. R. (1965). *J. Inst. Maths. Applics.*, **1**, 309.

THE NUMERICAL STUDY OF TOPOLOGICAL DEFECTS

E. P. S. Shellard

Department of Applied Mathematics and Theoretical Physics, University of Cambridge, Cambridge, UK

Abstract. We consider the numerical study of topological defects arising in relativistic field theories, with a particular emphasis on their cosmological significance. The basic properties and implications of defects — domain walls, cosmic strings, monopoles and textures — are briefly introduced. We then discuss numerical approaches for analysing defect interactions and evolution with either the full underlying field theory or, alternatively, a low-energy effective theory. For cosmic strings, we give specific examples of the problems encountered in both approaches, notably with the Goldstone and abelian-Higgs models and with the Nambu action in an expanding universe.

1 INTRODUCTION

Key to many recent developments in particle physics are the notions of spontaneous symmetry breaking and symmetry restoration at high temperature. These ideas are closely linked with unification, by which it is hoped that a description of all the known particle interactions can be brought together; in theories of grand unification (GUTs) all but the gravitational force are subsumed in a large Lie group G. Placed in a cosmological context, these compelling ideas imply that the hot early universe passed through a succession of phase transitions until eventually reaching the standard model

$$G \to H \to ... \to SU(3)_C \times SU(2)_L \times U(1)_Y \to SU(3)_C \times U(1)_Q.$$

The important point to emphasize here is that the formation of topological defects of one type or another — domain walls, cosmic strings, monopoles or textures — at any such phase transition is inevitable. On the one hand, the degenerate vacuum manifold $M = G/H$ will invariably have a non-trivial topology with which we can associate defect solutions. On the other, causality restricts correlations to be smaller than the particle horizon, so causally separated regions of space will choose different random values in M. This results in the formation of defects at the intersections of such regions — an effect known as the Kibble mechanism.

Topological defects, then, play an integral role in cosmology if we accept these central tenets of modern particle physics. Their potential implications must be confronted if there is to be any hope of pushing our understanding of the early universe further back towards the moment of creation. Already, for example, at electroweak symmetry breaking, $t \sim 10^{-11}s$, the phase transition is associated with rather subtle — albeit transient — topological effects when sphalerons and local textures appear. These cause baryon-number violating processes and may thereby have created the baryon asymmetry of the universe. Here, however, the context will generally be earlier GUT phase transitions at which very massive defects form. These may have had important gravitational effects more in keeping with the emphasis of this meeting. The following is not intended to be a comprehensive review of all numerical work on the cosmology of topological defects. Instead, using strings as the central example, the discussion stresses the necessity of numerical approaches in the study of defects, while delineating the essential issues which must be addressed.

2 TOPOLOGICAL DEFECTS

2.1 A brief history of topological defects

It is worthwhile at this point, however, to briefly discuss the long history of interest in topological defects and the gradual development of an understanding of their role in quantum field theory. The first defects to be studied in relativistic field theories were one-dimensional solitons, such as the sine-Gordon soliton and the ϕ^4-kink. Many workers in the 1950's — amongst them, Perring, Enz, Finkelstein and Misner — hoped to exploit the particle-like nature of these solutions to explain the known spectrum of particle excitations. The credibility of this idea was considerably enhanced when Skyrme presented the first three-dimensional defect solution, a texture stabilized against collapse by higher order derivative terms. These objects, known as skyrmions, are still the subject of much interest as a phenomenological model describing nucleons. However, the actual role of topological defects in quantum field theory was not in providing an explanation of the particle spectrum, but in completing it with additional excitations of great diversity. Nambu was one of the first to recognize this as an inevitable consequence of the non-trivial vacuum topology arising when symmetries are spontaneously broken. He even suggested a potential cosmological role for defects in a prescient remark, "If my view is correct, the universe may have a kind of domain structure. In one part of the universe you may have one preferred direction of the axis; in another part, the direction of the axis may be different."

It was not until the 1970's, however, that these ideas began to be taken more seriously. The two important developments being the discovery of finite-energy defect solutions in renormalizable field theories and the idea of symmetry restoration at high temperature, both of which carried directly over from condensed matter physics. These provided both the defects and the inevitable formation mechanism discussed previously. In Kibble's watershed paper where these ideas were detailed, he also pointed out a concise classification scheme for defects using homotopy groups.

Domain walls were the first defects to be a recognized cosmological catastrophe because they quickly dominate the universe and destroy its homogeneity unless they are very light. The well-known 'monopole problem' of GUT models was the next suggested consequence of defects, since in any symmetry breaking $G \to H \times U(1)$ the formation of monopoles is an inevitable result of the topology of the vacuum manifold M. Like domain walls, monopoles can be removed through the formation of strings at a subsequent phase transition — making a transient hybrid network — but this problem was also a significant motivation in the development of inflation. Finally, in the early 1980's topological defects began to be seen in a more positive light when it was suggested by Zeldovich and Vilenkin that cosmic strings might have provided the primordial fluctuations about which galaxies and other large-scale structure formed. Since then, a wide variety of other cosmological consequences for defects have been postulated, ranging from baryogenesis through to the observable electromagnetic effects of superconducting strings. The study of defects remains an area of active interest, which will continue unabated into the future because of its sound theoretical basis. References for further reading on this topic, as well as the remainder of section 2, can be found in the bibliography.

2.2 Basic properties
(i) Domain walls
Solitons in one-dimension are the most familiar topological defects. The best-known is probably the sine-Gordon soliton with the analytic solution,

$$\phi(x) = 4\eta \tan^{-1} \exp\left(\sqrt{\lambda}\,\eta x\right) \tag{1}$$

arising from the periodic potential,

$$V(\phi) = 2\lambda\eta^4\left(1 - \cos\left(\phi/\eta\right)\right). \tag{2}$$

The classical stability of (1) can be interpreted as a balance between dispersion and a non-linear self-interaction. However, this is to overlook the true topological origin of the stability; the potential has an infinite set of degenerate minima (labelled by the integers Z) and soliton solutions correspond to an interpolation between these. The field of the soliton climbs over the top of the potential, so the configuration has non-zero energy. To remove the soliton entails the infinite cost of pulling the field on one side (out to ∞) over this barrier. The soliton width $\delta \sim (\sqrt{\lambda}\,\eta)^{-1}$ is found by balancing the gradient $(\nabla\phi)^2$ and potential $V(\phi)$ terms in the energy functional. The surface energy density is typically enormous, $\sigma \sim \sqrt{\lambda}\,\eta^3$, for realistic models.

As pointed out already, domain walls have a disastrous effect on the homogeneity of the universe unless the symmetry breaking scale is very low, $\eta \lesssim 10\,MeV$. Even a domain wall network that loses energy sufficiently rapidly to remain scale-invariant relative to the Hubble radius — in a 'scaling' solution — must eventually dominate the universe because its energy density falls as $\rho_{DW} \sim t^{-1}$, in contrast to the background energy density $\rho \sim t^{-2}$. This rules out high energy models in which domain walls appear, unless there is some mechanism by which they can be rapidly removed at an early stage.

(ii) Cosmic strings

Domain walls are associated with a disconnected vacuum manifold M, whereas string solutions arise when M is not simply connected, that is, when there are unshrinkable loops in M. The simplest illustration is provided by the spontaneous breaking of a global $U(1)$ symmetry in the Goldstone model,

$$L = \tfrac{1}{2} \left(\partial_\mu \phi^* \right) \left(\partial^\mu \phi \right) - V(\phi) \tag{3}$$

where ϕ is a complex scalar field and we take a quartic potential,

$$V(\phi) = \frac{\lambda}{4} \left(\phi^* \phi - \eta^2 \right)^2 . \tag{4}$$

This is the "Mexican hat" potential with a degenerate circle of ground states, $M = S^1$, which is illustrated in figure 1(a). Field configurations for a vortex-string correspond to a non-trivial winding in the phase α of the complex field $\phi = |\phi| e^{i\alpha}$. The cross-section of an $n = 1$ is shown in figure 1(b); as we encircle the string in physical space, the field winds non-trivially about M. To resolve this phase twist continuously the field must approach $|\phi| = 0$ at the vortex core. This gives a non-zero core energy per unit length $\mu \sim \eta^2$, but there is an additional logarithmically divergent contribution due to phase gradient terms which must be cut-off by the inter-string distance.

Quantitatively, it proves easier to study the finite energy Nielsen-Olesen solutions arising in the abelian-Higgs model. Here, we introduce the vector field A^μ in the local-gauged Langrangian,

$$L = D_\mu \phi^* D^\mu \phi - V(\phi) - \tfrac{1}{4} F_{\mu\nu} F^{\mu\nu} \tag{5}$$

where the covariant derivative is given by $D_\mu = \partial_\mu - ieA_\mu$, the antisymmetric field strength tensor is $F_{\mu\nu} = \partial_\mu A_\nu - \partial_\nu A_\mu$ and e is the gauge coupling (the global $U(1)$-model is simply the $e \to 0$ limit). In an appropriate gauge, the Higgs field ϕ takes the same form as for the global string at large distances from the core,

$$\phi \approx \eta \, e^{in\theta} , \tag{6}$$

where n is an integer, the string winding number. The gauge field asymptotically approaches

$$A_\mu \approx \frac{1}{ie} \partial_\mu \ln \phi . \tag{7}$$

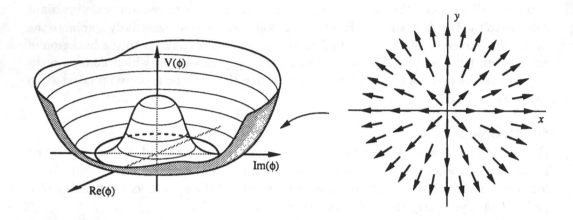

Figure 1. Field configuration for a vortex string. A loop in physical space is mapped non-trivially into the degenerate circle of minima in the "Mexican hat" potential.

In this apparently "pure gauge" form, both $D_\mu \phi$ and $F_{\mu\nu}$ exponentially approach zero away from the core, and so μ is finite. By Stokes theorem, there is also a non-zero magnetic flux flowing along the string, $\Phi_B = 2\pi n/e$. For GUT-scale strings, the energy per unit length is enormous, $\mu \sim \eta^2 \sim 10^{22} \, g/cm$, and so we can expect significant gravitational effects.

(iii) Monopoles

If the vacuum manifold is generalized from a circle S^1 to a two-sphere S^2, say by breaking $SO(3)$ with a three component vector $\vec{\phi}$, then there will be non-trivial field configurations corresponding to point-like defects or monopoles. Far from the core the monopole takes a "hedgehog" configuration,

$$\vec{\phi} = \eta \hat{r} \tag{8}$$

where $\vec{\phi}$ points radially outward in the direction of the unit vector $\hat{r}$. Such monopole configurations can be either local (finite energy with a physical magnetic flux) or global (requiring a long-distance cut-off). Monopoles are usually associated with an unbroken $U(1)$ symmetry, so grand unified models which incorporate the $U(1)_{em}$ of electromagnetism must confront a 'monopole' problem in the early universe.

(iv) Textures

A vacuum manifold corresponding to a three-sphere S^3 can give rise to textures. These are sometimes called non-singular solitons because the field is nowhere constrained to rise from the vacuum manifold M. The simplest analogue occurs in one dimension with a broken global $U(1)$ symmetry as in equation (1); the complex scalar ϕ can have a non-trivial winding about M with identical boundary values $\phi(\infty) = \phi(-\infty)$ while remaining in M, $|\phi| = \eta$. Textures are well-known in a condensed matter context, notably, in superfluid ^3He and in liquid crystals.

(v) Classification
The topological classification of defects is most succinctly achieved with the n homotopy groups. If the vacuum manifold is disconnected then $\pi_0(M) \neq I$ is non-trivial, if it is not simply-connected, that is, there are unshrinkable loops, then $\pi_1(M) \neq I$, and so forth. The correspondence with defect solutions can be summarized in the following table:

Topological Defect	Dimension	Classification
Domain Walls	2	$\pi_0(M)$
Strings	1	$\pi_1(M)$
Monopoles	0	$\pi_2(M)$
Textures	—	$\pi_3(M)$

Topology, however, only provides sufficiency conditions for the existence of stable defect solutions. Textures in three dimensions, for example, are unstable to collapse. Moreover, hybrid defects formed when there are a variety of symmetry breakings can prove to be very transient; monopoles which are connected by strings at a subsequent phase transition will annihilate very rapidly, as will domain walls bounded by strings.

2.3 Gravitational effects of defects
Defects have peculiar gravitational effects because they have a region of false vacuum trapped in their interior. Like the inflationary universe, though for a lower dimensionality, defects will have an equation of state with negative pressure. In consequence, domain walls, for example, are gravitationally repulsive. Spacetimes with domain walls are time dependent and are somewhat analogous to a three dimensional de Sitter space. The spacetime about a straight cosmic string is equivalent to Minkowski space with a wedge removed, that is, to a conical spacetime with a non-zero deficit angle. Although the string bends light, it exerts no gravitational force on a particle at rest. This metric has some peculiar possibilities, such as Gott's two-string spacetime which possesses closed timelike curves, but it is difficult to envisage situations where these could be physically realized. The metric for a string arising from the breaking of a global symmetry is different because it exhibits a singularity at a finite distance from the core — fortunately, beyond the present Hubble radius for a GUT-scale global string. The global monopole metric is analogous to the string metric except that it

has deficit *solid* angle. Finally, textures are unstable and approach a scale-invariant solution during collapse with a corresponding time-dependent spherically symmetric metric.

These gravitational effects are important if defects are to be observed. Cosmic strings, for example, can be expected to create double images of distant galaxies and quasars. Additionally, they should leave a "stringy" signature in the cosmic microwave background which may be resolved on small angular scales. Light or 'soft' domain walls, on the other hand, can be expected to create inhomogeneities on the largest angular scales. This is also the appropriate regime to search for textures which would create large spherical distortions in the CBR during collapse. There is also the possibility of the detection of a primordial background of gravitational waves due to the decay of defects. For strings, the comparison of predictions from the decay of oscillating loops with timings of the millisecond pulsar has provided a fairly stringent constraint on the energy scale of symmetry breaking $\eta \lesssim 5 \times 10^{16} GeV$.

One of the strongest motivations for the study of topological defects is the possibility that they provide the density fluctuations for large-scale structure formation. Zeldovich pointed out the difficulty of creating fluctuations on large scales because physical processes could only act on length-scales below the causal horizon. Phase transitions before $10^{-4}s$, for example, can only create fluctuations of about one *parsec*, some eight orders of magnitude below the largest currently observed structures which are over $100\,Mpc$ in extent. Three possible mechanisms have been postulated to avoid assuming an initial fluctuation spectrum in the boundary conditions of the universe: In order of popularity, these are inflation, topological defects, and explosions. Inflation has pole position because it simultaneously circumvents several other shortcomings of the standard cosmology. Indeed, it has achieved the status of orthodoxy in the astrophysical community and one of its variants has become known as the "standard" cold dark matter model. The difficulty with inflation, however, is that it remains a compelling paradigm in search of a realistic model. The original motivation with supercooling in a false vacuum state at a GUT-scale phase transition has encountered severe fine-tuning problems in the Higgs sector. Instead, in current models an exceedingly weakly-coupled scalar field, the 'inflaton,' is drawn from the magician's hat, along with some assumptions about the initial state of the universe. Moreover, inflation seems most appropriate as the universe emerges from the Planck epoch, $t \sim m_{\rm pl}^{-1} \approx 10^{-43}s$, as for chaotic inflation models. This potentially begs a number of quantum gravity issues about which we remain in ignorance.

In contrast, the theoretical foundations for topological defects are considerably more sound and involve relatively minimal extrapolations from known physics. They give rise to isocurvature fluctuations, unlike the adiabatic fluctuations from inflation. From the time of formation at a phase transition, the defect is compensated by an underdensity in radiation; only when a specific scale crosses the Hubble radius can the defect and the radiation underdensity begin to separate, allowing the accretion of matter to begin. Currently, three models show varying degrees of viability and provide an approximately scale-invariant spectrum: cosmic strings (in a variety of guises), global monopoles and global textures. Each alternative has potentially desirable features, chiefly related to the non-gaussian nature of the fluctuations. Further development, however, is required in developing a detailed picture of structure formation. Note also that textures appear to be in serious trouble with the recent COBE microwave background results, if the normalization on galactic scales is to be believed.

The final structure formation possibility is that matter collects on expanding shells, the aftermath of violent early explosions. Generally, the seeds for these explosions are primordial in origin, such as superconducting strings, so it is really the inverse of the defect mechanism. The difficulty is that models for explosive galaxy formation invariably cause unacceptable distortions in the near-perfect black body spectrum of the microwave background.

2.4 Other applications

Topological defects did not originate in relativistic field theories, they have been well-known in condensed matter physics for a considerable time. There are the domains in ferromagnetic materials, the vortices in superfluids and superconductors, the line and point defects in crystalline substances, and a wide variety of defect phenomena in liquid crystals. In a higher energy context, there are also the defects anticipated in the superfluid and superconducting interiors of neutron stars. Irrespective of their physical context, however, defects of the same dimensionality exhibit remarkably similar properties — whether, for example, it is the reconnection of cosmic strings in the hot early universe or vortices in superfluid helium-4 in the low temperature laboratory. Advances in an understanding of defects in one discipline, therefore, can often find much broader application.

Recent mathematical advances in the study of topological defect scattering have also spurred considerable interest. At special parameter values for a variety of field theories — known as the Bogomol'nyi limit — the inter-defect potential vanishes and a multi-dimensional moduli space of static solutions exists. At low velocity, motion can be approximated by geodesics on this moduli space and is determined by the geometry. This reduction greatly simplifies such interaction problems and, in some circumstances of special symmetry, renders them analytically tractable. The approach thus provides powerful geometric insight into non-linear dynamics.

2.5 Numerical approaches

Few studies of non-linear phenomena avoid reliance on numerical methods, and topological defects are no exception. Indeed, many of the most important developments in
the study of solitons and defects have derived from numerical studies. With only a few
notable exceptions, one can do little more analytically than develop a heuristic picture
of defect interactions and evolution. In a cosmological context, such arguments provide a valuable guide, but for determining critical quantitative properties, such as the
cosmic string density, past experience indicates that these approaches have difficulty
obtaining even the correct order of magnitude. Numerical simulations, therefore, have
been vital both in understanding qualitative topological defect interaction properties
and in developing quantitative models for their large-scale evolution. This happens
to be an appropriate division for describing numerical approaches to the study of the
cosmological implications of defects. On the one hand, the actual defect solutions are
studied by evolving the full field theoretic equations of motion. Where it is numerically feasible, this provides a detailed picture of defect interactions, but the limited
dynamic range restricts the usefulness of this approach for studying the evolution
of a large array of defects, such as a string network in an expanding universe. On
the other hand, therefore, it is appropriate to replace the field theory with a low-
energy effective theory with a greatly reduced number of degrees of freedom, such
as the Nambu action for strings. This allows the characterization of defect dynamics at vastly lower computational expense. Numerical approaches are also vital in
secondary analyses to determine the cosmological consequences of an evolving defect
network, whether this be the production of gravity waves or studies of the growth
of density perturbations with N-body codes. In the following two sections we shall
consider both these approaches in more detail.

3 FIELD THEORETIC APPROACH

3.1 The physical problem

There are a variety of motivations for attempting to understand defect interactions
and dynamics by adopting a brute force approach and tackling the full non-linear
field theory. For cosmic strings, for example, it was essential to know whether or not
reconnection or the "exchange of partners" occurs at string crossing. This determines
their cosmological significance since, if heavy strings either become entangled or even
simply pass through each other, they would soon come to dominate the universe —
this would preclude loop formation by the network which is the only effective energy
loss mechanism. Although there were intuitive reasons for believing that strings
would reconnect, until this issue was settled numerically [1-3], a serious question

mark remained about the viability of the cosmic string scenario. Another example was presented by ambiguities in analysing global strings and the radiation spectrum of emitted Goldstone bosons [4], as well as other issues in axion models [5]. Similar questions apply to all the other topological defects in determining their cosmological implications and many remain to be numerically resolved.

There are also mathematical questions related to the defect moduli space approach discussed in section 2.4. While this has proved to be a powerful and successful approach which enormously simplifies the characterization of defect dynamics in special cases, it is still without a rigorous mathematical justification. Instead, there is a heavy dependence on numerical results in guiding heuristic arguments to uncover the geometry of the defect moduli space. Indeed, confirmation by direct comparison with numerical simulations has been particularly important in establishing the veracity of this approach.

At this point, it is important to mention a caveat due to quantum corrections. In the following, defect simulations using the classical field equations will be described. This is a valid approximation only in what is known as the weak-coupling limit, when the additional correction terms are suppressed by powers of a small coupling constant. As a concrete example, for the abelian-Higgs model (5), the weak-coupling limit corresponds to the parameter values, $1 \gg \lambda \gg e^2$. It is mathematically interesting to explore other parameter regions (the Bogomol'nyi limit $\lambda = 2e^2$, for example), but if we wish to draw direct physical conclusions this caveat must be kept in mind.

Finally, it is worthwhile noting that current computational limitations restrict the numerical study of defects to relatively simple field theories such as $U(1)$ and $SU(2)$ models. Memory constraints are particularly acute for local-gauged field theories because of the multiplicity of vector fields which must be stored and evolved. Undoubtedly, more complicated models will become numerically accessible in the not-too-distant future, but for the time being we must infer the properties of 'realistic' defects (say, in GUT models) from more simple models. This is not unreasonable because the appearance of many defects, due to symmetry breaking in a large group G, can often be attributed to a direct embedding of either $U(1)$ or $SU(2)$ in G. As an illustration of the number of fields involved, one of the more realistic GUT models (in which strings appear) is based on $Spin(10)$ being broken in the complex **126** representation, that is, a model possessing over 250 field components in the Higgs sector — before gauge fields and subsequent symmetry breakings are even considered.

3.2 Field equations and evolution algorithms
Issues pertinent to the numerical solution of field theories in which defects arise are

best illustrated with specific examples. The abelian-Higgs model (5) is appropriate in this regard because it is the simplest model possessing gauge fields. The field equations are a second order system of hyperbolic partial differential equations,

$$(\partial^\mu - ieA^\mu)(\partial_\mu - ieA_\mu)\phi - \lambda\phi(\phi\phi^* - \eta^2) = 0 \tag{9}$$

$$\partial_\mu F^{\mu\nu} = e\text{Im}[\phi(\partial^\nu + ieA^\nu)\phi^*]. \tag{10}$$

To discretize this system for numerical solution, it is usual to reduce the evolution equations to first order form before rewriting as a specific finite difference scheme. In some special cases, it is possible to solve the equations directly in second order form, such as in the global U(1)-model.

Standard leap-frog algorithms are amongst the most efficient for the solution of multi-dimensional hyperbolic problems. For the system $\partial u/\partial t = \partial F(u)/\partial x + \partial G(u)/\partial y$, a scheme which is second-order accurate in both time and space takes the simple form,

$$u_{j,k}^{n+1} = u_{j,k}^{n-1} + \left(\frac{\Delta t}{\Delta x}\right)[F_{j+1,k}^n - F_{j-1,k}^n] + \left(\frac{\Delta t}{\Delta y}\right)[G_{j,k+1}^n - G_{j,k-1}^n] \tag{11}$$

where u, F, G are n-component vectors. Additional terms deriving from the potential term in (10) are suitably spatially averaged. The timestep is limited by the appropriate Courant condition for stability. It is fairly restrictive as for all explicit schemes; for (9-10) in first order form and in d-dimensions, this is simply $\Delta t \leq \Delta x/\sqrt{d}$. This leapfrog approach has been successfully adapted for domain wall interactions [6], global strings [1,5,7,8], the abelian-Higgs model [3,9] and a wide variety of other topological defect simulations, such as σ-model lumps [10] and global monopoles [11]. It is clear from (11), that data for the leapfrog scheme must be stored at two time levels. If memory restrictions are of paramount importance, one can use algorithms which only require one level of data, such as the second-order Lax-Wendroff or Crank-Nicholson schemes employed in references [2,12].

(i) Gauge constraints

Like problems in general relativity, there is the freedom to choose an appropriate gauge. For the equations (9-10), some simplification derives from choosing the Lorentz gauge of electromagnetism,

$$\partial_\mu A^\mu = 0, \tag{12}$$

but it is also a convenient gauge in which to boost static solutions to relativistic velocities as initial data. The gauge condition (12) then corresponds to an auxiliary constraint which must remain satisfied as (9-10) are evolved forward in time. The presence of such a gauge freedom has often been regarded as a serious numerical difficulty and, in the past, methods were developed to ensure there was no drift from the chosen gauge by ensuring that it was enforced at every timestep. Such concerns, however, appear to be unfounded. The evolution equations (9-10) should accurately preserve the original gauge choice and any 'gauge drift' should be regarded as a shortcoming of the numerical algorithm employed (such a scheme would be better abandoned, rather than artificially patched). This is borne out by extensive experience which, for appropriate numerical schemes, has yet to uncover serious problems with gauge choice. Instead, the redundancy evident in the constraint equation (12) can be regarded as a very positive feature because, along with energy conservation, it provides an important check on the growth of numerical errors.

There is a significant exception to the foregoing, when the dynamics of the system can be described with a hamiltonian formalism using techniques taken from lattice gauge field theory. This gauge-invariant approach was adapted by Moriarty *et. al.* [3] in a study of vortex motion in the abelian-Higgs model. Here, as a consequence of the symmetries of the discretized hamiltonian, the system automatically obeys the constraint equations which correspond to a lattice version of Gauss' law. Quoted figures for numerical accuracy are impressive and it is certainly an elegant approach which may become more widely adopted. However, there can be little doubt that these methods are considerably more complicated to implement and they also appear to be more computationally expensive.

(ii) Higher order algorithms and pseudo-spectral techniques
As noted in section 3.1, if the underlying symmetry group G is enlarged, the number of scalar and gauge field components increases to prohibitive levels. A gauged $SU(2)$ model already pushes currently available numerical capabilities to the limit, both in the memory required to represent topological defects and in cpu usage. Under such circumstances, especially in two or more dimensions, there is generally a substantial payback from the extra effort involved in implementing higher order schemes. This is due to the lower number of points required to represent the same data and, although the allowed CFL timestep Δt is generally more restrictive in relative terms, the actual timestep increases because the spatial step-size Δx can be considerably larger. For the same accuracy, therefore, one anticipates savings of up to several orders of magnitude in computer memory and time.

A fairly minimal higher-order extension to the leapfrog scheme (11) is the well-known Kriess-Oliger scheme [13], and is obtained with two additional correction terms

$$u_{j,k}^{n+1} = u_{j,k}^{n-1} + \left(\frac{\Delta t}{\Delta x}\right)[-\tfrac{1}{6}F_{j+2,k}^n + \tfrac{4}{3}F_{j+1,k}^n - \tfrac{4}{3}F_{j-1,k}^n + \tfrac{1}{6}F_{j-2,k}^n)]$$
$$+ \left(\frac{\Delta t}{\Delta y}\right)[-\tfrac{1}{6}G_{j,k+2}^n + \tfrac{4}{3}G_{j,k+1}^n - \tfrac{4}{3}G_{j,k-1}^n + \tfrac{1}{6}G_{j,k-2}^n)].$$

$$(13)$$

This scheme, which now entails a 5×5 stencil, is again second-order accurate in time but is spatially fourth order accurate. The stability of the scheme, and so the maximal timestep, can be considerably enhanced with further corrections which involve few additional operations [14].

The most impressive gains, however, are anticipated with pseudospectral and related techniques which are discussed elsewhere in this volume. Preliminary work with topological defect models confirms these expectations and they have proved effective in earlier studies of one-dimensional soliton interactions, as for the Korteweg–de Vries equation [15]. The pseudospectral method can be regarded as the limit of higher-order finite difference schemes of increasing accuracy. It is based on the trigonometric interpolation of data which, in practice, means performing a periodic discrete convolution using a fast Fourier transform. The disadvantage of this method is that it is best suited for problems which are naturally periodic, and limitations are apparent in models which develop shocks and contact discontinuities. Given the non-trivial boundary conditions associated with the presence of a topological defect, the imposition of periodicity appears to restrict the set of possible initial configurations, but there are ways by which this can be circumvented.

3.3 Initial data and boundary conditions
(i) Relaxation and multi-grid methods
For static defect solutions possessing either cylindrical or spherical symmetry, the substitution of an ansatz like (6-7) into the field equations yields a set of ordinary differential equations in one variable. Numerical solution is straightforward given appropriate boundary conditions, with one-dimensional relaxation techniques being far superior to shooting methods because of the high sensitivity of the solution to the initial data. Solutions can be obtained with standard NAG or IMSL routines, but rapidly convergent code for the two point boundary value problem can also be adapted from reference [16].

Where the static defect solutions do not possess special symmetries, the solution must be found by relaxation techniques in higher dimensions using appropriate boundary conditions. Examples are monopole solutions with winding number $n > 1$ and any general multi-defect configuration. Relaxation for non-linear elliptic boundary value problems in more than one dimension is often hampered by slow convergence. For repetitive calculations it is generally worthwhile implementing efficient multigrid methods [17]. These recursively apply error corrections from coarser grids where convergence is faster. Multigrid methods have been applied to the study of defects in references [3,18].

(ii) The Abrikosov ansatz

The Lorentz invariance of the field theories under consideration, implies that any static solution can be boosted up to an arbitrary velocity with an appropriate Lorentz transformation. To consider interactions of several defect solutions we can employ an appropriate variant of the Abrikosov ansatz for vortices in the abelian-Higgs model. Here, assuming we have the vortex solution $(\phi(\mathbf{x}), A_\mu(\mathbf{x}))$ we can obtain an approximate multi-vortex configuration by using

$$\tilde{\phi}(\mathbf{x}) = \prod_i \phi(\mathbf{x} - \mathbf{x}_i), \qquad \tilde{A}^\mu(\mathbf{x}) = \sum_i A^\mu(\mathbf{x} - \mathbf{x}_i), \qquad (14)$$

where the $\{\mathbf{x}_i\}$ are the positions of the vortex centres. Given the exponential fall-off of the inter-vortex potential, the ansatz (14) will be very accurate if all the vortices are widely separated, $|\mathbf{x}_i - \mathbf{x}_j| \gg \delta$. Even in other models where there is a power-law fall-off, an ansatz analogous to (14) provides initial data of sufficient accuracy for dynamical simulations.

(iii) Periodic and reflective boundary conditions

In two and three dimensions, memory limitations necessitate the imposition of artificial boundary conditions. Reflective boundary conditions — generally specifying that the field derivatives normal to the boundary vanish (Neumann) — have been the most widely employed in studying defect interactions. They can be imposed for any initial configuration, but their disadvantages are particularly evident in models with long-distance forces; any defect will be attracted towards its image anti-defect, ultimately annihilating at the boundary. For this reason, where applicable, periodic boundary conditions are to be preferred. A particularly nice implementation of periodic boundary conditions in two dimensions is presented in [3], where they are imposed with an appropriate gauge transformation for abelian-Higgs vortex configurations of non-zero winding. However, in three-dimensional simulations, either set of boundary conditions can be expected to influence the interior dynamics because of their close proximity.

(iv) Absorbing boundary conditions

There are several alternative means by which to minimize the effect of these boundary conditions [19]: (a) By changing variables, an infinite or very large region can be mapped into a much smaller domain, thus removing the boundaries to an arbitrary distance from the centrally located defect interactions. However, short wavelength modes will not propagate accurately into regions where the grid-size is changing rapidly, so there must be a smooth transition over a considerable number of grid points. (b) Another method is to impose absorbing layers around the interior regions by adding dissipative terms to the finite difference scheme. This ensures that energy is not reflected but, to be successful, this dissipation must be smoothly switched on, again 'wasting' a large number of exterior grid-points. (c) More sophisticated boundary conditions can be devised by changing the finite difference scheme at the boundary in order to match to asymptotic solutions which are exclusively outgoing modes; energy, then, is allowed to propagate out of, but not into, the computational domain. This is generally the most successful approach if the fields near the boundary can be accurately treated by a linearized version of the field equations. For simulations of Goldstone boson radiation from global strings (3-4) this entailed simply solving the outgoing massless wave equation at the boundary, for example, $\alpha_t = -\alpha_x$ on the outer face normal to the positive x-axis [4]. This lowest order boundary condition, corresponding to a type (0,0) Padé approximant, is remarkably effective at absorbing radiation for all but very small incident angles. Excellent absorption can be achieved with the (2,0) or (2,2) Padé approximants [20], though implementation is a considerably more complicated procedure.

3.4 Example 1. Vortex scattering in the abelian-Higgs model

Useful illustrations of the foregoing discussion can be made with some specific examples. Vortex interactions in the abelian-Higgs model have been extensively studied both numerically [2,3,9] and analytically [21,22] (with some numerical input). Qualitative topological questions concerning string reconnection provided the original motivation, but later more emphasis was placed on quantitative geometrical properties underlying vortex scattering in two dimensions [9,22,3]. Simulations of slow-motion vortex scattering, such as that illustrated in figure 2, employed the leapfrog algorithm (11) and reflective boundary conditions with the maximum memory available on a large μVax (only the central region of the grid is shown). Initial data was created by using boosted cylindrically symmetric solutions, found with a 1D relaxation technique, which were multiplied together using the Abrikosov ansatz (14).

For critical parameter values $\lambda = 2e^2$ (the Bogomol'nyi limit), these dynamical results allowed an approximate determination of the metric on the two-vortex moduli space (recall the discussion in section 2.4). This is the snub-nosed cone illustrated in figure 3, with the radial coordinate defining the separation of the vortices and the angular coordinate is identified by π instead of 2π. A geodesic passing over the top of the cone corresponds to right-angled vortex scattering in a head-on collision. One can argue that this phenomena underlies string reconnection in three dimensions [23]. A more accurate determination of this metric is not dynamical and, instead, requires a direct examination of static two-vortex solutions at arbitrary separation using relaxation techniques — simultaneous over-correction [22] and multigrid methods [23].

3.5 Example 2: Global strings and relativistic superfluids

A variety of fascinating non-linear defect physics is exhibited by the global strings of

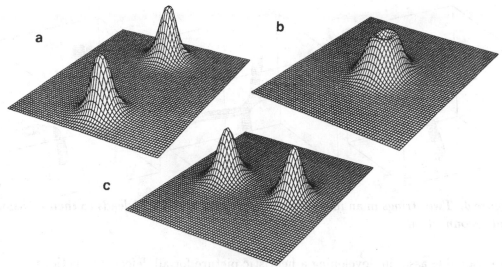

Figure 2. Energy density plots for two vortices in a head-on collision [9].

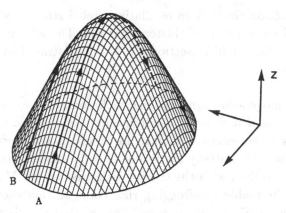

Figure 3. Approximate embedding in R^3 of the metric on the two-vortex moduli space, also showing two geodesic paths [9].

the simple Goldstone model (3-4), ranging from reconnection and radiation through to relativistic superfluid phenomena. The outcome of a string crossing is illustrated in figure 4. The stabilized leapfrog algorithm employed proved to be remarkably robust, despite the energetic aftermath of the interaction. The highly relativistic collapse and annihilation of the circular loop in figure 5(a) provided an even more dramatic test — note the Lorentz contraction of the string. Here, three-dimensional contours for a particular value of $|\phi|$ are plotted using a graphics package specifically developed with defect interactions in mind (by Dr W. Boucher, University of Cambridge). These interactions can also be animated for capture on video. Such animations prove to be

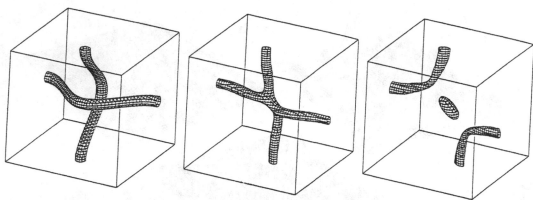

Figure 4. Two strings in an initially static configurations which leads to their collision and reconnection.

an invaluable asset in developing a heuristic picture for all defect interactions.

In figure 6, a cross-section through an oscillating global string is shown to illustrate radiative losses to phase waves or Goldstone bosons. The efficacy of the absorbing boundary conditions, Padé (0,0) in section 3.3, is demonstrated by the near-absence of reflected radiation.

The addition of a homogeneous time-varying background, $\alpha = \omega t$, implies that the global string behaves as if it were a vortex-line in a relativistic superfluid [25]. In particular, the string experiences a Magnus force which acts perpendicular to the direction of fluid flow. An initially static vortex ring configuration will develop a velocity at which it is stable, with the tension balanced by the Magnus force. Even more remarkable is the stable leapfrogging ring configuration shown in figure 5(b) — periodic boundary conditions are imposed here so that the rings can continue to propagate through the box faces [26]. Beyond being a relativistic analogue of super-fluid vortex-lines, these correspond directly to excitations which would be expected in the interior of soliton or Q-stars (which are also discussed in this volume).

3.6 Example 3. Tricks for cosmological evolution
It is straightforward to generalize the string field equations (9) to an isotropic and expanding background (again dispensing with gauge fields),

$$\ddot{\phi} - 2\frac{\dot{a}}{a}\frac{1}{\tau}\dot{\phi} - \nabla^2\phi - a^2\lambda(\phi\phi^* - \eta^2) = 0 , \qquad (15)$$

where $a(\tau)$ is the cosmological scale factor, τ is the conformal time $d\tau = dt/a$, and with time derivative $\dot{\phi} \equiv \partial\phi/\partial\tau$. The key modification is the second Hubble damping

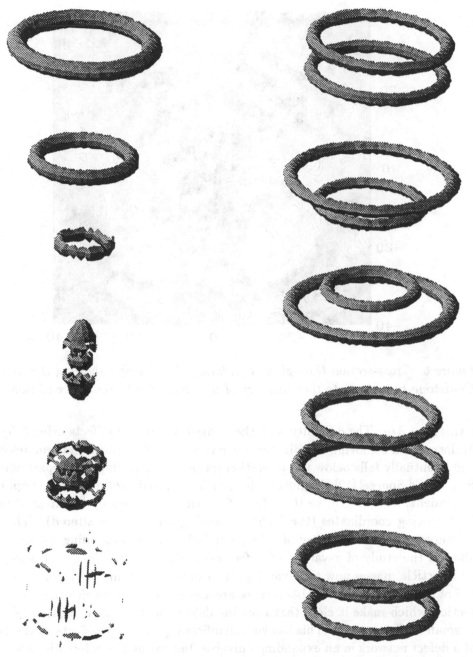

Figure 5. (a) The collapse and annihilation of an initially static global string in vacuum [25]. (b) The addition of a homogeneous time-varying background, makes the strings behave like a vortex-line in a relativistic superfluid. Above a minimum size, vortex rings are stable.

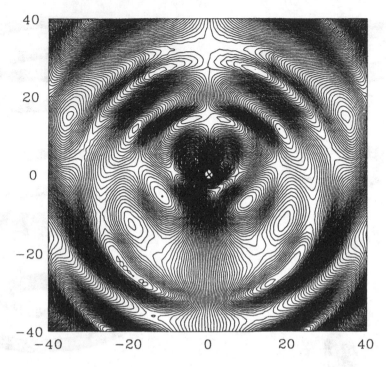

Figure 6. Cross-section through an oscillating global string showing the radiation of Goldstone bosons. Note the efficiency of the absorbing boundary conditions.

term, $H = \dot{a}/a$. The difficulty with the numerical study of defects using (15) is that, in the natural comoving coordinates, the physical width of the defect remains constant and eventually falls below the resolution set by the comoving grid separation. Press, Ryden and Spergel [12] circumvent this problem by making the quartic coupling time-dependent, $\lambda = a^{-2}(\tau)$, so that the defect width $\delta \sim (\sqrt{\lambda}\eta)^{-1}$ remains "constant" in comoving coordinates (the Hubble damping term is also altered). This greatly enhances the dynamic range of defect simulations in an expanding universe and has allowed the study of a variety of defect networks. Nevertheless, with the potential and particle masses now time-varying, the exact relationship to the original problem is not entirely obvious. Also serious are the considerations discussed in the next section which make it clear that evolving defects accumulate a substantial quantity of small-scale structure. This has very significant quantitative effects on the evolution of a defect network in an expanding universe. Increasing the defect width will tend to "wash out" substructure below this comoving scale. This may be acceptable in cases where there is a high degree of radiative damping or for textures where the defect size is not λ-dependent. However, it is an approach worthy of further investigation.

4 LOW-ENERGY EFFECTIVE ACTIONS

4.1 Reduction to an effective theory

As noted previously, a reduction in the number of degrees of freedom is essential if we are to numerically study the stochastic properties of a large system of defects. Such a low energy effective theory can usually be derived from the full defect field theory by integrating out massive modes associated with the defect cross-section. Here, we shall concentrate on the evolution of a string network in an expanding universe because this has been studied in some detail, though the following discussion can be applied to domain walls and hybrid defect systems, as well as strings in other contexts.

The Nambu action for strings can be derived from the abelian-Higgs model (5) if we assume that the scale of perturbations along the string is much larger than its width δ [27]. Essentially, we can regard the string as sweeping out a two-dimensional world-sheet $x^\mu(\sigma, \tau)$ in space-time, with τ a timelike parameter and σ spacelike, labelling points along the string. The Nambu action is then

$$S = \mu \int \sqrt{-\gamma}\, d\tau\, d\sigma \qquad (16)$$

where $\gamma_{\alpha\beta}$ (α, $\beta = 1, 2$) is the worldsheet metric and $\gamma = \det(\gamma_{\alpha\beta})$. The action (16) has a much more general model-independent motivation since it corresponds to the area of the worldsheet. This is analogous to the action of a relativistic point particle which is proportional to the particle worldline.

In an expanding universe with the conformally flat metric $ds^2 = a^2(\tau)(d\tau^2 - d\vec{x}^2)$, we can vary the action (16) to obtain the equations of motion

$$\ddot{\vec{x}} + 2\frac{\dot{a}}{a}\dot{\vec{x}}\left(1 - \dot{\vec{x}}^2\right) = \left(\frac{1}{\epsilon}\right)\left(\frac{\vec{x}'}{\epsilon}\right)'$$

$$\dot{\epsilon} = -2\frac{\dot{a}}{a}\epsilon\dot{\vec{x}}^2 \qquad (17)$$

where ϵ is proportional to the energy density along the string

$$\epsilon = \left(\vec{x}'^2 / 1 - \dot{\vec{x}}^2\right)^{\frac{1}{2}}. \qquad (18)$$

Note that there are residual gauge freedoms which have been removed by the identification of the worldsheet time with conformal time τ, and by the gauge condition

$$\dot{\vec{x}} \cdot \vec{x}' = 0 \qquad (19)$$

which makes the velocity $\dot{\vec{x}}$ perpendicular to the string's direction $\vec{x}'$. Though this system of equations has been tackled directly numerically, it is more usual to consider them in first order form by adopting the right and left-moving coordinates

$$\vec{\alpha} = \vec{x}' - \epsilon \dot{\vec{x}} \tag{20}$$

$$\vec{\beta} = \vec{x}' + \epsilon \dot{\vec{x}} . \tag{21}$$

Equation (17) then becomes

$$\dot{\vec{\alpha}} = -\left(\frac{\vec{\alpha}}{\epsilon}\right)' - \frac{\dot{a}}{a}(\vec{\beta} - \vec{\alpha}) \tag{22}$$

$$\dot{\vec{\beta}} = \left(\frac{\vec{\beta}}{\epsilon}\right)' - \frac{\dot{a}}{a}(\vec{\alpha} - \vec{\beta}) \tag{23}$$

$$\dot{\epsilon} = -\frac{\dot{a}}{a}\epsilon^{-1}(\epsilon^2 - \vec{\alpha}.\vec{\beta}) \tag{24}$$

$$\dot{\vec{x}} = \frac{1}{2\epsilon}(\vec{\beta} - \vec{\alpha}) . \tag{25}$$

With a slow expansion rate, these equations are effectively in conservation form.

4.2 Evolution algorithms

Even in flat space it is not trivial to numerically integrate the equations (22-25) for an interacting string network. Every time strings reconnect, four kinks are created; these are discontinuities in $\dot{\vec{x}}$ and $\vec{x}'$ which travel along the string at the speed of light. Moreover, reconnection entails an unequal partitioning of energies ϵ between adjacent grid points, so accurate evolution must be maintained despite this spatial ϵ-variation, as well as the presence of kinks. Standard finite-difference techniques, however, are flawed because contact discontinuities are progressively smeared; in N timesteps, the width of the transition typically grows as $N^{1/(R+1)}$, where R is the order of accuracy of the numerical scheme [28]. There are also problems from oscillations created by overshoots across the discontinuity, particularly for schemes with $R > 1$. Such oscillations can adversely affect accuracy, introducing non-linear instabilities which must be dealt with by adding artificial viscosity or pre-processing (spreading out) at the price of further smearing — approaches used for strings in references [29] and [30], respectively. An alternative is to attempt the complicated procedure of "shock fronting" by tracking the position of each propagating kink [31].

On the other hand, the technique used in references [32,33] by the present author was a hyperbolic conservation method designed for solving problems involving shocks. This was chosen after extensive trials with many algorithms ranging from spectral methods to conservation finite-difference schemes. The actual high-resolution total variation non-increasing (TVNI) algorithm [34] is related to artificial compression methods which use analytic estimates of the smearing rate of discontinuities to apply appropriate non-linear corrections, thus avoiding the use of viscosity [35]. The scheme, which employs a five-point wide stencil, can be proven to be second-order accurate through comparison with the one-step Lax–Wendroff scheme. Finally, the updating of string position $\vec{x}$ with (9) was performed with an inter-leaved two-step Runge–Kutta algorithm. Numerical tests showed that kinks spread to between three to five grid points in the first few timesteps and then remained at this width for thousands of loop oscillations. The scheme also accurately reproduced simple analytical solutions in an expanding universe. The redundancy due to the gauge conditions in the evolution equations provides an important check on the accuracy of the numerical scheme. Each was preserved to within a few percent during the course of the simulations. For example, energy conservation for small non-interacting loops was the poorest, but for the key long strings this energy constraint always remained above 97% for thousands of timesteps.

In order to satisfy the Courant stability condition $\Delta t < \epsilon \Delta \sigma$, a string loop was broken up into regions of variable length to isolate regions with small values of ϵ. Each of these regions was then evolved forward using the necessary number of timesteps satisfying the Courant condition. Such flexible time-stepping is considerably more efficient than straight evolution with the minimal Δt.

4.3 Additional numerical difficulties

Realistic initial conditions have to be generated for a string configuration in an expanding universe. This is generally achieved using the Monte-Carlo method first employed for static networks by Vachaspati and Vilenkin [36]. The strings are effectively random walks with a characteristic length scale ξ_0 with boundary conditions set by taking the numerical box to be periodic in comoving coordinates. Faster relaxation to the 'scaling' solution (see below) could be achieved by initially boosting the strings with uncorrelated random velocities. The numerical results showed some sensitivity to the number of points N_0 per initial correlation length ξ_0 and the minimum allowed loop size $L_{\min}$. Below a ratio of 2:1 with $N_0 > 30$ this dependence was observed to flatten — this is related to the typical wavelength of small-scale structure which builds up on the strings.

A particularly time-consuming part of a string simulation is the location of string crossings after each timestep. A naive routine comparing all points takes an enormous $n(n-1)/2$ computations where n is the number of points — this, or a simple variant, was the chief bottleneck in early numerical codes [29] and [30]. Vast improvements can be made upon this by sorting the string points into linked lists of cells with a size set by the maximum inter-point distance. This radix sort is an $n\ln n$ algorithm — most cpu time remains devoted to the accurate time evolution of the network.

Each segment is then compared with others in its own and neighbouring cells using a succession of increasingly more stringent tests to determine if a crossing took place. The final test was suggested in [30] and is associated with the changing sign of the tetrahedron volume defined by the two segments. The actual reconnection is achieved by evolving the associated string segments back to the intersection point and then 'exchanging partners' by locally reparametrizing the strings [32].

The final ingredient in the large code — well over ten thousand lines in length — is the analysis of string properties. To this end there are many additional routines which monitor processes such as intersections and loop creation and others which measure properties of the network such as the fractal dimension [33]. Recent additions include fourier transform routines which allow the rate of gravitational radiation to be estimated, though at this stage without backreaction [34].

4.4 Cosmic string networks
The notion of 'scaling' underlies the viability of cosmic string models in the early universe. This requires that the string network remain self-similar relative to the Hubble radius in order that it not dominate the energy density of the universe. Basically, the energy density of strings must remain constant relative to the background, potentially also providing the source of the scale-invariant fluctuations for galaxy formation.

The typical correlation of the network at 'scaling' grows in proportion to the horizon

$$\xi(t) = \gamma t \tag{26}$$

where γ is a constant, while the energy of such a Brownian array of strings is simply

$$\rho_S = \frac{\mu}{\xi^2}. \tag{27}$$

Evidence that the string network does approach a 'scaling' density is provided in figure 7 for a string network in the radiation era [32] — the solution is an 'attractor.' While the approach to a 'scaling' solution was supported by all groups, the actual value of this density was a matter of continuing controversy. However, the latest independent high-resolution simulations [31,33] have converged and are consistent with the relative density, $\rho/\mu t^2 = 14 \pm 3$.

Figure 8 illustrates an evolved cosmic string network in a box with a side-length about one quarter of the horizon [32]. The importance of high resolution numerical techniques is immediately apparent. There is a preponderance of small loops, kinks and other small wavelength modes on scales below the overall long string correlation length which is roughly comparable to the horizon, $\xi \sim t$. This 'fractal' substructure

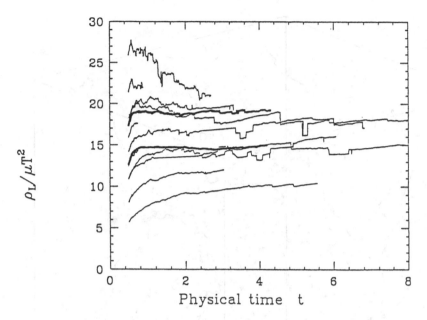

Figure 7. Approach of the relative density of a cosmic string network toward 'scaling' [32].

evolves somewhat during string evolution but its overall relative energy per correlation length remains constant. The characterization of this small-scale structure has proved resistant to analytic approaches and remains an outstanding issue numerically. While probably not significant for the overall long string density, it has important consequences for gravitational radiation estimates because it determines small loop creation sizes and thereby the peak radiation wavelength of the network. Animations of the string network are also of great assistance in observing and distinguishing the dominant physical processes at work.

5 CONCLUSION

The numerical study of topological defects — whether directly with the underlying field theory or, alternatively, with an effective low energy theory — will be area of considerable activity into the foreseeable future. Many outstanding issues remain in particle physics, cosmology, general relativity and also in a condensed matter context, with advances in one field potentially having a much broader significance. Even the first progress towards an analytic characterization of non-linear dynamics has relied heavily on numerical results as an adjunct and for verification. In the cosmological realm, the study of the inevitable role of topological defects remains at an early stage. It is an area ripe for further development.

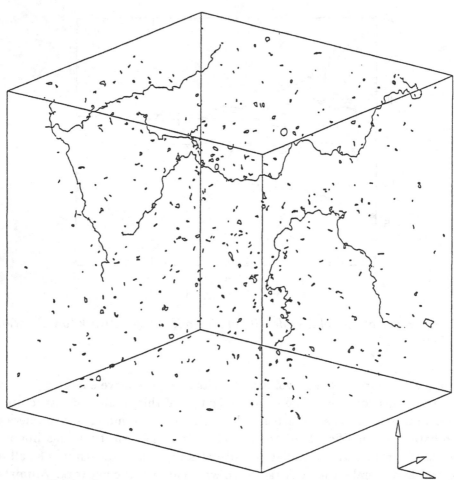

Figure 8. Evolved cosmic string network in an expanding universe, revealing two length scales [32]: the long string correlation length and that associated with the significant small-scale structure. The horizon is approximately 3.5 times the box-size.

6 ACKNOWLEDGEMENTS

I am grateful to the organizers of this numerical relativity conference for their invitation and the opportunity it presented of discussing many shared numerical difficulties and techniques. Section 4 is largely based on previously published work from an ongoing collaboration with Bruce Allen (University of Wisconsin, Milwaukee). I benefitted greatly from a graduate lecture course in numerical analysis given by N.F. Trefethen at MIT. The graphics package for the figures in section 3 was developed by Dr W. Boucher for the representation of topological defects and other three-dimensional objects. I am also grateful for discussions with John Stewart.

BIBLIOGRAPHY
(The following references are recommended for further reading on the general issues related to topological defects which were raised in the Introduction).

Topological defect review: Vilenkin, A. (1985). *Phys. Rep.*, **121**, 263.

Defect formation: Kibble, T. W. B. (1976). *J. Phys. A*, **9**, 1387; (1980). *Phys. Rep.*, **67**, 183.

Domain walls: As above and Hill, C., Fry J. and Schramm, D. (1989). *Phys. Rev. D*, **39**, 3571.

Cosmic strings: As above and Witten, E. (1985). *Nucl. Phys. B*, **249**, 557.

Monopoles: Preskill, J. (1984). *Ann. Rev. Nucl. Part. Sci.*, **34**, 461; Barriola, M. and A. Vilenkin (1989). *Phys. Rev. Lett.*, **63**, 341.

Textures: Turok, N. (1991). *Physica Scripta T*, **36**, 135.

REFERENCES
[1] Shellard, E. P. S. (1987). *Nucl. Phys. B*, **283**, 624.

[2] Matzner, R. A. (1988). *Computers in Physics*, 51.

[3] Moriarty, K. J. M., Myers, E. and Rebbi, C. (1988). *Phys. Lett. B*, **207**, 411. Moriarty, K. J. M., Myers, E. and Rebbi, C. (1991). Princeton preprint.

[4] Davis, R. L. and Shellard, E. P. S. (1989). *Nucl. Phys. B*, **324**, 167.

[5] Shellard, E. P. S. (1986). In *Proceedings of the 26th Liege International Colloquium on "The Origin and Early History of the Universe,"* 173. Shellard, E. P. S. (1990). In *Proceedings of the Cambridge Workshop on "Cosmic Strings,"* G. W. Gibbons, S. W. Hawking and T. Vashaspati (eds.), Cambridge University Press, Cambridge.

[6] Hawking, S. W., Moss, I. G. and Stewart, J. M. (1983). *Phys. Rev. D*, **26**, 347.

[7] Hecht, M. W. and DeGrand, T. A. (1990). *Phys. Rev. D*, **42**, 519.

[8] Perivolaropoulos, L. (1991). Brown preprint, Brown-HET-802.

[9] Shellard, E. P. S. and Ruback, P. J. (1988). *Phys. Lett. B*, **209**, 262.

[10] Leese, R. A., Peyrard M. and Zakrzewski, W. J. (1990). *Nonlinearity*, **3**, 387.

[11] Bennett, D. and S. Rhie (1990). *Phys. Rev. Lett.*, **65**, 1709.

[12] Press, W., Ryden, B. and Spergel, D. (1989). Harvard-Smithsonian NSF-ITP-89-51.

[13] Kriess, H. O. and Oliger, J. (1973). *GARP Publ. Ser. No. 10*.

[14] Abarbanel, S. and Gottlieb, D. (1980). *SIAM J. Sci. Stat. Comput.*, **1**, 426.

[15] Fornberg, B. (1987). *Geophysics*, **52**, 483.

[16] Press, W. H., Flannery, B. P., Teukolsky, S. A. and Vetterling, W. T. (1986). *Numerical Recipes*, Cambridge University Press, Cambridge.

[17] Stuben, K. and Trottenberg, V. (1982). *Multigrid Methods*, A. Dold and B. Eckman (eds.), Spring-Verlag, Berlin.

[18] Leese, R. A. and T. Samols, (1992). DAMTP preprint.

[19] Trefethen, L. N. (1988). Graduate lecture notes, MIT.

[20] Trefethen, L. N. and Halpern, L. (1986). *Math. Comp.*, **47**, 421.

[21] Ruback, P. J. (1988). *Nucl. Phys. B*, **296**, 669.

[22] Samols, T. *Comm. Math. Phys.*, to appear.

[23] Shellard, E. P. S. (1988). In *Proceedings of the Yale Workshop on Cosmic strings: The Current Status*, F. Accetta and L. Krauss (eds.), World Scientific, Singapore.

[24] Shellard, E. P. S. In preparation.

[25] Davis, R. L. and Shellard, E. P. S. (1989). *Phys. Rev. Lett.*, **63**, 2021.

[26] Shellard, E. P. S. and Davis, R. L. (1990). MIT preprint, CTP No. 1760. Shellard, E. P. S. In preparation.

[27] D. Forster. (1974). *Nucl. Phys. B*, **81**, 84.

[28] Harten, A. (1977). *Comm. Pure and Appl. Math.*, **XXX**, 611.

[29] Albrecht, A. and Turok, N. (1985). *Phys. Rev. Lett.*, **54**, 1868. Albrecht, A. and Turok, N. (1989). *Phys. Rev. D*, **40**, 973.

[30] Bennett, D. P. and Bouchet, F. R. (1988). *Phys. Rev. Lett.*, **60**, 257.

[31] Bennett, D. P. and Bouchet, F. R. (1990). In *Proceedings of the Cambridge Workshop on "Cosmic Strings,"* G. W. Gibbons, S. W. Hawking and T. Vashaspati (eds.), Cambridge University Press, Cambridge.

[32] Allen, B. and Shellard, E. P. S. (1990). *Phys. Rev. Lett.*, **64**, 119.

[33] Shellard, E. P. S. and Allen, B. (1990). In *Proceedings of the Cambridge Workshop on "Cosmic Strings,"* G. W. Gibbons, S. W. Hawking and T. Vachaspati (eds.), Cambridge University Press, Cambridge.

[34] Harten, A. (1983). *J. Comput. Phys.*, **49**, 357.

[35] Sod, G. (1985). *Numerical Methods in Fluid Dynamics*, Cambridge University Press, Cambridge.

[36] Vachaspati, T. and Vilenkin, A. (1984). *Phys. Rev. D*, **30**, 2036.

[37] Allen, B. and Shellard, E. P. S. (1992). *Phys. Rev. D*, to appear.

COMPUTATIONS OF BUBBLE GROWTH DURING THE COSMOLOGICAL QUARK-HADRON TRANSITION

J. C. Miller

Trieste Astronomical Observatory, Trieste, Italy

O. Pantano

Physics Department, University of Padua, Padua, Italy

Abstract. An outline is given of a scheme being used for making computations of the growth of single hadronic bubbles during the cosmological quark-hadron transition. The code uses a standard Lagrangian finite-difference scheme for flow within the bulk of each phase together with continuous tracking of the phase interface across the grid by means of a characteristic method with iterative solution of junction conditions.

1 INTRODUCTION

In view of the subject of this meeting, our emphasis here will be on the computational aspects of our study of the cosmological quark-hadron transition (Miller & Pantano 1989, 1990; Pantano 1989). However, as a preliminary, it is good to recall some fundamental points of the physics lying behind the calculations.

According to present ideas, hadrons are composed of quarks which move freely within a hadron but are strongly constrained from leaving. A phenomenological description of this is provided by the MIT bag model (Chodos et al 1974) where the region occupied by the quarks is associated with a false vacuum state characterized by a uniform vacuum energy density B and an associated negative pressure $-B$. If normal hadronic matter were compressed to high enough density, the individual hadrons would overlap and the quarks would become free to move within the entire interior region, giving rise to a quark-gluon plasma. Heavy-ion collision experiments at CERN and Brookhaven are aiming to create transient plasma in the course of collisions and to look for signatures of its decay. The universe had the right conditions for quark-gluon plasma to exist at times before ten microseconds after the big bang and subsequent expansion and cooling would then have led to a *confinement transition* producing hadrons. If, as seems quite possible (although it is somewhat disfavoured by current lattice gauge computations), this was a first order phase transition, then there could be interesting consequences for astrophysics and cosmology particularly concerning dark matter problems, the formation of structure and possible effects on cosmological nucleosynthesis. The interest of these possible consequences is sufficiently great that it seems worthwhile to investigate the picture corresponding to a first order transition for as long as this has not been definitively ruled out as a possibility.

We consider here the scenario in which the universe cools to slightly below the critical temperature for the transition T_c while remaining in the quark phase and bubbles of the new hadron phase are then nucleated within the slightly supercooled plasma (see Witten 1984; Kajantie & Kurki-Suonio 1986). Each bubble then starts to grow, restrained at first by surface tension, with the phase interface behaving as a subsonic deflagration front and pushing a compression wave out ahead of it. Subsequently, compression waves from adjacent bubbles meet, producing a region of disturbed fluid, and then the bubbles themselves meet and coalesce giving rise to disconnected quark regions which proceed to shrink.

In addition to strongly interacting particles there are also photons, leptons and neutrinos present in each phase and these play an important role in determining how the transition proceeds. The compound media are *relativistic* and are composed of particles which interact on *widely differing length scales* (strong, electromagnetic and weak) and *surface tension* is important at several stages of the process. This is a detailed problem of relativistic hydrodynamics with particle physics input. (Since the energy density in each phase is provided almost entirely by relativistic particles, it is necessary to use relativistic hydrodynamics for calculations of this process even in the limit of small bulk velocities.)

At some stages, the situation may tend to a *similarity solution* (see Kurki-Suonio 1985) in which case the system of partial differential equations governing the fluid motion reduces to a system of ordinary differential equations and it is possible to make analytic progress in obtaining a solution for the flow pattern. In general, however, it is necessary to solve the set of partial differential equations as an initial value problem in order to follow the progress of the transition and this involves quite extensive numerical calculations. We have been carrying out an investigation of this kind and we here give a description of the methodology used and present some results.

2 THE METHOD OF SOLUTION

Our present code uses spherical symmetry (which is suitable for considering the growth of isolated bubbles) and is fully relativistic. In fact, full *general* relativity has been used since it does not introduce serious additional complications beyond those already present in a *special* relativistic formulation and it allows us to include the background expansion of the universe in a natural way. The fluids are at present being treated as perfect. For the equation of state of the quark matter, we have used both the MIT bag model and also an expression proposed by Bonometto & Sokołowski (1985) as a fit to results from lattice QCD calculations, while the hadronic matter has been treated as a gas of massless pointlike pions. The other components in each phase are included as ideal relativistic gases.

The picture being considered consists of a finite hadronic region surrounded by quark-gluon plasma extending out to infinity, with the two media being separated by a phase interface having an associated surface energy and surface tension. An explicit Lagrangian finite difference scheme has been used for the bulk of each phase while the phase interface has been treated as an exact discontinuity surface and tracked continuously through the finite difference grid using a characteristic scheme with iterative solution of junction conditions.

We write the space-time metric as

$$ds^2 = -a^2 dt^2 + b^2 d\mu^2 + R^2 \left(d\theta^2 + sin^2\theta \, d\phi^2\right), \tag{1}$$

with μ being a comoving radial coordinate having its origin at the centre of symmetry. (In this work, we use units for which $c = \hbar = k = 1$). For small net baryon number, it is a reasonable approximation to take the energy density e and the pressure p as depending only on temperature so that $p = p(e)$. The system of hydrodynamical equations can be written as

$$u_t = -a \left[4\pi R^2 \frac{\Gamma}{w} p_\mu + G\left(\frac{m}{R^2} + 4\pi pR\right)\right], \tag{2}$$

$$R_t = au, \tag{3}$$

$$\frac{(\rho R^2)_t}{\rho R^2} = -a\left(\frac{u_\mu}{R_\mu}\right), \tag{4}$$

$$e_t = w\rho_t, \tag{5}$$

$$\frac{(aw)_\mu}{aw} = \frac{(e_\mu - w\rho_\mu)}{\rho w}, \tag{6}$$

$$m_\mu = 4\pi R^2 e R_\mu, \tag{7}$$

$$\Gamma = 4\pi \rho R^2 R_\mu = \left(1 + u^2 - 2Gm/R\right)^{1/2}, \tag{8}$$

$$b = \left(4\pi R^2 \rho\right)^{-1}, \tag{9}$$

$$p = p(e). \tag{10}$$

Here u is the radial component of fluid four-velocity in the associated Schwarzschild frame, R is the Schwarzschild circumference coordinate, Γ is the general relativistic analogue of the Lorentz factor, ρ is the relative compression factor, w is the specific enthalpy $(= (e + p)/\rho)$ and the subscripts denote standard partial derivatives. The mass function m can also be calculated using the alternative equation

$$m_t = -4\pi R^2 apu.$$ (11)

The time evolution equations in the above set can be rewritten as a set of ordinary differential equations along the *characteristic directions* (which are the paths of ingoing and outgoing sound waves and advective flow lines). Doing this is very useful in connection with understanding the structure of hydrodynamic flow near to a phase interface as well as playing a key role in our method of calculation. The characteristic equations are

$$du \pm \left(\frac{\Gamma}{\rho w c_s}\right) dp + \left[G\left(\frac{m}{R^2} + 4\pi pR\right) \pm \frac{2u\Gamma c_s}{R}\right] a \, dt = 0$$

$$\text{along} \quad d\mu = \pm 4\pi R^2 a\rho c_s \, dt,$$ (12)

$$\left\{\begin{array}{l} d\rho = (wc_s{}^2)^{-1} \, dp \\ dR = au \, dt \\ dm = -4\pi R^2 apu \, dt \end{array}\right\} \qquad \text{along} \quad d\mu = 0.$$ (13)

Here, c_s is the sound speed given by $c_s^2 = (\partial p/\partial e)$.

It is reasonable to treat the phase interface as an exact discontinuity with the fluid variables on either side being linked by suitable junction conditions. Unlike a shock, a transition front separates two different media and surface effects need to be taken into account. The required junction conditions for energy-momentum conservation may be derived with the aid of the Gauss-Codazzi formalism (Israel 1966). For the particular case where the surface tension σ is independent of temperature one obtains:

$$[eb^2\dot{\mu}_s^2 + pa^2]^\pm = -\frac{\sigma}{2}f^2\left\{\frac{1}{ab}\frac{d}{dt}\left(\frac{b^2\dot{\mu}_s}{f}\right) + \frac{f_\mu}{ab} + \frac{2}{fR}(b\dot{\mu}_s u + a\Gamma)\right\}^\pm,$$ (14)

and

$$[ab(e + p)]^\pm = 0.$$ (15)

where $\mu_s(t)$ is the interface location, $\dot{\mu}_s = d\mu_s/dt$, $f = (a^2 - b^2\dot{\mu}_s^2)^{1/2}$, and $[A]^\pm = A^+ - A^-$, $\{A\}^\pm = A^+ + A^-$ with the superscripts $\pm$ indicating quantities immediately ahead of and behind the interface. The comoving derivative is taken along the world-line of the interface. There are also three metric junction conditions coming from continuity across the interface of R, dR/dt and ds:

$$[R]^\pm = 0$$ (16)

$$[au + b\dot{\mu}_s\Gamma]^{\pm} = 0 \tag{17}$$

$$[a^2 - b^2\dot{\mu}_s^2]^{\pm} = 0 \tag{18}$$

The mass function m receives a contribution from the surface energy. At the time of nucleation of a bubble, conditions are essentially Newtonian so that

$$[m]^{\pm} = 4\pi R^2 \sigma, \tag{19}$$

and the subsequent time evolution is given by

$$\frac{d}{dt}[m]^{\pm} = 4\pi R^2 [b\dot{\mu}_s\Gamma e - apu]^{\pm}. \tag{20}$$

Next, we turn to the nature of the solution near to the phase interface. In our numerical scheme, we use the characteristic equations only for calculating quantities adjacent to the interface but for understanding how the solution works, it is very helpful to consider what would happen if the whole calculation were made in this way. Suppose that we knew the complete solution everywhere on a base time level t and wanted to calculate the new solution at a general point X on time level $t + \Delta t$ (see figure 1). In order to find this, we would need to use data at points L, M and N where the forward, backward and advective characteristics intersect the base time level. Equations (12) would be solved along the forward and backward characteristics to give u and p; ρ, R and m would be calculated from (13) along the advective characteristic while a would be given by (6) which has to be integrated across the time slice. The boundary condition for a, which may be imposed at any suitable point, determines the scaling of the time coordinate t.

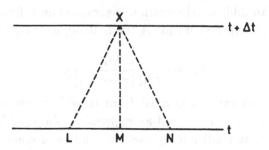

Figure 1. *Illustration of the characteristic solution at a general point. XM is the advective characteristic; XL and XN are the forward and backward characteristics respectively.*

In the vicinity of a phase interface, the situation becomes more complicated. There are two possible types of solution for the transition front: *detonations*, which move supersonically relative to the medium ahead, and *deflagrations* which are subsonic.

It is almost certainly the deflagration solutions which are relevant in the present context but it is nevertheless interesting to compare the properties of the two types of solution. In the theory of detonations, a central role is played by Chapman-Jouguet processes, for which the front moves at the sound speed relative to the medium behind (see Courant & Friedrichs 1948), and so we will use this for our comparison. However, similar consequences hold also for strong detonations which move subsonically relative to the medium behind. Figure 2 shows the characteristics for points immediately ahead of and behind the phase interface for the cases of a Chapman-Jouguet detonation (a) and a deflagration (b). Since the detonation is moving supersonically relative to the medium ahead, all three characteristics may be drawn from the point just ahead back to the base time level and so all of the hydrodynamical equations can be solved for this point as also can the a equation whose boundary condition may be fixed in the medium ahead of the interface. This means that the state of the fluid ahead is completely determined by initial data in the same phase and so it is not affected by the presence of the transition front. In the medium behind, only the forward characteristic can be drawn up to the fluid element just behind the interface. However, the junction conditions have to be satisfied across the interface and these together with the forward characteristic equation are exactly sufficient to determine completely the velocity of the transition surface and the values of u, p, ρ, a, m and R just behind it, once the state of the fluid ahead is known. For the deflagration, which moves subsonically relative to the medium ahead, it is no longer possible to draw a forward characteristic up to the point just ahead from within the same phase and so the medium ahead *is* now influenced by the presence of the transition front. The lack of this forward characteristic also has the consequence that the number of equations is now one less than the number of unknowns. In order to close the system, an additional equation is required and, from a physical point of view, this should be an expression linking the hydrodynamical energy flux across the interface

$$F_H = \frac{aw\dot\mu_s}{4\pi R_s^2 (a^2 - b^2 \dot\mu_s^2)},$$

(21)

with the transformation rate as derived from considerations of the details of the transformation process. We have used an expression obtained by following a line of argument similar to that traditionally used in studies of classical bubble dynamics (for example, studies of the expansion of vapour bubbles in water — see Theofanous et al. 1969). The procedure there is to work in the new (vapour) phase and use kinetic theory, calculating from the Maxwellian the ideal flux away from the interface $\Phi(T_l)$ and the ideal flux towards the interface $\Phi(T_v)$ where T_l and T_v are the temperatures of the liquid and the vapour respectively, adjacent to the interface. The net flux is

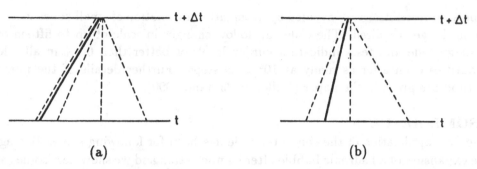

Figure 2. *The characteristic structure of a detonation solution (a) and a deflagration solution (b). The characteristics are shown with dashed lines and the path of the interface is marked by the heavy line.*

then written as

$$F_{l \to v} = \alpha \left[\Phi(T_l) - \Phi(T_v) \right], \tag{22}$$

where α is an accommodation coefficient ($0 \le \alpha \le 1$) which takes account of deviations away from the ideal situation. In the present case, we use an analogous expression but with the appropriate radiative fluxes:

$$F_{q \to h} = \frac{1}{4} \alpha g_h \left(\frac{\pi^2}{30} \right) \left(T_q^4 - T_h^4 \right), \tag{23}$$

where g_h is the number of relativistic degrees of freedom in the hadron phase and T_q and T_h refer to temperatures in the quark and hadron phases respectively, adjacent to the interface. For small initial supercooling, the temperatures remain close to T_c and (23) becomes

$$F_{q \to h} \to \alpha g_h \left(\frac{\pi^2}{30} \right) T_c^4 \left(\frac{\Delta T}{T_c} \right), \tag{24}$$

where $\Delta T = (T_q - T_h)$.

The overall strategy of the calculation, then, is to use the system of equations (2) – (10) to determine the flow within the bulk of each phase, by means of a standard finite difference method, while the interface is tracked continuously through the finite difference grid by making a simultaneous solution of the relevant characteristic equations together with the junction conditions and the transformation rate equation (given by equating the expressions for F_H and $F_{q \to h}$). The interface is an internal boundary for the problem and the state of the fluid on either side of it provides boundary conditions for the standard finite difference solution within the bulk of each of the two phases. In order to follow large changes in scale, the grid is chosen to have exponentially increasing spacing between nodes and unnecessarily stringent time-step restrictions are avoided by automatic cancellation of small zones when they are no longer required. The code can follow changes in scale of up to fifteen orders of magnitude and tests indicate accuracy levels of better than 0.1% in all relevant quantities even after as many as 10^5 time steps. Further details of the numerical method are given in the paper (Miller & Pantano 1990).

3 SOME RESULTS

The first application of the computer code has been for following the early stages in the expansion of a hadronic bubble after its nucleation and we show here some sample results from a simplified calculation in which only strongly interacting particles are considered. Full details of these calculations, including a discussion of how the results depend on the values of the input parameters, are contained in the paper mentioned above.

For the calculation whose results are illustrated here, the quark medium was represented by the bag model with two relativistic quark flavours and the critical temperature was taken as 150 MeV. Figures 3 and 4 show the velocity and energy density profiles in the fluid as functions of radius R at various times during the bubble expansion (for nucleation temperature $T_N = 0.98 \, T_c$, $\alpha = 1$ and $\sigma = T_c^3$). Initially, surface tension dominates but its role progressively decreases during the expansion until eventually its effect becomes negligible and an asymptotic regime is reached where the interface velocity is almost constant. This happens when the bubble radius has increased by about two orders of magnitude above its nucleation value. The hadronic material inside the bubble is then essentially at rest and the compression wave expanding out through the quark medium is becoming self-similar. There is no evidence for any shock forming at the leading edge of the compression wave but this is not surprising in view of the fact that the maximum velocity attained by the bubble surface ($0.103 \, c$ in this case) is small compared with the sound speed ($c/\sqrt{3}$). Most of the latent heat released in the transition does, in fact, go into thermal energy of matter in the compression wave but this is nevertheless a very small perturbation of the pre-existing quark-gluon energy density and the velocity profile is very similar to that for an incompressible fluid.

4 CONCLUSION

Calculations involving only strongly interacting particles are a reasonable approximation for the first part of the bubble expansion but this ceases to be the case when the bubble radius begins to get near to the mean free path of the electromagnetically interacting particles which will also be present ($\lambda_e \simeq 10^3 \, fm$). First, these particles start to provide a significant long-range energy and momentum transport mechanism, causing deviations away from the previous solution. Later, when the characteristic

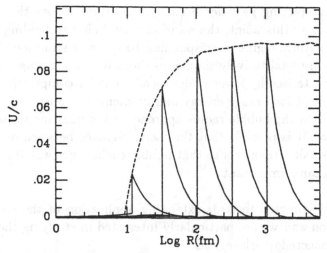

Figure 3. Fluid velocity at successive times (continuous lines), is plotted against radius R. (See text for the parameters of the run.) The dashed line shows the behaviour of the fluid velocity just ahead of the interface as the bubble expands.

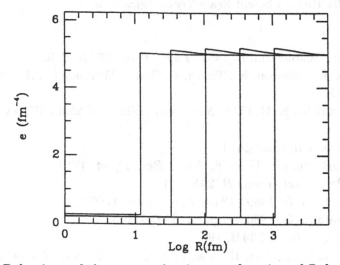

Figure 4. Behaviour of the energy density as a function of R for the same run as in figure 3.

length scales of the problem become large compared with λ_e, they may be taken as being in thermal equilibrium with the strongly interacting matter and moving together with it as a single fluid. Perfect fluid hydrodynamic equations may then be used as before except that the additional degrees of freedom must be included in the equations of state. For the parameters used in figures 3 and 4, the contribution of the electromagnetically interacting particles becomes important before the asymptotic regime is reached and, at this point, the solution would change tending eventually towards another asymptotic regime corresponding to the new situation. These new asymptotic solutions are qualitatively similar to those for only strongly interacting particles but are characterised by lower velocities and smaller compressions. At even larger scales, the effect of long-range energy and momentum transport by neutrinos becomes important when the bubble radius approaches the neutrino mean free path ($\lambda_\nu \simeq 1\,cm$). However, it is possible that the mean distance between bubble nucleation sites might be smaller than this so that bubble coalescence would occur before neutrino transport became important.

Present work is being directed towards extending application of the code to later stages of the transition and we are particularly interested in studying the final contraction of the disconnected quark regions.

ACKNOWLEDGEMENTS
This research is being carried out with financial support from the Italian Ministero dell'Università e della Ricerca Scientifica e Tecnologica.

REFERENCES
Bonometto, S. A. and Sokołowski, L. 1985. *Phys. Lett.*, **107A**, 210.

Chodos, A., Jaffe, R. L., Johnson, K., Thorn, C. B. and Weisskopf, V.F. 1974. *Phys. Rev. D*, **9**, 3471.

Courant, R. and Friedrichs, K. O. 1948. *Supersonic Flow and Shock Waves*, Springer-Verlag.

Israel, W. 1966. *Nuovo Cimento*, **44**, 1.

Kajantie, K. and Kurki-Suonio, H. 1986. *Phys. Rev. D*, **34**, 1719.

Kurki-Suonio, H. 1985. *Nucl. Phys. B*, **255**, 231.

Miller, J. C. and Pantano, O. 1989. *Phys. Rev. D*, **40**, 1789.

Miller, J. C. and Pantano, O. 1990. *Phys. Rev. D*, **42**, 3334.

Pantano, O. 1989. *Phys. Lett.*, **224B**, 195.

Theofanous, T. G., Biasi, L., Isbin, H. S. and Fauske, H. K. 1969. *Chem. Eng. Sci.*, **24**, 885.

Witten, E. 1984. *Phys. Rev. D*, **30**, 272.

INITIAL DATA OF AXISYMMETRIC GRAVITATIONAL WAVES WITH A COSMOLOGICAL CONSTANT

Ken-Ichi Nakao

Department of Physics, Kyoto University, Kyoto, Japan

Kei-Ichi Maeda

Department of Physics, Waseda University, Tokyo, Japan

Takashi Nakamura

Yukawa Institute for Theoretical Physics, Kyoto University, Kyoto, Japan

Ken-Ichi Oohara

National Laboratory for High Energy Physics, Oho, Japan

Abstract. We investigate initial data for localized gravitational waves in space-times with a cosmological constant Λ. By choosing the appropriate extrinsic curvature, we find that the Hamiltonian and momentum constraints turn out to be the same as those of the time-symmetric initial value problem for vacuum space-times without Λ. As initial data, we consider Brill waves and discuss the cosmological apparent horizon. Just as with Brill waves in asymptotically flat space-time, the gravitational "mass" of these waves is positive. Waves with large gravitational mass cause a strong cosmic expansion. Hence, the large amount of gravitational waves do not seem to be an obstacle to the cosmic no-hair conjecture.

1 INTRODUCTION

The present isotropy and homogeneity of our universe is something of a mystery within the framework of the standard big bang scenario. The inflationary universe scenario, however, is one of the favourable models which may explain the so-called homogeneity problem [1]. In this scenario, when a phase transition of the vacuum occurs due to an inflaton scalar field and supercooling results, the vacuum energy of the scalar field plays the role of a cosmological constant and the space-time behaves like the de Sitter one with a rapid cosmic expansion. This phenomenon is called inflation. As a result, all inhomogeneities go outside the horizon by rapid cosmic expansion. After inflation, the vacuum energy of the scalar field decays into radiation and the standard big bang scenario is recovered.

However, there still remains a question in the above scenario. Even if inhomogeneities are very large before inflation, is the de Sitter-like rapid cosmic expansion realized as long as there is a vacuum energy? In connection with this matter, there is "the cosmic no-hair conjecture" which states that all space-times approach the de Sitter space-time if a positive cosmological constant exists [2]. If this conjecture is true, there is no doubt about the answer to the above question. However, in general, the inhomogeneities have energy and produce a gravitational field themselves. So when the inhomogeneities are very large, we can imagine that they are not homogenized by the cosmic expansion but, rather, could collapse into black holes or naked singularities.

We can imagine various kinds of inhomogeneities which are classified into two parts: one due to an inflaton scalar field and the other due to known fields. The latter type of inhomogeneity may be further divided into two classes: one is the inhomogeneity of the space-time itself due to the gravitational waves and the other is the inhomogeneity of ordinary matter. The inhomogeneity of the effective cosmological constant is crucial for the practical inflationary scenario and therefore several investigations have been performed by analytic and numerical approaches [3-6]. On the other hand, there is little work on the other type of inhomogeneities, which is the topic of this article.

Here, our attention is focused on initial data of axially symmetric gravitational waves in a universe with a positive cosmological constant Λ. Furthermore, we concentrate on the localized gravitational waves with an asymptotic Schwarzschild-de Sitter region. We also assume that the traceless part of the extrinsic curvature vanishes. When we adopt the constant-mean-curvature slicing, i.e. $TrK = -\sqrt{3\Lambda}$, the momentum constraint becomes trivial and the Hamiltonian constraint reduces to the same form as that of the time-symmetric initial value problem without Λ, namely

$$^{(3)}R = 0, \tag{1.1}$$

where $^{(3)}R$ is the scalar curvature of the three dimensional spacelike hypersurface. Following Brill, we can construct initial data for gravitational waves even for the case with non-zero Λ. Brill discussed the positivity of the mass of gravitational waves with an asymptotically flat region and showed that when the amplitude of a gravitational wave exceeds some critical value, the three dimensional spacelike hypersurface is not asymptotically flat but, rather, is closed by the energy of the gravitational wave. This feature is the same in our case.

This is same form as that for the time-symmetric gravitational waves discussed by Brill. Since there is a non-vanishing TrK, however, this data corresponds to a snapshot of gravitational waves on a uniformly expanding background space-time.

Hereafter, we assume that the topology of a three dimensional spacelike surface is R^3. Following Brill, we write the intrinsic metric of the three dimensional spacelike hypersurface in the following form

$$dl^2 = \psi^4(R,z)[e^{Aq(R,z)}(dR^2 + dz^2) + R^2 d\varphi^2], \tag{2.4}$$

where A is a constant which determines the amplitude of the gravitational waves, and $q(R,z)$ is an arbitrary function which satisfies

$$q = 0 = \partial_R q \qquad\qquad \text{at } R = 0, \tag{2.5a}$$
$$q \to O(1/r^2) \qquad \text{or faster for } r \to +\infty, \tag{2.5b}$$

where $r = \sqrt{R^2 + z^2}$. We adopt the following function $q(R,z)$

$$q(R,z) = \left(\frac{R}{r_0}\right)^2 \exp\left(-\frac{r^2}{2r_0^2}\right), \tag{2.6}$$

where r_0 is a constant, which corresponds to the width of the gravitational waves.

Then the Hamiltonian constraint is written as

$$\frac{\partial^2 \psi}{\partial R^2} + \frac{1}{R}\frac{\partial \psi}{\partial R} + \frac{\partial^2 \psi}{\partial z^2} + \frac{A}{8}U(R,z)\psi = 0, \tag{2.7}$$

where

$$U(R,z) = \left(\frac{2}{r_0^2} - \frac{6R^2}{r_0^4} + \frac{R^2 r^2}{r_0^6}\right)\exp\left(-\frac{r^2}{2r_0^2}\right). \tag{2.8}$$

Since we are considering the case of an asymptotic Schwarzschild-de Sitter region, we impose the following asymptotic behavior for the conformal factor ψ

$$\psi \to 1 + \frac{M}{2r} \qquad \text{for } r \to +\infty, \tag{2.9}$$

where M is a constant. Although the above condition is apparently the same as the asymptotic flatness condition, this condition corresponds to an asymptotic Schwarzschild-de Sitter boundary condition because $TrK = -\sqrt{3\Lambda}$. It is worth noting that M corresponds to the gravitational mass if the space-time is the Schwarzschild-de Sitter space-time, which is the case when $A = 0$ and there is a singularity at $r = 0$. Hence, it is the gravitational mass of the gravitational waves and, as shown by Brill, it is always positive for non-vanishing A.

However, in contrast with Brill waves, there is also a cosmic expansion effect in our case. This causes anti-trapped surfaces and a cosmological apparent horizon, which will be discussed in section 3. Since the expansion rate H_C is the inverse of the areal radius of the cosmological apparent horizon in a homogeneous and isotropic universe, we investigate H_C also in our inhomogeneous case and show that H_C is a monotonically increasing function with respect to the gravitational mass of the waves. Therefore, the large mass seems to lead to strong cosmic expansion. This is the same feature as the spherically symmetric case of a Schwarzschild-de Sitter asymptotic region. Thus, it is likely that large gravitational waves do not necessarily violate the cosmic no-hair conjecture.

2 INITIAL DATA OF AXIALLY SYMMETRIC GRAVITATIONAL WAVES WITH A COSMOLOGICAL CONSTANT

When we adopt the 3+1 formalism to solve the Einstein equations, we must solve the constraint equations in order to set up the initial data. The Hamiltonian and momentum constraint equations for a vacuum space-time with a cosmological constant Λ are given by

$$^{(3)}R - K_i^j K_j^i + TrK^2 = 6H_0^2, \qquad (2.1a)$$

$$D_j(K_i^j - \delta_i^j TrK) = 0, \qquad (2.1b)$$

with

$$H_0 \equiv \sqrt{\frac{\Lambda}{3}},$$

where K_i^j is the extrinsic curvature, $TrK \equiv K_i^i$ and D_j is the covariant derivative on the three dimensional spacelike hypersurface.

Even though Λ exists, we can obtain the initial data for the gravitational waves by a similar procedure to that of Brill [8] and of Eppley [9]. Brill considered the vacuum Einstein equations and discussed the initial data of gravitational waves. Brill waves are non-trivial axially symmetric vacuum solutions of the constraint equations with vanishing extrinsic curvature (time symmetry). It is considered to be a snapshot of axially symmetric gravitational waves at a moment on maximum expansion (or minimum contraction). Since our case contains a cosmological constant, we adopt the following initial data for the extrinsic curvature

$$K_i^j = -H_0\delta_i^j, \qquad (2.2)$$

instead of time-symmetric initial data. It is easy to see that the above data satisfy the momentum constraint trivially since the extrinsic curvature has only a constant trace part. The Hamiltonian constraint with (2.2) becomes

$$^{(3)}R = 0. \qquad (2.3)$$

From regularity and reflection symmetry with respect to the equatorial plane, the conformal factor ψ should satisfy the following boundary conditions

$$\partial_R\psi|_{R=0} = \partial_z\psi|_{z=0} = 0. \tag{2.10}$$

When A exceeds the critical value obtained by Brill, the three dimensional spacelike hypersurface is closed by the energy of the gravitational waves. However, here, we consider only cases with an asymptotic Schwarzschild-de Sitter asymptotic region, in which such a possibility is excluded.

3 THE COSMOLOGICAL APPARENT HORIZONS

When the gravitational waves are sufficiently localized and have large amplitude, then trapped surfaces and an apparent horizon are formed in an asymptotically flat space-time [9]. Also, in our case, *appropriately* large amounts of gravitational waves form trapped surfaces and an apparent horizon.

The cosmic expansion effect causes another kind of trapped surface. We call it an anti-trapped surface. The anti-trapped surface is a closed spacelike 2-surface such that both expansions of the future-directed *ingoing* and *outgoing* null geodesic congruence orthogonal to the surface are positive. So the area of a wavefront of light, which is emitted inward and orthogonal to an anti-trapped surface, does not decrease but, rather, increases at that moment. Here, we shall define the cosmological apparent horizon by the inner boundary of the anti-trapped surfaces, that is the outermost closed spacelike 2-surface with vanishing expansion of the future-directed ingoing null geodesic congruence orthogonal to the surface. In this article we focus on the cosmological apparent horizon.

It should be noted that such surfaces are observer-dependent, as can be seen in the example of the de Sitter universe. However, since we are interested in the cosmic expansion effect on gravitational waves, the observer consists of the gravitational waves themselves, which are localized near the origin. Thus we focus on the anti-trapped surfaces and the cosmological apparent horizon enclosing the origin.

Suppose we have some closed spacelike 2-surface S for our initial data with metric (2.4). Then the convergence ρ_{in} and ρ_{out} of the ingoing and outgoing null geodesic congruence orthogonal to S are given by

$$\rho_{in} = +\rho^{(TS)} - 2H_0, \tag{3.1}$$
$$\rho_{out} = -\rho^{(TS)} - 2H_0, \tag{3.2}$$

with

$$\rho^{(TS)} \equiv -D_k s^k, \tag{3.3}$$

where s^k is the inward spacelike unit normal vector of S. $\rho^{(TS)}$ is the convergence for the case of time-symmetric initial data and $2H_0$ in the equations for ρ_{in} and ρ_{out} corresponds to the cosmic expansion. Then anti-trapped surfaces are found when $\rho_{in} < 0$ and $\rho_{out} < 0$ and the cosmological apparent horizon is obtained when $\rho_{in} = 0$. In order to find these surfaces, we adopt the prescription proposed by Sasaki et al [10].

In figure 1, the cosmological horizons are depicted. The dashed line corresponds to the case of $A = 0$, that is de Sitter space-time, and the solid line to the case of $A = 4.3$ with the gravitational mass $M/r_0 = 2.93$, where $H_0 = 0.1/r_0$. In the case of Schwarzschild-de Sitter space-time, when the gravitational mass of the black hole exceeds the critical value $M_{crit} = 1/\sqrt{27H_0^2}$, the cosmological apparent horizon vanishes. On the other hand, there always exists a cosmological apparent horizon in the case of gravitational waves, since the topology of the spacelike hypersurface is R^3, while the spacelike hypersurface of Schwarzschild-de Sitter space-time has the topology of $R \times S^2$. However, it is difficult to find the cosmological apparent horizons for gravitational waves with $A > 4.3$ since the shape of the cosmological horizons are highly deformed.

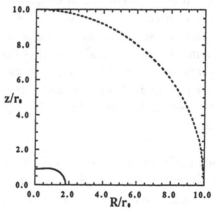

Figure 1.   *The cosmological apparent horizons with $H_0 = 0.1/r_0$ depicted in the cylindrical coordinates $(R/r_0, z/r_0)$. The dashed line is the cosmological horizon with $A = 0$ which corresponds to de Sitter space-time. The solid line is the case $A = 4.3$, with gravitational mass $M = 2.93r_0$.*

Let S_C to be the area of the cosmological apparent horizon. Then the areal radius of the cosmological horizon r_C is defined by $S_C = 4\pi r_C^2$ and its inverse is denoted by H_C, i.e.

$$H_C \equiv 1/r_C. \tag{3.4}$$

It should be noted that H_C is equal to the expansion rate H_0 in the case of the de Sitter universe H_0. Hence, we regard H_C as a cosmic expansion rate.

In figure 2, we show the relation between H_C normalized by H_0 and the gravitational mass M. The dashed line is H_C/H_0 for Schwarzschild-de Sitter space-time with $M = M_{crit} < 1/\sqrt{27H_0^2} \simeq 1.92$ and the squares correspond to our initial data. As expected, when the gravitational mass is small, H_C is virtually the same for the two cases. As can be seen from this figure, H_C monotonically increases with respect to M. Gravitational waves with large mass seem to cause strong cosmic expansion. This feature agrees with a spherically symmetric space-time with an asymptotic Schwarzschild-de Sitter region. The simplest example is the Oppenheimer-Snyder space-time with a cosmological constant, which describes the motion of a homogeneous dust sphere. This example shows that a dust sphere with large gravitational mass exceeding the critical value M_{crit} can not recollapse but must expand and approach the de Sitter space-time asymptotically if it is initially expanding [7]. Hence we show that gravitational waves with large gravitational mass do not seem to violate the cosmic no-hair conjecture.

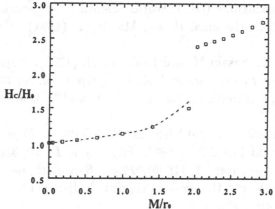

Figure 2. H_C/H_0 *is depicted as a function of the gravitational mass M/r_0. The dashed line is H_C/H_0 for Schwarzschild-de Sitter space-time. On the other hand, H_C/H_0 for the gravitational waves are depicted by squares. For small masses, H_C/H_0 of the gravitational waves agree with that of Schwarzschild-de Sitter space-time.*

4 CONCLUSION

We have investigated the initial data of axisymmetric gravitational waves with a cosmological constant Λ, which correspond to snap shots of gravitational waves on a uniformly expanding background space. The effect of the cosmic expansion causes an anti-trapped region. We have introduced the concept of cosmological apparent horizon defined by the inner boundary of the anti-trapped region. Since the inverse of the areal radius of the cosmological apparent horizon H_C is the cosmic expansion rate in the case of the de Sitter universe, we have considered H_C of our initial data to see the effect of the gravitational mass of waves on cosmic expansion. We have shown that H_C is a monotonically increasing function with respect to the gravitational mass M of the gravitational waves. Hence, the gravitational waves with large M seem to cause a strong cosmic expansion effect and do not seem to be an obstacle to the cosmic no-hair conjecture. This feature is essentially the same as in a spherically symmetric space-time with an asymptotic Schwarzschild-de Sitter region [7].

REFERENCES

[1] Guth, A. H. (1981). *Phys. Rev. D*, **23**, 347.
     Sato, K. (1981) *Mon. Not. Roy. Astron. Soc.*, **195**, 467.
     Albrecht, A. and Steinhardt, P. J. (1982). *Phys. Rev. Lett.*, **48**, 1220.
     Linde, A. D. (1982). *Phys. Lett. B*, **108**, 389.
[2] Gibbons G. W. and Hawking, S. W. (1977). *Phys. Rev. D*, **15**, 2738.
     Hawking S. W. and Moss, I. G. (1982). *Phys. Lett. B*, **110**, 35.
     Maeda, K. (1989). In *Proceedings of 5th Marcel Grossmann Meeting*, D. G. Blair
        and M. J. Buckingham (eds.), World Scientific, Singapore, 145.
[3] Sato, K., Sasaki, M., Kodama, H. and Maeda, T. (1981). *Prog. Theor. Phys.*,
        65, 1443.
     Maeda, K., Sato, K., Sasaki M. and Kodama, H. (1982). *Phys. Lett.*, **108B**, 98.
     Sato, K. (1987). In *Proceedings of I. A. U. Symposium*, No.130.
[4] Kurki-Suonio, H., Centrella, J., Matzner, R. A. and Wilson, J. R. (1987). *Phys.
        Rev. D*, **35**, 435.
[5] Holcomb, K. A., Park, S. J. and Vishniac, E. T. (1989). *Phys. Rev. D*, **39**, 1058.
[6] Goldwirth, D. S. and Piran, T. (1989). *Phys. Rev. D*, **40**, 3269.
     Goldwirth, D. S. and Piran, T. (1990). *Phys. Rev. Lett.*, **64**, 2852.
[7] Nakao, K. (1991). Kyoto University Preprint, KUNS-1112.
[8] Brill, D. (1959). *Ann. Phys.*, **7**, 466.
[9] Eppley, K. (1977). *Phys. Rev. D*, **16**, 1609.
[10] Sasaki, M., Maeda, K., Miyama S. and Nakamura, T. (1980). *Prog. Theor. Phys.
        Lett.*, **63**, 1051.

PANEL DISCUSSION

PANEL DISCUSSION

The workshop concluded with a panel discussion, chaired by Chris Clarke, with John Miller, Silvano Bonazzola and Matt Choptuik comprising the panel. The session was recorded by (what turned out to be) a rather inadequate tape recorder. I have tried on the one hand to make sense in the transcription of the passages which were unclear, and on the other to edit out some of the more repetitious moments, but I must apologise if I have ended up misreporting any of the participants. (Ed.)

Chris Clarke:
I would like the panel to start off by saying what they think are the main highlights of the meeting, the things which have struck them most about the results which have been presented here and also, if possible, what they think are the main problems of where we're going.

John Miller:
A particular impression which I have got from listening to the contributions in this meeting is that the situation now is really very different from what it was only a few years ago in similar sorts of meeting. At that time there was really one sort of method that everybody used. There were experiments with other sorts of method but there was a very strong brand leader. This is still true, but the brand leader has much stronger competition now and I have been very interested to hear during this meeting about how the various competing methods are coming along. I will write up on the blackboard what I think is some sort of structure for the different methods which are around now. The brand leader is clearly the standard single grid 3+1 finite difference method. These have been used for a long time, there is a lot of experience in the use of them and a lot of sophistication has been developed. We have heard here in this meeting also a couple of applications of characteristic based methods, where one is focussing on the causal structure. This can be very important when looking at shock solutions or other solutions with discontinuities, like phase-transition surfaces. I think that these methods come alongside the standard single grid 3+1 finite difference methods, in that one is just looking at the causal structure in a more detailed way, but it is within that context. Characteristic based methods, of course, are not at all new, they go right back into the early days of numerical analysis. Also, under this heading, I'll put the multi-grid methods which are also doing standard 3+1 finite differences but not on a single grid, using the ability to control errors by using arrangements of grids of different sizes. So all of those I've put together in one group.

The second group, which we actually haven't heard much about in this meeting, are the finite element methods which I'll link together with Regge calculus. The fundamental difference between these and the first group, is that the first group deal with line elements whereas the second deals with volume elements. Regge calculus is the "fundamental finite element method" for GR. One can argue about the extent to which it is a finite element method in the normal sense but I will place it here.

Aside from these sorts of method, another philosophy is provided by the spectral methods and the multiquadrics that we have been hearing about. Here, we are talking about expanding everything in terms of basis functions with support over the whole solution domain and this is a quite different approach from what is going on with the methods which I was talking about previously. Also, in a completely different category, there are the 2+2 methods, which is another area of considerable activity. So, whereas a few years ago, we would have all been here talking almost entirely about the 3+1 finite difference methods, now there is a whole range of techniques which are really in court and are serious competitors. It is not clear, in the end, that the standard method is going to be the best way to do the biggest front line calculations. It is certainly still a powerful competitor, because of all the wealth of experience that there is with it, but study of the other techniques is certainly a growth area.

Now the second thing that I want to say is that one of the bug-bears of numerical computing is that it is very easy to get answers which are rubbish, and one needs to know when this is happening. This is not always clear, particularly since the calculations that we are really going to be interested in over the next years are ones in multi-dimensions, where one is pushing up against the limits of computer power, of resolution and so on. In GR calculations one is having to introduce slicing conditions and coordinate systems and when the calculation has been finished, it can be quite hard to disentangle what it all means physically. I feel that we need to think very carefully about this, partly in the sense of knowing when we have got rubbish and partly in being able to make an interpretation of results that we have got that are alright. It is not enough just to have a mathematical solution of a problem. We want to know about the physics — what is really going on. Some people would say that all we are really concerned with is looking at how things seem as viewed from infinity, and, in particular, in the gravitational waves which are given out by the process. Therefore, one should not try too hard to think about the interpretation of what is going on in the middle. You should be sitting at infinity and, if everything is clear there, then it is alright. However, I think that is not enough; we are *not* only wanting to know what is happening as seen from infinity. We want to get some good idea of what is happening down there where the gravity is strong. So we need to have some good tools, conceptual tools, for understanding these things, and I think that in the coming years we need to pay attention to this.

So my overall impression of the meeting is that the subject has become more varied than it used to be. We now have a range of strongly competitive techniques and that is a healthy situation.

Silvano Bonazzola:

I agree with much of what John says. An important aspect of this field is that of validity implication, and the need to check results by different methods. We know that it is a very important problem, for example, to control the error due to the viscosity, either artificial or intrinsic. Let us not forget the intrinsic viscosity involved in any method. The Spanish group showed us a very simple example, namely even in the most simple case of the collapse of a spherical geometry, we can get a very different result if you use one method instead of another one. So this is very important, because gravitational radiation emitted from collapsing objects depends strongly on the viscocity. I would like to talk to you about a new concept of viscocity. This is the vanishing viscocity. It is well known that in the case of the spectral method the error vanishes as e^{-N}, where N is the number of degrees of freedom. The error due to the artificial vanishing viscosity has the same property and vanishes with the same law. This is a new concept and it could be important in the future.

I think John forgot one class of methods, namely SPH (smooth particle hydrodynamics) methods. In this conference, we have heard nothing about this method, whereas we heard a lot about it at the last conference. It's a very useful method when you have to treat problems without defined boundaries, for example. The Japanese group showed us their work on the coallescence of neutron stars computed with this method. The drawback of this method is that it has very high intrinsic viscocity. Of course, the intrinsic viscocity depends on the number of particles present, which is an important consideration given the current state of computers. In many ways, this is a good method for starting up a problem, and for getting some feeling for it. There are not many people using this method. In fact, it appears from this conference that the majority of people working in numerical relativity use explicit schemes. It would be wiser to use implicit schemes. For example, Joan Centrella spoke about using something like 200 hours of Cray time. Well, with implicit methods we can save a lot of time.

What should we do in the future? I am an astrophysicist and so I shall consider the question from this viewpoint. In the last month there has been a small revolution in astrophysics, and we are very excited about it. I don't know if you have heard about gamma ray bursts? Gamma ray bursts are something which were discovered

25 years ago with military satellites planned to control nuclear explosions around the earth. Instead of discovering nuclear explosions, they discovered some other emissions, which were short wave gamma ray emissions coming from space. Even only one month ago, we understood very little about these processes. Everybody, with one exception, considered these gamma ray bursts to be galactic in character. The model in fashion was that of a big neutron star undergoing nuclear explosions. The one person who did not share this idea was Bodhan Paczyński. He thought the sources were extra-galactic. Now the American experiment BATSE on board the GRO satellite was able to detect gamma ray bursts over the whole sky with very high sensitivity due to the big surface of the detector, and showed that they were extra-galactic. I mean, they have a very high isotropic angular distribution and they have $\log N / \log S$ curves similar to quasars. Many models were proposed. Paczyński suggested that they originated from the coallescence of neutron stars. The initial reaction of Brandon Carter, however, was that this could not be possible, because you cannot get gamma ray emissions from coallescing neutron stars, due to the matter screening the gamma ray bursts. At best you can have X-ray emissions. We did some statistics on the probability of having coallescences of neutron stars. We found that gamma rays could only be detected within one giga parsec. This is too rare to be compatible with the estimated rate of coallescences of neutron stars. We are out by a factor of a thousand.

When Paczyński sent us a preprint in which he showed clearly that the more conservative hypothesis is to consider that the gamma ray bursts are extra-galactic, we gave a copy to Brandon. This was a Friday afternoon, and on Monday morning Brandon arrived with a paper written proposing a new model. He was so excited that if anyone were to make an objection, he was ready to kill them! (Laughter). Ah, you know Brandon as well? What is his idea? You start of with a small black hole, 10^5 solar masses say, and with an ordinary star passing very close to the black hole. He defined a penetration parameter β, which is given in terms of the ratio of the Schwarzschild radius and the radius of the star. This can lead to a compression of the star by a factor of one thousand. The compressed star is then heated up to about 10^9 degrees, and produces gamma ray emmissions. My first objection (for which I survived) is that the observations reveal that the light curve of the bursts has spikes with very short rising time of the order of 10^{-3} seconds. How can we explain these spikes? Well, we were able to think of a way of doing it, and it leads to a prediction (namely that the rising time of the light is correlated with the total length of the bursts) which seems to work. This is a more complicated problem than black hole coallescences, but it is a nice problem for future work in numerical relativity.

Matt Choptuik:
Now for something completely different, which may be completely controversial and get some discussion going. Well, let me just sketch my view of numerical relativity. (Ed: Matt rubbed out everything written on the blackboard by the previous two speakers — laughter). In my experience, this thing has gone in three stages: some formalism, some implementation and out of this comes the output, which we hope isn't garbage, and the physics. Now I think that we should all agree as numerical relativists that the physics is the only thing we should be really interested in. As mathematicians, numerical analysts and computer scientists, it is fine to be worried about the other two stages, but if this was really a conference on numerical relativity then the high points should be the new physics. In that respect, I was particularly impressed by the work of Ken-Ichi Oohara and that by Mark Dubal. It think it is truly remarkable that Mark has come up with this initial data for two black holes in such a short period of time. It is a long standing problem. To me, those were the two high points. In my opinion, the first two things (formalism and implementation) really don't matter that much. I am really glad to see all these methods coming up, although I don't think there has been particularly a proliferation of methods in the past few years. To say that there is a dominant method in numerical relativity seems strange because there are so few people who do it anyway. So whoever is doing most work at one time is going to appear to be the most dominating in the field.

John Miller:
Certainly, the alternative methods go back in time, but there has been more talked about here.

Matt Choptuik:
Well that is because there are two or three people here talking about them. (Laughter). You can even make arguments on theoretical grounds that the formulation and implementation shouldn't matter. For example, Mark Dubal talked about an example where the multiquadric method performed very well on a problem involving the Laplacian on the unit cube. I took it as a challenge to show that you could do just as well with a standard finite difference method.

But the lesson in GR is that we are looking for classical solutions. There is a large class of solutions that we think will be smooth and quite regular and still have a lot of interesting physics. The fact that the solutions are classical means, in some sense, that when you get down to the resolution where the solutions are truly smooth on the scale of your mesh, you can expect an exponential increase in accuracy for a linear investment in extra computational work — which means your problem is completely solved. Now this is a sort of theoretical statement, but you also find it happening in practice. What happens is that methods tend to be "partially tuned" and then have computational complexities which differ by constant factors. Constant factors can't matter (fundamentally) because tomorrow's machines are going to be twice as fast, and in different parts of the world they are already running at different speeds; yet people do reasonable research in various places. And it is always easy, relatively speaking, to find another processor. What isn't easy is if you design a method where you start off with some resolution and then you want to double the resolution (number of algebraic unknowns) and the computing time goes up by a factor of four or something. Then, you will always be disappointed and never be satisfied by your answers, or at least very rarely. That, in a nutshell, is what is happening with lattice QCD, for example, where they face the problem that for a linear increase in accuracy, they basically have to invest a proportionately exponential amount of computing time. Therefore, I can state with some confidence, that in the year 2010, they will still be saying we need bigger computers, whereas we are now getting to the point where the biggest problem is in interpreting the physics. Within implementation, I include things such as designing a multiple mesh algorithm, graphics, and all sorts of other stuff which takes a lot of time. It takes most of graduate students' time when they are doing numerical relativity these days. Yet this is something that is not talked about in relativity text books at all, and in most cases has to be picked up sort of on the fly. So my hope is that some day that this stuff will be streamlined so that we can spend more time doing physics.

I also want to stress the notion that we *do* have ways of assessing accuracy of solutions and I will just show you one example. Again this is in the context of my one and only problem of the collapsing scalar field. Tsvi Piran and Dalia Goldwirth had a code based on a characteristic formulation of the problem and from my PhD thesis I had a standard single grid 3+1 code which I augmented using Richardson-extrapolation techniques. Now the results come from quite a strong-field-solution (Figure 1); if you increase the initial amplitude of the scalar field by a few per cent a black hole will form. We performed a numerical coordinate transformation (which was essentially another numerical code) to take the Cauchy results and put them into characteristic coordinates. Now since this is a strong field solution, there is no exact solution, which is why we were doing the problem. Figure 2 shows some plots of norms of differences between various solutions — both in the Cauchy and characteristic cases — and essentially analytic results. The "analytic" results were produced through multi-level extrapolation, but the point is that the vertical scales in the plots are logarithmic (log base 2). The black dots are the Cauchy results and the white squares are the characteristic results. What you can see, for example, if you look at the Cauchy results (Figure 2a), is that as you go from one dot to the next smaller dot, where there is an increase of a factor of 2 in resolution, the dots are basically separated

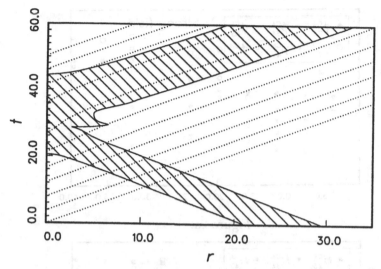

*Figure 1. Space-time plot showing principal features of a strong-field calculation performed in a comparison of characteristic (**CH**) and Cauchy (**CA**) codes for spherically symmetric massless scalar collapse. **CA** uses the (r, t) coordinate system described in Choptuik's contribution in this volume, and Richardson-extrapolative finite-difference techniques. **CH** uses an outgoing null coordinate, U, and an areal radial coordinate, R, and employs the method of characteristics as a discretization technique. The initial data is a single, ingoing Gaussian pulse of scalar radiation and the boundaries of the hatched region are $1/e$ contours of $|\phi|$. Dotted lines are the outgoing null geodesics, $U = [0.0, 3.6, 7.6, 11.6, 15.6, 19.6, 23.6, 27.6, 31.6, 35.6, 39.6]$, along which the quantities $\phi(R, U)$ and $H(R, U) \equiv (R\phi)_{,R}$ were compared. Note the "bending" of the null rays in the central region where the scalar field is most strongly self-gravitating.*

by four units all the way along. This manifestly demonstrates that the solution is converging with $O(h^4)$ error. For the characteristic results, what you see in the strong field interaction region, is that the spacing between the squares, which were initially 4 units apart, end up 2 units apart. So you can actually see the convergence of the code change there. You can also see that at late times the Cauchy results are more accurate than the characteristic results. So in this case we could generate more accurate results along characteristics by taking the Cauchy results and transforming them, than by actually solving along the characteristics. This is another statement of my claim that it really doesn't matter what coordinate system you use; what really matters is that you do the numerical analysis correctly. Figure 3 shows some actual point-wise results—what you get by subtracting a given solution from the base-line solution. Figure 3a shows the Cauchy result on a coarse mesh and two finer meshes at an early time, and you can see the solution converging and the same for the characteristic

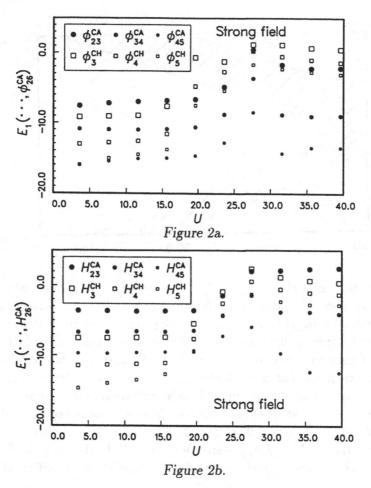

Figure 2a.

Figure 2b.

Figure 2. Results from convergence tests of **CH** and **CA** evolutions of the strong-field initial data. Plotted in each graph is a base-2 logarithmic measure of the percentage error in various numerical solutions, relative to high-accuracy reference solutions (ϕ_{26}^{CA}, H_{26}^{CA}) generated using multi-level Richardson extrapolation. (Thus, $E_1(\cdots) = 0.0$ or -10.0 corresponds to 1% or about 0.001% relative deviation, respectively.) Symbols of the same shape and shading, but having different sizes, correspond to calculations performed by a given code using different basic discretization scales (resolutions), h_ℓ, satisfying $h_\ell = 2\,h_{\ell+1}$. $O(h^q)$ convergence is signaled by a constant separation of q units between different-sized markers at a fixed U. $O(h^4)$ and $O(h^3)$ convergence is observed for ϕ^{CA} and H^{CA}, respectively, throughout the calculation. ϕ^{CH} and H^{CH} exhibit nearly $O(h^4)$ convergence at early retarded times. Once the pulse of scalar radiation reaches $R = 0$, the convergence of both **CH** quantities drops to $O(h^2)$, so that at late times, and at a given resolution, the **CA** results tend to be more accurate than the **CH** values.

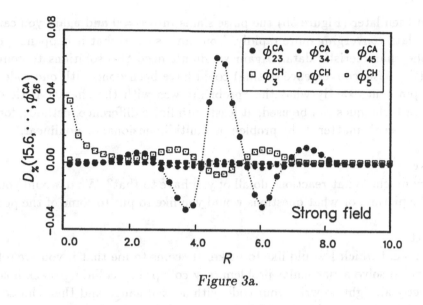

Figure 3a.

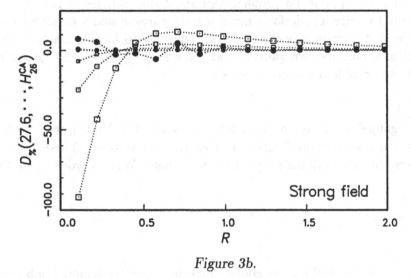

Figure 3b.

Figure 3.   Illustration of point-wise convergence of representative values from the
strong-field calculation. $D_{\%}(\cdots)$ is the percentage relative deviation of a given quantity from the high-accuracy reference values, and the different marker sizes again
indicate computations performed with different resolutions. In Fig 3a, rapid convergence of both ϕ^{CA} and ϕ^{CH} is evident; in Fig 3b, deterioration of the convergence of
H^{CH} near $R = 0$ is apparent. However, it seems clear that both codes are converging
to the same solution.

values. And then later (Figure 3b) the pulse slams into $r = 0$ and again you can see the Cauchy data converging quite rapidly. You can also see what is happening quite clearly in the characteristic data. Again we didn't need two solutions to compare — all this (the basic convergence testing) could have been done with *one* code. We could even pinpoint exactly what the "problem" was with the characteristic code. These sorts of techniques can be used, at least with finite difference methods, for any problems. It doesn't matter if the problem is multidimensional or nonlinear.

Chris Clarke:
Thanks very much. What reactions do all of you have to that? What would you like to add to the picture, or what questions would you like to put to some of the people?

John Barrett:
I have a comment which I would like to make. It seems to me that if you are setting up your code to solve a particular problem, say collapse or colliding black holes, it seems perfectly all right to write your code with a fixed gauge and then choose that gauge so that for that particular problem you avoid singularities as it evolves. But if you are trying to write a code for a more general purpose and you hope to run on anything, then that approach does not work. One has to have a method whereby the code itself can pick its own gauge, or at some stage you can change the gauge appropriate to the problem you are solving.

Matt Choptuik:
I always put "gauge" in quotes because my supervisor, Bill Unhru, used to bat me about the ears if I said "gauge" instead of coordinate system. These days codes already have coordinate systems deeply built into them. What is your suggestion?

John Barrett:
I'm just interested.

Chris Clarke:
Surely, shouldn't we be looking for codes which have a conspicuous knob on them outside that you can just twiddle to alter the gauge?

Matt Choptuik:
But then you have got to have a conspicuous set of routines to generate your equations.

Chris Clarke:
It's quite hard to imagine what such a thing would look like.

Matt Choptuik
Well I think the day of monolithic "codes" is coming to an end. I think in the future things should be more modular.

John Barrett:
Do you think there should be a different code for each different problem?

Matt Choptuik:
Well, I mean, you have compilers that generate machine language from Fortran. So there are some people working on code compilers that generate finite difference and spectral methods and everything else from high level specifications of the problem. So that is one option.

Chris Clarke:
You seem to be talking about adapting them.

Matt Choptuik;
In Austin, at Schlumberger Research, they have a working system which is being used.

Silvano Bonazzola:
Actually, I do not know if I agree with this philosophy, because I think if you don't see what the question is, then you cannot understand what to do. I prefer the other philosophy of people building faster computers with better chips.

John Miller:
I think this depends partly on how much you are worried about the efficiency of the code you produce. If you want something which is a black box which will decide for you how to do things, then what would probably happen is that you would get a code that is not as efficient as it could be. Although I know that computer power is increasing so that maybe there will come a time when people care less about the absolute efficiency of what they do, I think that time hasn't arrived yet. We do still care quite a lot about efficiency. We are up near the limits of what can be done with present computers and I think that until we get past that, probably one needs to have purpose built codes. It is clearly most efficient to have codes that are particularly made for the problem you are talking about. If one gets to the stage where that efficiency is not important, then one can go to black boxes, but I am a bit mistrustful of black boxes, because they are always written by somebody else and you never quite know how well they have done it.

Matt Choptuik:
A code generator is not a black box. I mean, do you mistrust the Cray assembly language that is generated by that for Fortran?

John Miller:
No.

Matt Choptuik:
Well, people used to. In the old days they would go in and check their assembly code.

John Barrett:
There comes a pont where you can't do everything yourself.

John Miller:
But where should you set that point?

John Barrett:
It has struck me that there seems to have been an awful lot of discussion throughout the whole conference about how to avoid coordinate singularities, how to get round them. That discussion seems to be at the end, after you have written your code whereas, it seems to me, you should really worry about this at the beginning.

James Vickers:
It is also related to the fact that, at the end of the day, you are going to get the answer in some coordinate system. Knowing what the metric is in some particular coordinate system doesn't tell you a tremendous amount about the physics. For example, one would like to know something about the geodesic structure in a strong gravitational region. One might also want to know about the stability of horizons. It seems to me, if one's thinking of doing numerical experiments which go to help one understand the structure of relativity, rather than help one understand astrophysical observations, that is, if one wants to use numerical relativity as a theorist rather than as an astrophysicist, then one would want to have information about rather different things in the future. None of these issues have yet been addressed. I can seen that they are not the ones which one would need immediately, but later on when really one wants to do experiments, one wants to have answers available which aren't just the metric but which tell you more about the space-time structure.

Matt Choptuik:
Well what sort of things are you thinking about? For example, in the scalar field problem, I can give you a black box which, basically, gives you metric coefficients and

then you can do whatever processing you want on them afterwards.

James Vickers:
Yes, you could give me the metric coefficients, but this would not enable me to solve the geodesic structure.

Matt Choptuik:
You just integrate them up.

James Vickers:
Well, I'm not going to do that by hand! And then I want some way of displaying them that helps me understand what is going on.

Chris Clarke:
There is a whole other area of numerical relativity which doesn't really exist yet, like questions of how the horizon is formed. How do you actually discover numerically where the horizon is, how smooth the horizon is and all that sort of thing?

James Vickers:
I am thinking about the stability of the horizon in the Kerr solution, for example. Some people looked at that theoretically from the point of view of perturbation theory, in a way which not everyone considers is satisfactory. If one has great confidence in the numerical code, which I'm not in a position to judge, then one could probably investigate that problem numerically. The answer that one would get numerically would enable one to formulate theorems which would be possible to prove analytically.

Matt Choptuik:
I am sure that will happen.

James Vickers:
Sure, but at the moment people don't seem to have looked at that sort of problem.

Matt Choptuik:
That is largely due to the fact that the codes that have been developed in numerical relativity are up against the boundary limits of computer time and so people don't get to make results of parameter space surveys. You can design systems nowadays which will integrate ODEs in real time in front of your eyes. So you really can twiddle knobs and see bifurcation patterns and things like that. There is no reason in principle why this won't happen. That is a while down the road.

James Vickers:
It seems to me that it is just about reaching the stage where that is something which is feasible and I would encourage people to pursue it.

John Miller:
This sort of problem is extremely difficult numerically. It probably will be done in the end, but it is quite formidable.

Ray d'Inverno:
I started off this workshop with some sort of personal overview of numerical relativity and I would like to return to a basic question I raised there. It seems to me that in numerical relativity there is the good news and the bad news. The good news is that there is a whole class of problems which one really wants to know about in GR, and this seems to be the only tool available for investigating them. The other good news is that it is a young field and yet has already produced a body of results which is impressive, especially when you think of how few people are involved in it. It is a good field in the sense that it is likely that observational results will soon be coming along which are crying out for explanation. Numerical Relativity is the only tool which can provide the answers. It's also a good field in that it can attract good research students to it and one can get them working on feasible problems almost from day one. Certainly they can do some work straightaway, whereas this is a problem in the more theoretical aspects of general relativity (at least we find it so), because it takes such a long time to tool students up and get them started. This, in turn, leads to real problems in finishing in a reasonable amount of time. Moreover, even if graduate students don't stay in the field, they are computer literate and that means they are employable. The bad news, it seems to me, is the status the field currently possesses in the main stream of general relativity. I don't know what the panel's opinion is. Do you feel that it doesn't quite possess the credibility it ought to have? In the opening talk, I described how computer algebra systems originally had this credibility problem too. They seem to have cracked it now as a result of different systems producing the same results, and hence, demonstrating reliability. So now computer algebra is just used as a tool. I get the impression that at international conferences on general relativity which include numerical relativity presentations, the pictures go up on the screen and people sit there politely. But when they go away to coffee, they mutter under their breath: is it reliable?

Chris Clarke:
It's only numerical! (Laughter).

Ray d'Inverno:
I take Matt's point about reliability in his case, and his investigation is quite important, but maybe reliability can only be shown for that particular problem.

Matt Choptuik:
Why is it only for that particular problem? That's all you can do in a problem. An experimenter doesn't say "I have measured this result and it is accurate and so is every other result that anybody makes."

Ray d'Inverno:
Yes, well that would give me confidence in believing the results you have obtained for that problem. But what about other problems where there haven't been attempts to do that?

Matt Choptuik:
Well, I am just saying you can do that with any problem. There is a general prescription. I can say no more. I can't make them do it. I can keep nattering at other people and eventually get a convert here and there.

Ray d'Inverno:
I have heard informally at this meeting about errors being discovered in reported work. This included a big effort to try and reproduce a well-known result, only to discover eventually that the result was wrong. The worry is that this situation is not uncommon.

Matt Choptuik:
But my claim is that if they use this convergence test they will know they are right. It also gives you confidence that you have got the right answers.

Chris Clarke:
I think that the analogy with experiment is the one to press, because now no experimenter would produce results where the error wasn't mentioned anywhere in the paper. Experimental training involves training in estimating errors, and that is part of the whole thing. I think if people start looking at numerical relativity as being what is now experimentation for the relativistic community, then they ought to take the same view. They shouldn't dream of producing results without some estimate of the error.

Silvano Bonazzola:
Yes, you speak for me on that too.

Ray d'Inverno:
Do members of the panel agree that we still have this credibility barrier?

Silvano Bonazzola:
Well, they will change their minds. (Laughter).

Matt Choptuik:
In North America, probably because there is a greater percentage of people doing numerical relativity now, it will gain credibility.

Silvano Bonazzola:
It reminds me of the problem with the four-colour theorem, which was demonstrated by computer. But for a mathematician, this is not a proof, only a demonstration. This is the same point of view.

Alan Rendall:
I think one way of gaining credibility would be if there was more interaction between numerics on the one hand and analytic theory on the other. To give you an example of something nice which happened, there were these solutions which were found numerically by Bartnik and McKinnon ((1988), *Phys. Rev. Lett.*, **61**, 141, Ed.) for the spherically symmetric Einstein-Yang-Mills equations and now it has been proved mathematically that they exist and that was a very nice piece of mathematics. So in the end everybody ends up gaining something. There the numerical calculation was actually a fairly trivial one which just involved solving ODEs, but it should be possible also to have this friendly interaction going on in other areas. If the theorists saw that something was coming out of numerical relativity which was obviously good for them, then they would tend more to take it seriously.

Silvano Bonazzola:
I think mathematicians and people working only in numerical relativity should concentrate on the physics of the problem. For example, why are people studying coalescences of black holes? Well, in part, it is because it is a challenge. But what are the major results we get from this work? If, with the next generation of wave detectors, we can observe the signature of the coallescence of two black holes, then we will be able to gain important information about the evolution of binary star systems. This is the importance of the work. So I think that nature has much more imagination than we do and proposes a lot more problems, which are not just academic. Moreover, I am in England and, since this is a pragmatic country, it is worth remarking that if you can justify your work as astrophysics you might be able to get much better funding support than you would otherwise!

John Miller:
Can I give my answer to Ray? I think this depends on the sort of cultural background from which one is looking at things. You are thinking of it from within a general relativity context, and I agree there is perhaps this difficulty of credibility with the mathematical relativists ...

Ray d'Inverno:
Who make up the bulk of the community.

John Miller:
Who make up the bulk of *this* community. They think that what is being done here by the numerical people is, perhaps, a little bit unreliable and question how it should be viewed. But there is another community also, which is the astronomy and cosmology community, and they keep asking other questions of people like us. They say "It's all very well relativists making their calculations, but what do they mean in terms of observations that we can make. That is the real world." Now for these people we are, perhaps, the acceptable end of relativity (laughter) because we are getting a bit nearer to experiments that can be made. So it depends on the background from which you ask the question.

Alan Rendall:
I'd like to say something else about coding errors. What should be contained in a paper which reports a numerical calculation? Sometimes, having looked at such papers, I've come away not thinking I have learnt very much. But it's not entirely clear to me, since I have never written such a paper myself, what should you ideally report. You can't include your program, that's obviously impossible. What information should be contained in a report of a numerical calculation?

Matt Choptuik:
Well the obvious answer is error estimates in principal quantities that you are presenting. There is always going to be some key graph or number that comes out of it.

John Barrett:
One of the most elementary mistakes is not to say what the axes in the graphs represent.

Chris Clarke:
To be specific, I've just stopped being a Deputy Editor of Classical and Quantum Gravity, and we were frequently being faced with the problem of the balance between

going into details of the numerics and going into details of the significance of the results. This here is a reflection of the division between the different communities. Do you need different sorts of papers addressing the two communities, or three communities, if you include the astrophysics community? Is there a sort of numerical relativity paper which can somehow get through to all three? What is the best publicity tactics — what sort of papers ought we to be writing?

John Miller:
Well this isn't quite an answer to the question, but I think it's very important, since you aren't going to be putting down listings of your code, to be very clear what the philosophy is behind the calculation that you are doing. It's this which, perhaps, can appeal to the various different communities. If one is very clear what are the ideas behind the calculation, the structure of it, why you are doing it, how reliable you think it is in terms of errors and so on, what the results of it are, what the applicability is; then, perhaps, you have some hope of addressing all these different communities at once. In the past, I think we have had some quarrel with the editors of Classical and Quantum Gravity, over this about exactly what the limits ought to be. There are technical papers in numerical relativity which are of interest to a particular audience, which is not necessarily the broader computing audience. It is particularly a numerical relativity audience, and it seems to me that Classical and Quantum Gravity is very well able to serve this. But I know of some papers (which are not mine!) where there have been comments like "This is too technical. We are not seeing results come out which are directly telling us something new about general relativity. They are talking about techniques and therefore why should they be in Classical and Quantum Gravity?" I think that is a pity; techniques ought to be published and Classical and Quantum Gravity is the right journal to be doing this.

James Vickers:
Yes, after all if it was some new formalism but you were not able to get a new result out of it, they wouldn't think twice about publishing it.

John Miller:
Sure, that is exactly the point. So I think it would be good if people saw the basic work in numerical relativity also in the same light. It is not necessarily the case that it has to give you a final answer that is earth shattering. If it's a technique which is useful for other numerical relativists for further development, then it's good if it's there.

Silvano Bonazzola:
Actually, I do not agree. There is a journal in which we can publish this sort of work,

namely the Journal of Computational Physics. However, it takes so long to publish. They ask for three referees and it can take two years before it is published. Yet you want your technique published.

Ray d'Inverno:

I think the point is that it is a different sort of discipline from the rest of standard theoretical GR and so it requires different sorts of papers. So probably the solution, in the longer term, is that we should have a journal devoted to this sort of activity. Then, even though it's a young science, the conventions will grow within it about the acceptable ways to present the various facets of techniques, results, etc.

John Barrett:

For a paper to be interesting to relativists, the results have to be delivered in such a way that relativists can read them. In particular, this means that the things which are displayed should have some geometrical significance. Too often, one sees a graph displayed of something like the (r, θ) component of the metric, followed by some explanation that this corresponds to something happening physically. Whereas to a relativist, this means absolutely nothing.

Matt Choptuik:

Maybe it will be of crucial importance to the next generation who are coming along and reproducing the results. Nobody makes a claim that it has any physical significance.

John Barrett:

From the numerical point of view that may be fine, but from the relativity point of view, to actually read the results ...

Matt Choptuik:

Why publish any equation — the equation may have no geometric significance?

John Barrett:

My point is: if you want to refer to a point of spherical symmetry, say, then you can actually measure the area of the sphere and that has a geometrical significance. Giving some coordinate doesn't. To write a paper that relativists can read, the results have to be displayed in that form. You have to talk about curvatures, areas, lengths, proper distances, how geodesics are behaving, and so on. That's something relativists are interested in.

Jörg Frauendiener:
But you could make the same criticism of someone working in exact solutions.

John Barrett:
Yes. The big advance in exact solutions came when people started talking about their global structure and drawing Penrose diagrams. Then you could see what the space-time means.

James Vickers:
That is close to another point: I would like somehow to use numerical techniques to actually help get a better understanding of global structure, even in the case of some space-times which one knows about analytically.

Ray d'Inverno:
We are probably not there yet, are we? I think James has a point. The challenge should be taken up. How many exact solutions do we have that insight of? Very, very few of them. That's an ultimate goal, to understand the solutions we have got. But that is a long way off. However, it doesn't stop intermediate results, namely exact solutions, being reported. In the same way, one shouldn't resent intermediate results involving codes which are grappling with the physics being reported. They too are in a stage where they are setting themselves up.

John Barrett:
I don't mean to grumble about it, but from the point of view of a relativist, one needs to go on a stage.

John Miller:
In fact what you are saying is related to what I was trying to say earlier, which is really the following. One would like to talk about things in invariant form, because that is what has real meaning. In order to make calculations you have to have coordinates. That's life. In order to make calculations involving the formation of horizons, you need slicing conditions that are of some suitable form, otherwise you can't carry on making your calculation. It is of interest to print out results that you get using your coordinate system, your slicing condition; one needs to have that. But then at the end you have to make some sense of it, and it is sometimes very difficult to do so. This is where one has to get back to conceptual understanding. We have to have conceptual understanding in the end, otherwise what are we doing? So I agree with you, but I think there is a role for publishing the (r, θ) component of the metric. It can be valuable to look at that, because in order to make calculations you have to have a coordinate system.

John Barrett:
I am fully in favour of publishing the line element and the exact solution as well. But there is another stage — it is not the whole story.

Nigel Bishop:
Could I introduce another subject? Computer technology is something which affects us all very much. I would like to make a couple of comments about it and see what people think. Computer power is growing and growing at the moment. We already have a gigaflop machine and there are prospects of having a terraflop machine in a few years. This is obviously going to affect the type of problem we can approach. There seems to be a corollary to that in that if we want to really get this sort of power, then one is looking at some form of parallel machine. Parallel machines are a fairly recent development, and we have tended not to write our codes in parallel. There are the prospects that, in the same way that vectorisation has become automatic in compilers, certain aspects of parallelization could also become automatic. So the prospect of parallelization is something which we should all be aware of, even if things are eventually done automatically, because it certainly helps to write things, having that idea in mind.

Matt Choptuik:
I don't agree. The codes we have will be among the easiest to parallelize, by virtue of the fact that they are easy to vectorize.

Chris Clarke:
Does this consideration effect the sort of language you use? Should we move more towards object-orientated languages, less declarative languages?

Matt Choptuik:
I think in ten years time everyone will be using Fortran 90. But that is not so much an object-orientated language. It is vector oriented. It's not clear you will need anything more here.

Chris Clarke:
Because the parallelization is more of a vector nature, you mean? Is that even the case with these sort of multigrid techniques, where it looks as though parallelization could well be more than just a vector aspect if you are looking at different grids? Maybe in that case parallelization means more than just vectorisation.

Matt Choptuik:
Yes.

Nigel Bishop:
It does effect, to some extent, the type of algorithm that one uses. Parallelization is natural on finite difference grids. I'm not so sure how easy it is to parallelize spectral methods. That may be a longer term approach.

John Miller:
I think that the future of top end computing is probably going in the direction that you say. It's going into multiprocessors. I know that you are an expert in this but, at the moment, the things are horribly difficult to use in practise.

Various voices:
That is not true.

Matt Choptuik:
I think at the next big general relativity meeting after GR13, you will see some interesting work coming out of it.

John Miller:
OK, I am not saying it is impossible to use the approach, but it is quite difficult. However, if one is talking about these massive transputer arrays, then one's into a different business.

Nigel Bishop:
I think you are partially right. It was true some years ago, but I think we are now moving to a situation where they are becoming easier to use. We are on our way.

Matt Choptuik:
Well, Connection machines have Fortran 90 compilers. In Texas, for example, they assigned an undergraduate project to parallelize a string code. That was a reasonable project, and seems to have worked quite well, quite naturally.

Chris Clarke:
I think we are at an intermediate phase, aren't we? Our experiences at Southampton are that we are working with a machine where the compilers are not yet good enough.

Matt Choptuik:
I think the Connection machine compilers are probably significantly better. They have got a lot of software engineers now.

Silvano Bonazzola:
But how much computation power do we need? For example, people working in incompressible hydrodynamics carry out simulations with spectral methods using a 256 cube in three dimensions. I have seen movies of it; it is fantastic. You see vortices, and you can enter the vortices. They now have simulations with Reynolds number of the order of 300. Do we need to produce simulations with higher Reynolds number? Maybe not.

Chris Clarke:
You don't have to go for the most difficult physics straight off. There is still a lot of comparatively easy physics which needs to be done.

Marquina:
From another point of view, I think a major problem is to find new and simple algorithms. I think when we use numerical methods in a scheme, the problem is to try to imitate the physics of the problem, or the qualities of the solution you are looking for. This requires new methods in numerical analysis and non-linear theory. Sometimes, you can make big savings with a smart method, with a simple method, though not in every case. The problem is how hard you have to look at numerical schemes in order to solve your equations. Another problem is to use numerical analysis to get the right formulation of the equations. Not every formulation is good to apply to a numerical scheme. Experimentation helps to find the correct formulation as well as the characteristics of the problem you are looking for, and even to discover new ones which you were not originally looking for.

Silvano Bonazzola:
This is another problem. What do we really want to be able to do? Let us suppose we have a supercomputer and we are able to model the sun. We put everything we know about in — nuclear reactions and so on. Then you end up with a nice model of the sun. What do you learn about the sun? Nothing. This is because you need to understand what are the most important physical processes occurring. So sometimes, instead of trying to consider all the problems at once, it is much better to study a small physical phenomenon and see just how important it is.

Chris Clarke:
Well, perhaps that is a good point at which to stop and thank everyone, especially the panelists.

John Miller:
Can I say a final thing? I think that we have all enjoyed this meeting very much.

Perhaps some of us felt a little bit grudging at the beginning; we thought that in the week before Christmas time, oughtn't we to have been doing other things rather than going to a relativity meeting? But as it has gone along, I think that we have all enjoyed it very much and so we ought to thank the organizers for all of their work on it.